META-ALGORITHMICS

META-ALGORITHMICS

PATTERNS FOR ROBUST, LOW-COST, HIGH-QUALITY SYSTEMS

Steven J. Simske

HP Labs, Colorado, USA

IEEE PRESS

WILEY

Registered office
John Wiley & Sons Ltd, The Atrium, Southern Gate, Chichester, West Sussex, PO19 8SQ, United Kingdom

For details of our global editorial offices, for customer services and for information about how to apply for permission to reuse the copyright material in this book please see our website at www.wiley.com.

Library of Congress Cataloging-in-Publication Data

Simske, Steven J.
  Meta-algorithmics : patterns for robust, low-cost, high-quality systems / Dr. Steven J. Simske, Hewlett-Packard Labs.
     pages cm
  ISBN 978-1-118-34336-4 (hardback)
  1. Computer algorithms.   2. Parallel algorithms.   3. Heuristic programming.   4. Computer systems–Costs.
5. Computer systems–Quality control.   I. Title.
  QA76.9.A43S543 2013
  005.1–dc23
                                                                                          2013004488

A catalogue record for this book is available from the British Library.

ISBN: 9781118343364

Typeset in 10/12pt Times by Aptara Inc., New Delhi, India

Printed in Singapore by Ho Printing Singapore Pte Ltd

Contents

Acknowledgments

The goals of this book were ambitious—perhaps too ambitious—both in breadth (domains addressed) and depth (number and variety of parallel processing and meta-algorithmic patterns). The book represents, or at least builds on, the work of many previous engineers, scientists and knowledge workers. Undoubtedly most, if not all, of the approaches in this book have been elaborated elsewhere, either overtly or in some disguise. One contribution of this book is to bring these disparate approaches together in one place, systematizing the design of intelligent parallel systems. In progressing from the design of parallel systems using traditional by-component and by-task approaches to meta-algorithmic parallelism, the advantages of hybridization for system accuracy, robustness and cost were shown.

I have a lot of people to thank for making this book a reality. First and foremost, I'd like to thank my wonderful family—Tess, Kieran and Dallen—for putting up with a year's worth of weekends and late nights spent designing and running the many "throw away" experiments necessary to illustrate the application of meta-algorithmics (not to mention that little thing of actually writing). Their patience and support made all the difference! Thanks, your presence in my life makes writing this book worthwhile.

I also thank Hewlett Packard (HP), my employer, for the go-ahead to write this book. While my "day job" work load was in no way lightened during the writing of the book (for one example, as I write this I have 150% more direct reports than I did when the contract to write this book was signed nearly a year and a half ago!), HP did streamline the contract and in no way micromanaged the process. Sometimes the biggest help is just getting out of the way, and I appreciate it. Special thanks to Yan Liu, Qin Zhu Lippert, Keith Moore and Eric Hanson for their support along the (HP) way.

The editorial staff at John Wiley & Sons has been tremendous. In particular Alex King, my main editor and the person who convinced me to write this book in the first place, has been a delight. Baljinder Kaur has been tremendous in finding typographical and logical errors during the editorial process. Her sharp eye and wit have both made the process a snap. I'd also like to thank Genna Manaog, Richard Davies, Liz Wingett, and Claire Bailey for their help.

Most of the photos and all of the tables and diagrams in the book were my creation, with one notable exception. There is an excellent scan of a 1996 brochure for the Cheyenne Mountain Zoo which I have used extensively to illustrate mixed-region segmentation. Thanks to Erica Meyer for providing copyright permission, not to mention for helping run one of the country's most unique attractions.

Have you found a mistake or two in reading the book? If not, you, like me, have Jason Aronoff to thank. Jason read each chapter after I finished the first draft, and his excellent and very timely feedback allowed me to send a second draft to Wiley (before deadline! I've heard that "never happens" from my friend and fellow author Bob Ulichney), where otherwise a very sloppy, choppy first draft would have had to suffice. Thanks to Marie Vans for her extra pair of eyes on the proofs.

On the science of meta-algorithmics, big thanks go to Sherif Yacoub, who framed out several of the patterns with me more than a decade ago. His analytical and design expertise greatly affected Chapter 7 in particular. I'd also like to thank Xiaofan Lin for excellent collaboration on various meta-algorithmic experiments (part of speech tagging and OCR, for example), not to mention his great leadership on voting patterns. My friend and colleague Igor Boyko worked with me on early meta-algorithmic search approaches. Yan Xiong also worked on several of the original experiments, and in particular discovered hybrid ways to perform journal splitting. John Burns led the team comprising all these *übersmart* researchers, and was tremendously supportive of early work.

I would be remiss at best to fail to mention Doug Heins, my friend and confidant, who has the most meta-algorithmic mind of anyone I know. That's right, it has improved accuracy, robustness and cost (yes cost—I owe a lot to him, but to date he has not charged me!). My deep thanks also to Dave Wright, who has extended meta-algorithmics to fantasy football and other areas. In addition to his great insights during the kernel meta-algorithmic text classification work, Dave continues to be a source of wisdom and perspective for me.

I can only begin to thank all my wonderful collaborators in the various domains—imaging to security to biometrics to speech analysis—covered in part in this book. Particular mention, however, goes to Guy Adams, Stephen Pollard, Reed Ayers, Henry Sang, Dave Isaacson, Marv Luttges, David Auter, Dalong Li and Matt Gaubatz. I wish to separately thank Margaret Sturgill, with whom I have collaborated for 18 years in various hybrid system architecture, classification and imaging projects.

Finally, a huge thanks to my many supportive friends, including Dave Barry (the man of positive energy), Jay Veazey (the wise mentor and font of insight), Joost van Der Water, Dave Klaus, Mick Keyes, Helen Balinsky, Gary Dispoto and Ellis Gayles, who have encouraged me throughout the book creation process. I hope this does not disappoint!

If you perform a Tessellation and Recombination pattern on the above paragraphs, the output would be quite obvious. I am a lucky man indeed. Thanks so much!

Steve Simske
17 April 2013

1

Introduction and Overview

Plus ça change, plus c'est la meme chose.

—Jean-Baptiste Alphonse Karr (1849)

There's even exponential growth in the rate of exponential growth.

—Ray Kurzweil (2001)

1.1 Introduction

Services, businesses, analytics, and other types of data services have moved from workstations, local area networks (LANs), and in-house IT infrastructure to the Internet, mobile devices, and more recently "the Cloud." This has broad implications in the privacy, security, versioning, and ultimately the long-term fate of data. These changes, however, provide a welcome opportunity for reconsidering the manner in which intelligent systems are designed, built and tested, deployed, and optimized during deployment. With the advent of powerful machine learning capabilities in the past two decades, it has become clear to the research community that learning algorithms, and systems based in all or part on these algorithms, are not only possible but also *essential* for modern business. However, the combined impact of mobile devices, ubiquitous services, and the cloud comprise a fundamental change in how systems themselves can—and I argue should—be designed. Services themselves can be transformed into learning systems, adaptive not just in terms of the specific parameters of their applications and algorithms but also in the repertoire (or set) and relationship (or architecture) between multiple applications and algorithms in the service.

With the nearly unlimited computing and data hosting possibilities now feasible, hitherto processor-bound and memory-bound applications, services, and decision-making (actionable analytics) approaches are now freed from many of their limitations. In fact, the cloud- and graphical processing unit (GPU)-based computation have made possible parallel processing on a grand scale. In recognition of this new reality, this book focuses on the algorithmic, analytic, and system patterns that can be used to better take advantage of this new norm of parallelism, and will help to move the fields of machine learning, analytics, inference, and classification to more squarely align with this new norm.

Meta-algorithmics: Patterns for Robust, Low-Cost, High-Quality Systems, First Edition. Steven J. Simske.
© 2013 John Wiley & Sons, Ltd. Published 2013 by John Wiley & Sons, Ltd.

In this chapter, I overview at an often high, but thematic, level the broad fields of machine intelligence, artificial intelligence, data mining, classification, recognition, and systems-based analysis. Standing on the shoulders of giants who have pioneered these fields before this book, the intent is to highlight the salient differences in multiple approaches to useful solutions in each of these arenas. Through this approach, I intend to engage all interested readers—from the interested newcomer to the field of intelligent systems design to the expert with far deeper experience than myself in one or more of these arenas—in the central themes of this book. In short, these themes are:

1. Instead of finding the best possible intelligent algorithm, system, or engine for a task, the system architect should look for the best combination of algorithms, systems, or engines to provide the best accuracy, robustness, adaptability, cost, and so on.
2. Parallel approaches to machine-intelligence-driven tasks such as data mining, classification, and recognition naturally lead to parallelism by task, parallelism by component, and eventually parallelism by meta-algorithmics.
3. Meta-algorithmic approaches and patterns provide a toolbox of potential solutions for intelligent systems design and deployment—accommodating architects of widely varying domain expertise, widely varying mathematical background, and widely varying experience with system design.

1.2   Why Is This Book Important?

Jean-Baptiste Alphonse Karr, right after the 1848 Revolutions rocked Europe, made the famous observation that "the more that things change, the more they stay the same." This statement anticipated Darwin's treatise on the Origin of Species by a decade, and is germane to this day. In the fast-changing world of the twenty-first century, in which Ray Kurzweil's musing on the rapidly increasing growth in the rate of growth is nearly cliché and Luddite musings on humanity losing control of data are *de rigueur*, perhaps it may be time to reconsider how large systems are architected. Designing a system to be robust to change—to anticipate change—may also be the right path to designing a system that is optimized for accuracy, cost, and other important performance parameters. One objective of this book is to provide a straightforward means of designing and building intelligent systems that are optimized for changing system requirements (adaptability), optimized for changing system input (robustness), and optimized for one or more other important system parameters (e.g., accuracy, efficiency, and cost). If such an objective can be achieved, then rather than being insensitive to change, the system will *benefit* from change. This is important because more and more, every system of value is actually an *intelligent* system.

The vision of this book is to provide a practical, systems-oriented, statistically driven approach to parallel—and specifically meta-algorithmics-driven—machine intelligence, with a particular emphasis on classification and recognition. Three primary types of parallelism will be considered: (1) parallelism by task—that is, the assignment of multiple, usually different tasks to parallel pipelines that would otherwise be performed sequentially by the same processor; (2) parallelism by component—wherein a larger machine intelligence task is assigned to a set of parallel pipelines, each performing the same task but on a different data set; and (3) parallelism by meta-algorithmics. This last topic—parallelism by meta-algorithmics—is in practice far more open to art as it is still both art and science. In this book, I will show how meta-algorithmics extend the more traditional forms of parallelism and, as such, can complement the other forms of parallelism to create better systems.

1.3   Organization of the Book

The book is organized in 11 chapters. In this first chapter, I provide the aims of the book and connect the material to the long, impressive history of research in other fields salient to intelligent systems. This is accomplished by reviewing this material in light of the book's perspective. In Chapter 2, I provide an overview of parallelism, especially considering the impact of GPUs, multi-core processors, virtualism, and cloud computing on the fundamental approaches for intelligent algorithm, system and service design. I complete the overview chapters of the book in Chapter 3, wherein I review the application domains within which I will be applying the different forms of parallelism in later chapters. This includes primary domains of focus selected to illustrate the depth of the approaches, and secondary domains to illustrate more fully the breadth. The primary domains are (1) document understanding, (2) image understanding, (3) biometrics, and (4) security printing. The secondary domains are (1) image segmentation, (2) speech recognition, (3) medical signal processing, (4) medical imaging, (5) natural language processing (NLP), (6) surveillance, (7) optical character recognition (OCR), and (8) security analytics. Of these primary and secondary domains, I end in each case with the security-related topics, as they provide perhaps the broadest, most interdisciplinary needs, thus affording an excellent opportunity to illustrate the design and development of complex systems.

In the next three chapters, the three broad types of parallelism are described and applied to the domains described in Chapter 3. Chapter 4 will address Parallelism by Task, which focuses on the use of multiple instances of (usually the same) data being analyzed in parallel by different algorithms, services, or intelligent engines. This chapter will also outline the advantages that cloud computing brings to this type of parallelism—namely, the ability to produce actionable output limited by the throughput of the slowest process. Chapter 5 then addresses Parallelism by Component, in which different partitions of the same data set are processed in parallel. The advances provided by GPUs will be highlighted in this chapter. The third and final broad category of parallelism—Parallelism by Meta-algorithm—will be introduced in Chapter 6. Because of their importance to the rest of the book, these approaches will be elaborated as belonging to three different classes—first-, second-, and third-order meta-algorithms—each with a specific set of design patterns for application. These patterns will be introduced in Chapter 6 before they are then applied to the domains of interest in the three chapters that follow.

In Chapter 7, the first-order meta-algorithmic patterns are explored. These relatively simple means of combining two or more sources of knowledge generation—algorithms, engines, systems, and so on—are shown to be generally applicable even when the combined generators are known only at the level of black box (input and output only). One pattern, Tessellation and Recombination, is shown to be especially useful for creating correct results even when none of the individual generators produces a correct result—a process called emergence. This pattern bridges us to the potentially more powerful patterns of Chapters 8 and 9. In Chapter 8, the second-order meta-algorithms are described. A very powerful tool for temporal and series-parallel design of meta-algorithmic systems, the confusion matrix, is explored in full.

In Chapter 9, third-order meta-algorithmic patterns—generally focused on feedback from the output to input and system-level machine learning—are overviewed.

The book concludes with Chapter 10—elaborating how parallelism by task, component, and meta-algorithm lead to more accurate, cost-sensitive, and/or robust systems—and Chapter 11, which looks to the future of intelligent systems design in light of the previous chapters.

It is clear that, in addition to the more straightforward parallel processing approaches (by task and by component), this book focuses on meta-algorithmics—or pattern-driven means of combining two or more algorithms, classification engines, or other systems. The value of specific patterns for meta-algorithmic systems stems from their ability to stand on the shoulders of the giants of intelligent systems; in particular, the giants in informatics, machine learning, data mining, and knowledge discovery. This book will cover some new theory in order to expostulate the meta-algorithmic approaches, but it is intended to be a practical, engineering-focused book—enough theory will be provided to make the academic reader comfortable with the systems eventually crafted using the meta-algorithmics and other parallelism approaches.

Building big systems for intelligence—knowledge discovery, classification, actionable analytics, and so on—relies on the interplay of many components. Meta-algorithmics position the system architect squarely in the "post-cloud" era, in which processing, storage, analysis, and other traditionally limited computing resources are much less scarce.

In the sections that follow, the background science to which this book owes its existence will be occasionally interpreted in light of meta-algorithmics, which themselves are not introduced until Section 2.5 or fully developed until Chapter 6. This should not be an impediment to reading this chapter, but concerned readers may feel free to look ahead at those sections if they so wish. More importantly, this background science will be reinterpreted in light of the needs of the so-called parallel forms of parallelism that comprise this book. We start with informatics.

1.4   Informatics

Informatics, like the term "analytics," is a broad field of knowledge creation with a plethora of definitions. In keeping with Dreyfus (1962), I herein consider informatics to include the study of algorithms, behavior, interactions, and structure of man-made systems that access, communicate, process, and store/archive information. Informatics is concerned with the timely delivery of the right information to the right person/people at the right time. Informatics, therefore, is innately amenable to parallelism. Figure 1.1 illustrates this in simplified form. Two sets of (one or more) *algorithms* are provided. The internals of these algorithms are not important to the overall *behavior* of the system, which is the function mapping the inputs to the outputs. The *interactions* between the two algorithmic subsystems are also a "black box" to the inputs and outputs, and the overall *structure* of the system is the set of all inputs, outputs, algorithms, and interactions.

Informatics, therefore, is the science of useful transformation of information. This implies that the outputs in Figure 1.1 are more sophisticated than—that is, contain information of increased value in comparison to—the inputs. A simple but still useful example of an informatics system based on the architecture shown in Figure 1.1 is given in Figure 1.2.

In Figure 1.2, a parallel architecture is used even though the task is simple enough to perform with a sequential design. The advantage of the parallel design is that subsections of the original

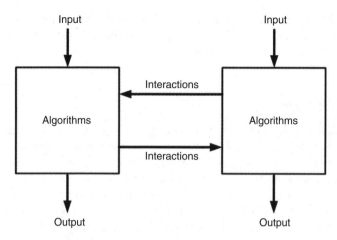

Figure 1.1 Simplified informatics system illustrating algorithms, behavior (mapping from input to output), interactions (data exchange between algorithm blocks), and structure (overall architecture)

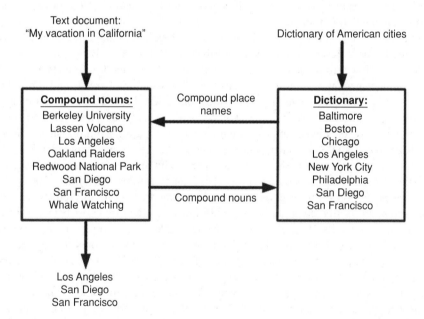

Figure 1.2 A simple system to pull compound place names from a document. The algorithm box to the left, "Compound nouns," extracts compound nouns from the input document. The algorithm box to the right, "Dictionary," inputs a dictionary of terms to search for in its input (which happens to be the output of the compound noun extractor). The terms occurring in each set are returned to the text document algorithm box and output as the set of compound place nouns: "Los Angeles," "San Diego," and "San Francisco"

document after being processed by the left (or "Compound nouns") algorithm, can be input to the right (or "Dictionary") algorithm and processed in parallel to the next subsection being processed by the "Compound nouns" algorithm.

1.5   Ensemble Learning

Informatics-based systems are thus a very general type of intelligent system. In this section, ensemble learning, which focuses on the handling of the output of two or more intelligent systems in parallel, is considered. Two reviews of ensemble learning are of particular utility—those provided by Berk (2004) and Sewell (2007). In Berk (2004), ensemble methods are defined as "bundled fits produced by a stochastic algorithm, the output of which is some combination of a large number of passes through the data." This bundling or combining of the fitted values from a number of fitting attempts is considered an algorithmic approach (Hothorn, 2003). In Berk (2004), classification and regression trees (CART), introduced in Breiman *et al.* (1984), are used to bridge from traditional modeling (e.g., mixture models, manifold-based systems, and others) to algorithmic approaches. Partitioning of the input is used to create subclasses of the input space, which correlate well with one among a plurality of classes. However, partitioning quickly leads to overfitting of the data and concomitant degradation of performance on test data when compared to training data.

To avoid this problem, ensemble methods are used. Bagging, random forests, and boosting are the three primary ensemble methods described in Berk (2004). Bagging, or "bootstrap aggregation," is shown to be definable as a simple algorithm: random samples are drawn N times with replacement and nonpruned classification (decision) trees are created. This process is repeated many times, after which the classification for each case in the overall data set is decided by majority voting. Overfitting is avoided by this "averaging" effect, but perhaps even more importantly by selecting an appropriate margin for the majority voting. This means some cases will go unclassified, but since multiple trees are created, these samples will likely be classified through another case. Should any samples be unassigned, they can be assigned by nearest neighbor or other decisioning approaches. Random forests (Breiman, 2001) further the randomness introduced by bagging via selecting a random subset of predictors to create the node splits during tree creation. They are designed to allow trade-off between bias and variance in the fitted value, with some success (Berk, 2004). Boosting (Schapire, 1999), on the other hand, is derived from a different learning approach, even though it may result in a very similar interpretative ability to that of random forests (Berk, 2004). Boosting is, generally speaking, the process by which the misclassified cases are more highly weighted after each iteration. It is argued that this approach avoids overfitting, and its famous incarnation, the AdaBoost (Freund and Schapire, 1996; Schapire, 1999), has certainly proven accurate in a number of machine learning problems. However, there are some concerns with the approach: the stopping criterion—usually the error value during training—is not always effective, and convergence is not guaranteed.

In Jain, Duin, and Mao (2000), 18 classifier combination schemes are overviewed. Among them are ensemble methods bagging and boosting, voting, and class set reduction. Interestingly, this article mentions the possibility of having individual classifiers use different feature sets and/or operate on different subsets of the input; for example, the random subspace method. This approach lays some of the groundwork for meta-algorithmics.

In Sewell (2007), ensemble learning is defined as an approach combining multiple learners. This review focuses on bagging, boosting, stacked generalization, and the random subset method. Here, Sewell refers to bootstrap aggregating, or bagging, as a "meta-algorithm," which is a special case of model averaging. Viewed this way, the bagging approach can be seen as an incipient form of the Voting meta-algorithmic pattern described in Section 6.2.3. It can be applied to classification or regression. However, as opposed to meta-algorithmic patterns, bagging operates on multiple related algorithms, such as decision stumps, and not on independently derived algorithms. Boosting is also described as a "meta-algorithm" that can be viewed directly as a model averaging approach. It, too, can be used for regression or classification. Boosting's value is in generating strong classifiers from a set of weak learners. This approach is an important part of the rationale for meta-algorithmics in general, as we will see in Chapters 6–9.

Stacked generalization (Wolpert, 1992) extends the training + validation approach to a plurality of base learners. This is a multiple model approach in that rather than implementing the base learner with the highest accuracy during validation, the base learners are combined, often nonlinearly, to create the "meta-learner." This paves the path for meta-algorithmic patterns such as Weighted Voting (Section 6.2.3), although stacked generalization is focused on combining weak learners, whereas meta-algorithmics are focused on combining strong learners, engines or intelligent systems.

The final ensemble method that introduces some of the conceptual framework for meta-algorithmics is the random subspace method (Ho, 1998), in which the original training set input space is partitioned into random subspaces. Separate learning machines are then trained on the subspaces and the meta-model combines the output of the models, usually through majority voting. This shares much in common with the mixture of experts approach (Jacobs *et al.*, 1991), which differs in that it has different components model the distribution in different regions of the input space and the gating function decides how to use these experts. The random subspace method leads to a single model—capable of classification or regression analysis—that can provide high accuracy even in the face of a highly nonlinear input space. Both the random subspace and mixture of experts approaches are analogous in some ways to the Predictive Selection meta-algorithmic approach (Section 6.2.4) and related meta-algorithmic patterns. However, as with the rest of the ensemble methods, these models stop at providing an improved *single model* for data analysis. Meta-algorithmics, on the other hand—as we will see in much of the rest of the book—use the output of ensemble methods, other classifiers, and other intelligent systems as their *starting points*. Meta-algorithmics combine multiple models to make better decisions, meaning that, for example, bagging, boosting, stacked generalization, and random subspace methods, could all be used together to create a more accurate, more robust, and/or more cost-effective system.

1.6 Machine Learning/Intelligence

The distinction between machine learning/intelligence and artificial intelligence is somewhat arbitrary. Here, I have decided to term those approaches that result in a readily interpretable, visible set of coefficients, equations, procedures, and/or system components as "machine learning." Those in which the details of how the system works are hidden are considered "artificial intelligence" systems. For the former, the intelligence is not "artificial," but rather based on an expert system, formula, or algorithm in line with human reasoning; for the latter, it simply works with an "intelligence" that is not immediately obvious. The distinction is not particularly important other than to allow me to collect some thoughts on these broad topics in a (relatively) structured manner.

For this machine learning overview, I primarily consulted Bishop (2006), Hastie, Tibshirani, and Friedman (2009), and Marsland (2009). While there are many other texts on machine learning, the combination of these three was appealingly broad and deep. In providing different foci, they were just the right balance between rigor and heuristics, and I refer the reader to these books for a far more in-depth coverage of these topics than I can provide in this section. Additional background texts consulted, which helped frame the development of machine and artificial intelligence in the decade leading up to support vector and ensemble methods, included Fogel (1995), Goldberg (1989), Leondes (1998), and Tveter (1998): these provided rich context for the more current state of machine intelligence and also highlighted for me the need for parallel and other hybrid methods.

1.6.1 Regression and Entropy

Regression is perhaps the simplest form of machine intelligence. Linear regression is often the introduction to least squares error, since this is the most common choice of loss function in regression (since least squares estimates have the smallest variance among all linear unbiased estimates). Introductory statistics courses teach the student how to perform linear regression and in so doing determine a least squares best fit model that is described in its entirety by the $\{slope, intercept\}$, or $\{\beta_1, \beta_0\}$. This fits precisely one definition of machine learning above: a readily interpretable, visible set of coefficients. The machine has learned that

$$\hat{y} = \beta_0 + \beta_1 X.$$

That is, the dependent variable y, as well as its estimate $\hat{y}$, is predicted by the independent variable X using only the model coefficients.

One of the more interesting applications of regression is in determining the complexity of the overall data. I define complexity here by the combination of the order of the regression, the residual variability determined by $1.0 - r^2$, and the entropy of the residual variability. Table 1.1 illustrates the residual variability as a ratio of the initial variability and the entropy

Table 1.1 Residual variability as a percentage of the original variability and entropy, $H(\Delta y)$, of the residual variability determined from $H(\Delta y) = -\sum p(\Delta y) \ln p(\Delta y)$, where $p(\Delta y)$ are computed on a histogram of 100 bins (maximum entropy $= 4.605$)

Order of Regression	Input Space A		Input Space B	
	Residual Variability	Residual Entropy	Residual Variability	Residual Entropy
1	0.432	2.377	0.345	2.561
2	0.167	3.398	0.156	2.856
3	0.055	4.101	0.087	3.019

of the residuals, $\Delta y = |y - y|$, for first-, second-, and third-order regression models of two sets of data, Input Spaces A and B.

The entropy for Table 1.1 is taken for 1% subranges of the overall range of X. The entropy is computed from

$$H(\Delta y) = -\sum p(\Delta y) \ln p(\Delta y),$$

where the maximum entropy is, therefore, $-\ln(0.01) = 4.605$. For Input Space A, the residual variability decreases as the order of regression increases. The entropy increases significantly, as well. These data indicate that much of the variability in the data is explained by a third-order regression: only 5.5% of the variability remains and the remaining error values have high entropy.

For Input Space B, however, the residual variability drops less, to 8.7% after a third-order fit. The entropy also does not increase significantly from second- to third-order best fit, indicating that there is still structure (i.e., a fourth-order or higher fit is appropriate) in the residual error values. Interpreting the data in Table 1.1 in this way is appropriate, since the simple first-, second-, or third-order regression models are a linear combination of fixed basis functions. Since reduced entropy indicates nonrandom structure remains in the data, it indicates that the regression model does not contain sufficient dimensionality, or else that a different model for the regression is needed altogether. This latter possibility is important, as it may mean we need to rethink the approach taken and instead use, for example, a support vector machine (SVM) with a kernel to transform the input space into one in which simpler regression models suffice.

Regardless, this simple example illustrates an important connection between regression and other intelligent system operations, such as classification (Section 1.9) or alternative models (Section 1.7). For example, one may conclude that Input Space A represents a third-order data set, and so gear its classification model accordingly. Input Space B, on the other hand, will require a more complicated classification model. This type of upfront data analysis is critical to any of the three forms of parallelism that are the main topic of this book.

1.6.2   SVMs and Kernels

SVMs are two-class, or "binary," classifiers that are designed to provide simplified decision boundaries between the two classes. Support vectors create boundaries for which the margin between the two classes is maximized, creating what is termed optimal separation. It is obvious that such an approach is highly sensitive to noise for small- and medium-sized data sets, since the only relevant subset of input data—the support vectors—are used to define the boundary and its margin, or spacing to either side of the decision boundary. An example of a support vector for a two-dimensional (2D), two-class data set is given in Figure 1.3.

SVMs (Cortes and Vapnik, 1995; Vapnik, 1995) focus on transforming the input space into a higher-order dimensionality, in which a linear decision boundary is crafted. There is art, and mathematical prowess, involved in creating a decision boundary, analogous to the creation of classification manifolds, as in Belkin and Niyogi (2003), Grimes and Donoho (2005), and elsewhere. As noted in Bishop (2006), direct solution of the optimization problem created by the search for an optimum margin is highly complex, and some classification engineers may prefer genetic, near-exhaustive, and/or artificial neural network (ANN) approaches to the

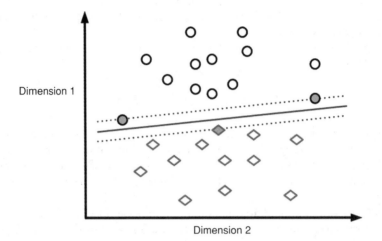

Figure 1.3 Example decision boundary (solid line) with margin to either side (dotted lines) as defined by the support vector (filled circles and diamond). The margin is the maximum width possible for the given data set in which classification errors are entirely eliminated (all circles at least one margin above the decision boundary, and all diamonds at least one margin below the decision boundary)

mathematically precise approach. However, some of this complexity is obviated by reducing the SVM optimization equation to what is known as the canonical representation of the decision hyperplane. This means the optimization is now recrafted as a familiar quadratic programming problem—a second-order function is optimized subject to a set of first-order inequality constraints.

The relationship between an SVM and regression models, such as described in the previous section, is subtle, as shall be illustrated in the next few paragraphs. The SVM is designed to maximize the margin of the decision boundary, as shown in Figure 1.3. As the size of the two populations increases, however, the percentage of points in each population that are being considered in forming the support vector drops. This is analogous to the much lower surface area to volume ratio of, say, an elephant in comparison to a mouse. It is clear that the margin *can* be optimized, but does that mean that it should? The answer to this somewhat troubling, but very important, classification question is, of course, "it depends."

To address this, I will push the SVM ramifications in another direction, trying to connect the decision boundary to regression and principle component analysis. Let us suppose that the margins around the decision boundary define a range of acceptable decision boundary slopes. This is shown in Figure 1.4 for the support vector introduced in Figure 1.3.

Figure 1.4 illustrates one means of using all of the samples in both of the classes to define the optimal decision boundary, and not just the support vectors. The decision boundary is allowed to be redefined such that none of the samples defining the support vector are misclassified, even though the slope of the decision boundary may change. The perpendiculars to the range limits of this new decision boundary are also shown.

In order to connect the range from line C to line D in Figure 1.4, a linear transformation of the data in Figures 1.3 and 1.4 is performed. In Figure 1.5, this transformation is performed by elongating the input space in a direction perpendicular to the original decision boundary. This

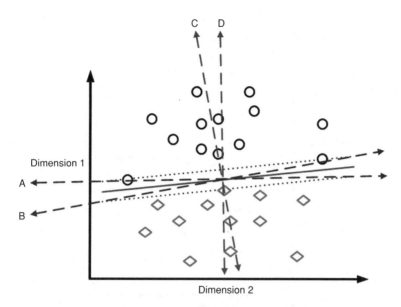

Figure 1.4   Support vector example from Figure 1.3 in which the Dimension 2 margins of the support vector (the "decision zone," or space within a margin of the optimal decision boundary) are used to define the range of slopes for the decision boundary. Line A indicates the minimum slope defined by the upper left and lower right limits of the decision zone, and line B indicates the maximum slope. The perpendiculars to lines A and B are lines D and C, respectively, and delimit the allowable range of slopes for the line of best fit to the overall data set

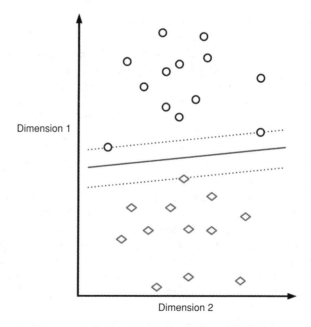

Figure 1.5   Support vector example from Figure 1.3 in which the data for the two classes (represented by circles and diamonds) are stretched vertically to ensure that the regression line of best fit and the principal component will both be roughly perpendicular to the decision boundary

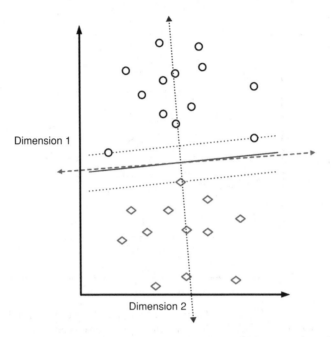

Dimension 1

Dimension 2

Figure 1.6 Redefinition of the decision boundary (dashed line, which passes through the centroid of the decision zone) based on defining it as perpendicular to the line of best fit to the transformed data ("dashed" and "dotted" line)

allows us to use regression and/or principal component analysis (PCA) to define the optimal decision boundary.

In this case, the decision boundary could be defined as the perpendicular to the line of best fit through the combined data. An illustration of this is given in Figure 1.6; the line of best fit is illustrated by the "dashed" and "dotted" line. This line of best fit is used to redefine the decision boundary as a perpendicular to it. In this example, the final system defined by the decision boundary is likely to be more robust than the original system because it minimizes the error of regression and yet is still compatible with the support vector decision zone.

Figures 1.3, 1.4, 1.5, and 1.6 provide some possibilities for modestly improving the support vector. However, it is more important to address less ideal systems in which the support vector cannot provide error-free classification. As noted in Figure 5.3 of the Marsland (2009) reference, one approach is to immediately assess each of the candidate margins for their errors: "If the classifier makes some errors, then the distance by which the points are over the border should be used to weight each error in order to decide how bad the classifier is." The equation for errors is itself turned into a quadratic programming problem, in order to simplify its computation. The error function to be minimized, however, incorporates an L1, or Manhattan, absolute difference, which may or may not be appropriate; for example, an L2 or Euclidean distance seems more appropriate based on the properties of least squares fitting described in Section 1.6.1.

The L2 distance is used in the example of Figures 1.7 and 1.8. In Figure 1.7, a problem in which the support vector decision boundary does not eliminate all classification errors is

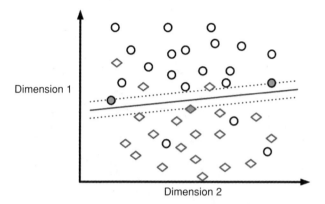

Figure 1.7 Decision boundary (solid line) with margin to either side (dotted lines) as defined by the support vector (filled circles and diamond). Here, there are six misclassifications

illustrated. In fact, 6 of 41 samples are misclassified. The support vector, however, provides a decision boundary and maximized margin.

Next, the entire set of data in Figure 1.7 is used in combination with the support vector decision zone to create a new decision boundary that has minimal squared error for the misclassified samples as a sum of their L2 values from the boundary. Figure 1.8 provides the optimum decision boundary that is compatible with the support vector range of allowable values as introduced in Figure 1.4. This can be compared with the decision boundary that minimizes the sum of L2 error distances overall in Figure 1.9.

Through this process, a boundary-based approach, the SVM, has been modified to be compatible with a population-based, "bottom-up," data-driven approach. Whether this approach provides a real advantage to SVM approaches compared to, for example, the kernel trick described next, remains to be seen. Certainly, it should be an area of research for the machine learning community. However, it does appear likely that such an approach, in minimizing the

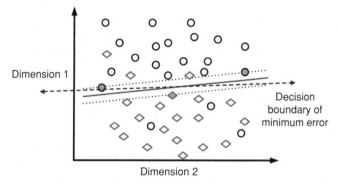

Figure 1.8 Decision boundary providing least squares error for the six misclassified samples of Figure 1.7 (dashed line), which does not deviate from the decision zone of the support vector

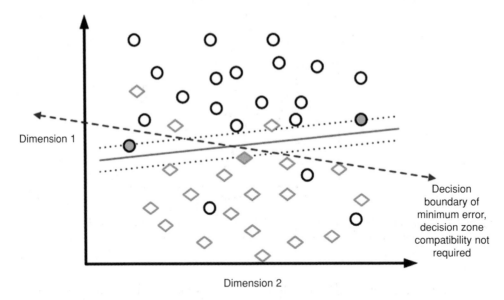

Figure 1.9 Decision boundary providing least squares error for the six misclassified samples of Figure 1.7 (dashed line), where it is allowed to deviated from the decision zone of the support vector

overall squared error and in maximizing the margin around the decision boundary, may be more robust.

Figures 1.7, 1.8, and 1.9 illustrate a coordinate space in which no linear separation of the class is possible. For linear separation, the following two equations must hold:

$$\text{If } (y > mx + b), y \in \text{Class}_A;$$

$$\text{If } (y < mx + b), y \in \text{Class}_B.$$

The kernel trick is introduced to find a linear decision boundary for separation of these two classes, since in real-world classification problems, the class separation is rarely as clean as shown in, for example, Figure 1.3. More commonly, the initial class separation, or decision boundary, is a Gerrymandered curve along the lines of that shown in Figure 1.10a. With the kernel trick we transform the decision boundary—albeit in a transformed coordinate system made possible by the increased dimensionality of the kernel—into a linear boundary as shown in Figure 1.10b.

Why is the transformation to a linear boundary important? My answer here is not based on SVM theory, as I am unaware of research to date that has addressed, let alone solved, this question. My interpretation is that, for one thing, it makes the boundary a minimal length pathway through the new dimension space, which minimizes the "error surface." For another, this single pathway boundary—when the right kernel is selected—minimizes overfitting, since the boundary is described by a surface the same order as the dimensionality of the class space.

In concluding this section on SVMs, I would like to point out their central role in current classification theory and practice. The kernel trick has, in some ways, allowed a classifier—the

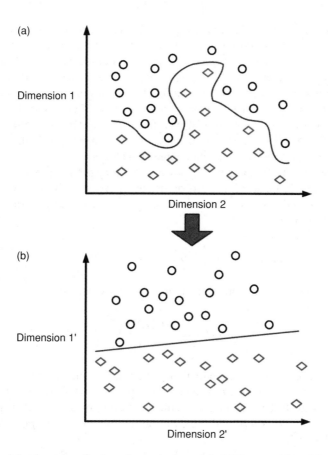

Figure 1.10 Decision boundary for two classes in the original 2D space (a) and in the transformed space (b) created within the kernel. In (b), the decision boundary is linear, which generally minimizes overfitting and differences between training and testing error rates

support vector—which might otherwise not be very effective for a given problem to become very effective when one or more kernels are utilized. Among these, determining the best results among the set of {linear, polynomial, radial basis, Fisher, Gaussian} kernels provide a good repertoire of choices that will generally result in one or more highly accurate SVM + kernel systems. For more on kernel engineering, please see Shawe-Taylor and Cristianini (2004).

1.6.3 Probability

Probability methods are important for a wide range of intelligent system tasks, including clustering, classification, and the determination of variance and covariance. Clustering is especially important to estimate the relative sizes and distribution of the classes of data, which in turn can be used to decide what, if any, transformation should be performed on each distribution before downstream analysis.

Probabilistic means of clustering data are as simple as the expectation maximization (EM)-based k-means clustering approach. The k-means clustering approach performs iterations consisting of two steps: updating the cluster assignment of the samples based on their location to the nearest cluster center, followed by updating the cluster centers based on the set of samples now belonging to them. Though simple, this algorithm can be readily modified to accept constraints, such as (1) having clusters with equal number of samples; (2) weighting different features used to determine the distances from each cluster center differently based on their predicted utility for later classification, regression, and so on; and (3) allowing predefined clusters to be accommodated concomitant to unknown clusters.

Another important set of approaches is based on joint distributions and conditional independence of features. In this case, the probability of an event A, given the occurrence of an event B, is the same as the probability of an event B, given the occurrence of an event A. The equation describing this is

$$p(A \mid B)^* p(B) = p(B \mid A)^* p(A),$$

from which the general form of Bayes' theorem is directly obtained:

$$p(A \mid B) = \frac{p(B \mid A)^* p(A)}{p(B)},$$

where, as noted above, $p(A \mid B)$ is the probability of event A given that the event B has already occurred. In many cases, we wish $p(A)$ and $p(B)$ to be very low (rarity of identifying event or events) but $p(A \mid B)$ to be very high (specificity of event or events). Rearranging Bayes' theorem to show the ratio of $p(A)$ and $p(B)$, we can see that the ratio of $p(A)/p(B)$ is equal to the ratio of $p(A \mid B)/(B \mid A)$:

$$\frac{p(A)}{p(B)} = \frac{p(A \mid B)}{p(B \mid A)}.$$

In many machine intelligence problems, we are concerned with events that occur with some probability when one or more of multiple other events occur. In fact, we may wish to find the probability of event A when considering all of a set of mutually exclusive events, denoted as B here. We thus rearrange to solve for $p(A)$:

$$p(A) = \frac{p(A \mid B)^* p(B)}{p(B \mid A)}.$$

Let us now consider the case where B can occur as one of N independent outcomes; that is, the sum of all $p(B_i) = 1.0$. We then obtain the following for $p(A)$:

$$p(A) = \frac{\sum\limits_{i=1,\ldots,N} p(A \mid B_i)^* p(B_i)}{\sum\limits_{i=1,\ldots,N} p(B_i \mid A)}.$$

This latter equation also governs the probability of event A for any subset of the outcomes of B; for example, under constrained situations in which one or more of the B_i is not allowed to, or cannot, occur. Why is this important? Because in many cases, we may not have a good estimate for the probability of event A. Event A, meanwhile, may be a triggering event for a specific downstream task, such as using a specific classifier tuned for the input data when the triggering event occurs. Thus, the rearranged generalized Bayesian equation allows us to compare different dependent events, here event A, for their overall probabilities against a mutually exclusive set of events, B.

1.6.4 Unsupervised Learning

Unsupervised learning techniques are concerned with clustering (aggregating like elements in a group), input space data distribution, or other operations that can be performed without training data. Training data is also referred to as ground-truthed data and as labeled data. With such unlabelled data, we have to rely on the structure of the data itself to infer patterns; that is, unsupervised learning.

One type of unsupervised learning is the k-means clustering approach described in Section 1.6.3. Related to the k-means clustering approach is the Gaussian mixture model (GMM), in which we know the number of classes that are in a data set, but have no labeled data. The GMM assumes that the data set is a function comprising the sum of multiple Gaussians, one each corresponding to the individual classes:

$$f(x) = \sum_{c=1}^{C} w_c G(x : \mu_c, X_c),$$

where x is the set of features, $f(x)$ is the output based on the set of features, C is the number of classes, w_c is the weight assigned to class c, and $G(x : \mu_c, X_c)$ is a C-order Gaussian function with mean μ_c and covariance matrix X_c. The classifier consists, after the model is built (also using an EM approach, per Section 9.2.2 of Bishop (2006)), of finding the maximum of the following:

$$\max_{k} \left\{ p(x_a \in k : k = 1, \ldots, C) = \frac{w_k G(x_a : \mu_k, X_k)}{\sum_{c=1}^{C} w_c G(x : \mu_c, X_c)} \right\}.$$

Weighting the classes is governed by the constraint

$$\sum_{k=1}^{C} w_k = 1.0.$$

For the EM approach to determining the GMM, the weights of the individual classes are proportional to the number of samples assigned to the individual classes. This is called the *average responsibility approach*, since the weighting is proportional to the class' responsibility in explaining the input data set. It should be noted that the individual features in the feature vector **x** can also be weighted differentially. There will be much more to say on this in later sections of the book.

Self-organizing feature maps (SOMs) (Kohonen, 1982) are an unsupervised learning approach to describe the topology of input data. This machine learning approach can be used to provide a lower-dimensional representation of the data in addition to its density estimation value. It should be noted that SOMs can be very useful at the front end of unsupervised clustering problems, as well. As will be discussed in detail in Chapter 8 and elsewhere in this book, one of the primary considerations in parallel system design is to decide how best to aggregate processes to run in parallel. For large intelligence systems, this may include deciding which aggregate classes to form first and from this determine the structure of a decision tree.

1.6.5   Dimensionality Reduction

SOMs, then, can be used for dimensionality reduction. Nearest neighbor techniques can also be used to reduce the dimensionality of a data set. This is important to classification problems too, since it is usually advantageous to eliminate isolated samples or even small clusters to avoid overtraining, the addition of ectopic classes, and other types of "output noise." Figure 1.11 illustrates how the 3-nearest neighbor approach is used to assign an unassigned sample to one of two classes. The process involved is the same as that used for assigning samples when the trained system is deployed.

The example of Figure 1.11 reduces the number of classifiers. More traditionally, however, *dimensionality reduction* is concerned with the removal of features from the feature space. In order to reduce the number of features, Marsland (2009) outlines three primary approaches:

1. Feature selection
2. Feature derivation
3. Clustering.

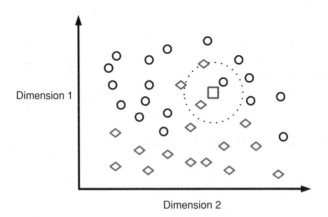

Figure 1.11   *k*-nearest neighbor classification applied to a formerly unclassified (square). The dotted circle has a radius sufficient to incorporate the nearest three neighbors. Based on the results, the square is assigned to the same class as the diamonds

Feature selection is a white box means of eliminating features inasmuch as we use our awareness of the relationship between a feature and the system output—for example, through correlation, confusion matrices, and so on—to decide whether or not to eliminate a feature. We can triage features through some relatively simple algorithms, including the following:

1. Select the feature with the lowest correlation with the output (alternatively, select the feature with the lowest accuracy output).
2. Remove the feature.
3. If system performance improves, drop the feature and return to Step 1.
4. If system performance does not improve, either terminate feature selection or jump to the next most poorly correlated (or accurate) feature and repeat Steps 1–3.

When employing Step 4 in this algorithm, we may choose to allow the "stack" to only get to a certain depth (say, 10% of the number of features) before terminating the search. Other forms of feature selection include the decision tree approaches and linear discriminant analysis (LDA). LDA is concerned with projecting to a lower-order dimension such that the samples in the higher-order dimension are optimally separated in the reduced dimension space. The approach is effectively half of an F-score approach: we wish to maximize the between-class scatter. The effect is similar to that discussed with regard to Figures 1.5 and 1.6. The "dashed" and "dotted" line in Figure 1.6, in fact, could be the output of an LDA, allowing the two dimensions in Figure 1.6 to be reduced to a single projected line.

Harkening back to the previous section, we observe that the LDA is performed on labeled data. However, there are several methods for dimensionality reduction that do not require labeled data. The first, PCA, is the process by which a matrix is transformed into its eigenvectors and eigenvalues; that is, orthogonal, uncorrelated components comprising a new set of dimensions in which the principal component is in the direction of maximum variance for the data set. Independent component analysis (ICA), on the other hand, assumes that the latent components are independent, and so strives to transform the input data space into a set of independent dimensions. Dimensionality reduction is readily achieved for PCA by dropping all of the eigenvectors whose eigenvalues are below a given threshold (Jolliffe, 1986), and for ICA based on the measurement of the mutual information of the components (Wang and Chang, 2006).

Feature derivation, a second method for dimensionality reduction, is concerned with transforming a (measured or original) feature set into another (derived or transformed) feature set. These new features can be combined features from the original space; for example, in imaging we may wish to combine red, green, and blue channel intensity into a *mean intensity* feature, which is simply the mathematical mean of the original three features. Note, however, that such transformations do not necessarily result in a reduced feature set immediately after transformation: in this example, we may replace the three features (red, green, and blue channel means) with three new ones (mean intensity, mean saturation, and mean hue) that provide us with a set better pruned using feature selection.

Both PCA and ICA, described above, perform feature derivation in addition to the feature selection. Each of them computes a principal component in the direction of maximum variance and derives the rest of the components based on orthogonality (PCA) or independence (ICA). Factor analysis (Dempster, Laird, and Rubin, 1977; Liu and Rubin, 1998) is another means of dimensionality reduction in which latent variables are discovered using an EM approach.

Multi-dimensional scaling (MDS) (Cox and Cox, 1994) and locally linear embedding (Roweis and Saul, 2000) are two other feature derivation processes.

The third method for dimensionality reduction, clustering, has been already addressed in light of k-means clustering, k-nearest neighbor (k-NN) clustering, and GMMs. As will be shown later in this chapter, a specific criterion, the F-score, can be used to decide whether or not to cluster.

Dimensionality reduction is important for several reasons. First off, in reducing the set of features on which regression, classification, and other machine learning tasks are based, it allows better compression of the data. Secondly, in reducing the correlation and/or dependency among the features, dimensionality reduction may make the system less sensitive to noise. Thirdly, dimensional reduction results in improved retrieval efficiency during querying.

1.6.6   Optimization and Search

Optimization is the process whereby, with a certain confidence, the best possible outcome for a system is obtained. Typically, a function of the output—for example, the cost of the system components, the cost of the processing, or the cost of correcting system errors—is to be optimized. For cost, the optimization is usually a minimization. Introductory calculus tells us that a smooth function can be effectively searched for its local optima using the gradient, or Jacobian operator. The steepest descent approach updates the vector $\mathbf{x}$ as follows:

$$\mathbf{x}_{k+1} = \mathbf{x}_k - \Delta_k \nabla f(\mathbf{x}_k),$$

where Δ_k is the distance to travel in the direction from $\mathbf{x}_k$ to $\mathbf{x}_{k+1}$ in order to reach a minimum. A related approach, which effectively uses the gradient descent method to create a trust region for a least squares approach, is the Levenberg–Marquardt algorithm (Levenberg, 1944). The conjugate gradient method is often a significant improvement on these methods, however, since it moves in conjugate directions sequentially—thereby avoiding consecutive small steps in the same direction.

For each of these approaches, one concern is that the gradient cannot move the vector $\mathbf{x}$ from the zone of a local optima. No method for searching other neighborhoods in the output space is provided. Thus, it is not unreasonable to combine such an optimization approach with a "location generator" that seeds an appropriate set of points in $\mathbf{x}$ to provide statistical confidence that an overall optimum has been achieved. This is illustrated in Figure 1.12, in which sufficient starting points exist to allow the finding of the overall optimum.

In later chapters of this book, first derivative approaches will largely focus on sensitivity analysis problems. However, it should be noted that the same concern applies to sensitivity analysis problems as do to the more general optimization problems. Namely, the input space must be seeded with a sufficient number of starting points to allow the overall optima to be found.

Another important machine intelligence system approach to optimization is search. The path to search is implicit in the grid pattern of seeded starting points in Figure 1.12. As the spacings between starting points, or nodes, in the grid become smaller, the odds of missing the overall optima also become smaller. Only optima with contours having effective radii of

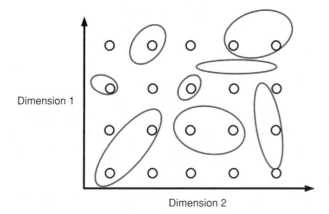

Dimension 1

Dimension 2

Figure 1.12   Example of an output space with sufficient seeding of starting points $\{x_j\}$. These are arranged in the grid (circles). The outer contours of regions for local optimization are shown in ovals. Half of the starting points belong to one of the local optima contours, and the average contour spans 1.25 of the starting points. It is highly unlikely that none of the starting points will reside inside the contour of the overall optimum

less than 70.7% of the internode distance could be unseeded. Assuming that the grid of seeded nodes is sufficient, therefore, the optima (or near-optima) can be found through an *exhaustive search* of the output values for all of the nodes. This search can be augmented by gradient methods as described above to find the optima around each of the nodes.

Another important use for search in machine intelligence is in optimizing pathways. Minimum pathways are important in network, traffic, and other multi-step processes. The benchmark for optimizing pathways, as for search, is exhaustive search, in which every possible pathway is attempted. Of course, this is hugely expensive computationally, even for a massive and massively parallel computing environment. The number of possible pathways, N_{PP}, for an N-node set of locations is given by

$$N_{\text{PP}} = \frac{(N-1)!}{2}.$$

As N increases to more than a trivial problem space, the number of pathways becomes unwieldy for the exhaustive search approach. As a consequence, effective means for searching a subset of the possible pathways must be selected. In Simske and Matthews (2004), several methods for selecting the next node in a pathway were given. One was termed the "lowest remaining distance" strategy, which is also known as the greedy search approach (Marsland, 2009). This approach often results in excessive distances for the last few node–node transitions. Another approach was termed the "centroid + clockwise traversal" method, which rotates around the centroid of all the nodes to the nearest—by angular value—node from the current node. This method is readily extended to a hybrid clustering/clockwise traversal approach that effectively introduces supernodes that are internally sequential. A final method mentioned in Marsland (2009) is the hill climbing approach, in which pairs of nodes are randomly swapped to see if improvements can be made over the best current set of pathways. This approach naturally leads to genetic algorithm approaches, which are described in Section 1.7.2.

1.7 Artificial Intelligence

In this overview chapter, artificial intelligence approaches are collectively viewed as machine reasoning systems in which the architecture of the connections is allowed to optimize without regard to a specific guiding formula or expert system rule set. Neural networks, which trade off architectural complexity with convergence speed, are overviewed first. Next, genetic algorithms are introduced and shown to be effective for searching a very large input space with a relatively small number of trials. We conclude this section with an overview of Markov methods, which are "memoryless" approaches in which the next state depends only on the current state.

1.7.1 Neural Networks

The term "neural network," or more correctly, "artificial neural network" (ANN), implies a system whose architecture is inspired by one or more physiological neural networks. While this may be the case in some instances, for the general case this is not an accurate representation of how ANNs are designed.

The human cerebral cortex, for example, is largely comprised six layers of neurons, with widely different relative layer thicknesses when comparing the different cortical lobes: frontal, parietal, temporal, and occipital (Kandel, Schwartz, and Jessell, 2000). This relatively structured architecture, however, is far different from that of the hippocampus—a brain structure associated with the incorporation of memory—or that of the cerebellum, a three-layered brain region containing more than half of the brain's neurons. The modal neuron in the brain makes thousands of synaptic connections to other neurons. These connections are characterized by the effects of released neurotransmitters on the postsynaptic neurons. These effects can be excitatory or inhibitory, synergistic or antagonistic, short-termed or long-termed, persistent or habituated, and even result in distinct effects on different postsynaptic neurons. Synaptic (and dendritic) projections can be local, into adjoining layers, or to distant brain regions. In other words, brain architecture is both complex and diverse.

Learning in the brain is also both complex and diverse. The efficacy of synapses, resulting in stronger or weaker influence on the postsynaptic neuron, can be altered through changes in neurotransmitter release, changes in neurotransmitter reuptake or removal from the synapse, changes in the morphology of the synapses, and changes in the number of synapses, depending on the strength and length of the reinforcement provided for a given connection. At a more macroscopic level, brain learning is also complex and diverse. Neural pathways are often carved rather than assembled—think of Michelangelo's *David* and not Picasso's *Baboon and Young*. This carving not only reinforces certain behaviors or preserves certain thoughts but also removes other behaviors and thoughts. This is in fact termed *neural sculpting*, and this too is not a uniform process throughout the brain; for example, it creates longer-lasting information pathways in the cerebral cortex than those in the hippocampus.

Let us compare these nature-made neural networks (NNN) to the artificial ones. For this comparison, the ANN of choice will be the multi-layer perceptron, or MLP, of Rumelhart,

Hinton, and McClelland (1986). This MLP connects an input layer to a hidden layer, and the hidden layer to the output layer, and comprises a three-layer network. The architecture of the MLP is far simpler than that of the NNNs described above in the following ways:

1. The ANN has a specific input and output layer. For an NNN, these would be functionally replaced by sensory (input) and motor (output) neurons, furthering the complexity of the NNN from that described above.
2. The ANN has only a single hidden layer that connects the input and output layers. The NNN, on the other hand, consists of three (cerebellum) or six (cerebral cortex) layers, or a much less structured, more highly interconnected architecture.
3. The number of connections in the MLP is very small. Typically, each input node, IN, connects to each hidden layer node, HLN, and each HLN in turn to each output node, ON. If the number of nodes in each of these three layers are N_{IN}, N_{HLN}, and N_{ON}, respectively, then the total number of connections, N_C, is

$$N_C = N_{HLN}(N_{IN} + N_{ON}),$$

for which the mean number of connections per node, $\mu_{C/N}$ is

$$\mu_{C/N} = \frac{N_{HLN}(N_{IN} + N_{ON})}{N_{IN} + N_{ON} + N_{HLN}}.$$

If the number of nodes at each of the three layers are equal, then

$$\mu_{C/N} = 0.67 N_{HLN}.$$

If, however, N_{IN} and N_{ON} are $\gg N_{HLN}$, as may be the case for imaging applications, then

$$\mu_{C/N} \approx N_{HLN}.$$

In other words, $\mu_{C/N}$ is $O(N)$. Unless the number of hidden nodes is abnormally large—the range of hidden nodes typically used is from roughly 20 to 100—this is approximately two orders of magnitude less than the mean number of connections per node in the human brain. As a better direct comparison, though, we should compare the ANN to a ganglion, that is, a peripheral nerve center. For this NNN, the mean number of connections per node—I should say per neuron here—is intermediate to that of the MLP and the human brain regions described above.

Regardless, we must keep in mind that the relationship between the mean number of connections and the complexity of a neural network—not to mention its capacity for learning—is nonlinear. Assuming that all synapses are independent and contribute relatively equally to learning and neural processing, the relationship between $\mu_{C/N}$ and ANN or NNN complexity is geometric.

4. The number of synapses is not equal to the number of connections in a living neural network, as it is in the ANN MLP. In an NNN, new synapses are grown during learning (Bailey and Kandel, 2008), allowing the same neuron not only to connect to new postsynaptic neurons but also to add new connections to existing postsynaptic connections.

5. Learning in an MLP consists of changing the weights on the connections between nodes. Learning in an NNN, as described above, is more complex. Here, learning includes the growth of new neurons, the growth of new synapses, and changes in the morphology and/or efficacy of synapses—along with the opposite of each of these in the case of neural sculpting.

6. Short-term and long-term memory also differ greatly when comparing the MLP and NNNs. In an MLP, the short-term learning is effectively the error feedback changing the weights of the connections. The long-term memory is the current state of the MLP, which is simply the set of weights on all of the connections. In an NNN, short-term and long-term memory are both quite complex. Synaptic plasticity, or the ability of chemical (neurotransmitter-based) synapses to change their efficacy, includes long-term potentiation (LTP) and increased synaptic interface surface area (ISISA). LTP is the process by which the amount of neurotransmitter released is altered—for example, greater amounts of excitatory neurotransmitter(s) are released into the synapse or lesser amounts of inhibitory neurotransmitter(s) are released into the synapse—or else the rate of removal of neurotransmitters from the synapse is altered. ISISA results in a greater relative postsynaptic potentials, enhancing the ability of the presynaptic neuron to affect the postsynaptic neuron. Both LTP and ISISA are associated with short-term learning, although ISISA can also be associated with long-term memory if the changes are significant—and permanent—enough. Of course, long-term memory in an NNN is even more complicated, since it includes the growth of new excitatory and/or inhibitory synapses, the projection of neuronal axons or dendrites into new regions, or even the growth of new neurons.

It is clear that ANNs such as the MLP are not as complex as the biological systems that (at least in theory) inspired them. It is certain that some ANN architecture will increase in complexity as their value for cognitive analysis is bettered, and better appreciated by the machine intelligence field. Even so, the general principles of the MLP are still quite likely to apply, and the various physiological attributes of NNNs are likely to be adopted slowly, and only in targeted applications.

The MLP is, in fact, simply an elaborate algorithm that is carried out on the structured sets of layers defined above. To initialize the MLP, an input vector is put into the INs. These nodes are arranged as a one-dimensional (1D) array, but 2D and 3D inputs are readily accommodated by simply mapping them to a 1D vector (e.g., the same way that 2D images are stored sequentially in memory, one raster line after the other). Weights along the connections are also randomly initiated, which contributes to the stochastic nature of ANN models. The input signals are fed forward through the network to the hidden layer and thence to the output layer. The state of the input signal, the weights along the connections, the sum of the weights at the next node, and the activation, or threshold, function (usually a sigmoid function) for determining whether the sum of weights causes firing of the hidden node, determine the set of states at the HLNs. The outputs of these nodes, the weights of the connections between HLNs and output layer nodes, and the same summing and activation approaches then determine the ON states.

Ground truthing or "target output" is then used to compute the error in the network. The expected, or targeted, output vector is compared to the actual output vector (the set of states of all ONs) and this error is then fed back into the network to update, in order, the hidden layer weights and then the input layer weights.

Each of these steps involves choosing certain system coefficients; for example, the activation function coefficient and the backward feedback coefficient. These coefficients are used to provide a trade-off between faster convergence time (preferred) and system convergence on a local optimum (not preferred). Increasing the learning rate (α) typically results in faster convergence, and also more likely results in the identification of a local optimum. Slower learning rates force the algorithm to inspect the topology at a higher resolution in exchange for slower convergence. This trade-off is, of course, problem dependent.

In this text, ANNs will be viewed as black boxes that produce a specific output, which we then use as input in a larger parallel processing intelligent system. For example, several neural networks can be run in parallel with different initial conditions to see if there is a preferred system convergence state or simply to provide multiple optima candidates. In this way, ANNs can be used in a method somewhat analogous to genetic algorithms, as described in the Section 1.7.2. Initial conditions based on an expert, or expert system, estimate may also converge faster and still provide an output that is optimal or very close to optimal. Regardless, ANNs are a very important type of algorithm to use in combination with other machine intelligence algorithms because they tend to work differently than many of the other common machine learning approaches, such as SVMs, boosting, genetic approaches, expert systems, and Bayesian systems. As such, they are a useful (and easy to implement) means of adding another intelligent system to a problem space. The advantages of this for the meta-algorithmic patterns outlined and elaborated in Chapters 6–9 will become obvious. The fact that there are a plethora of open source neural network software libraries is icing on the cake.

Note that in some ways, if I may reach a bit here, an ANN is analogous to the kernel method introduced in Section 1.6.2. The ANN—through the use of the hidden layer and a focus on optimizing connections between layers that are difficult, if not impossible, to map to any mathematical relationship between input and output—is able to solve problems without the solution architect having to really understand how the solution has been effected. In effect, a larger-dimensional solution space is created (the number of connections are much greater than the number of INs and ONs), which the MLP algorithm efficiently searches to create an intelligent system. The kernel trick also increases the dimensionality of the search space and in so doing allows a linear boundary to be found in that space that maps to a more complex—and more accurate—boundary in the lower-dimensional problem space.

1.7.2 Genetic Algorithms

Genetic algorithms (Goldberg, 1989) are efficient search approaches based on the principles and kinetics of natural selection and heredity. Specifically, genetic algorithms are based on the principle of a gene, which comprises a string of DNA or RNA triplets that code for a polypeptide (protein) chain or an RNA chain. For genetic algorithms, these triplets are represented with individual bits or loci, meaning that the sequences that can be optimized by the genetic algorithm can be binary in nature.

Any binary representation of a function to be searched can be handled by a genetic algorithm. In its simplest form, a genetic algorithm consists of the following elements and operations:

1. A set of strings that comprise a small subset of the possible search strings.
2. A means to populate this subset of search strings.
3. A fitness measurement for the search strings.

4. A means to select a next iteration of search strings, which includes replication, near-replication, crossover, mutation, and random selection.
5. A means of terminating the algorithm after a certain number of iterations of Steps 2–4.

These are relatively simple steps, and are well designed for searching over a specific output function.

As an example, suppose we wish to determine the best set of 10 query terms to include in a particular search. There are $2^{10} - 1 = 1023$ possible query sets (the null term query is disallowed), and we wish to get an acceptable recommended search query from just a small set of 50 search query candidates, called genes. We decided to use 10 randomly created genes, and then proceed from there with four rounds of gene shuffling. The initial set of 10 genes was determined by using a random number generator (RNG), with returned RNG values in the range [0.5, 1.0] being made 1s, and RNG values in the range [0.0, 0.5) being made 0s: example genes include {0001001110}, {1011011011}, and {0100001101}. Next, the fitness measurement was determined based on the relative success of finding the correct document using the search query based on each gene. The three genes shown have four, seven, and four terms, respectively (since the 1s indicate including the term). The fitness of the first and second queries (based on the genes) are weighted much higher than the third, and so these two are chosen—whereas the third is not—as the basis for the next set of 10 queries. A mutation rate of 5% and crossover rate of 90% is chosen. The first sequence {0001001110} is randomly assigned a crossover at the fourth bit with the second sequence {1011011011}, resulting in two new genes {0001011011} and {1011001110}. Next, one of these bits is mutated: the eighth bit of the second new gene, making it {1011001010}. Similar crossover and mutations led to an additional eight set of genes, for a total of 10 new genes in the second iteration. By the end of five such iterations, we have exhausted the 50 search query candidates, and the best quality search is, for example, 98.5% as good as the best single search out of all 1023 possible searches. Randomly selecting 50 queries would generally give a lower mean percentage of the best possible value, since the genetic algorithm preferentially reproduces the higher fitness genes, crossovers of two of them, and minor mutations on them.

Importantly, this set of steps, with only minor modification, can be used for other types of search, optimization, and classification. In a previous work, a genetic algorithm was used for a variant of the traveling salesman problem (TSP), in which a path through N locations is to be minimized (Simske and Matthews, 2004). In this case, the five elements and operations described above are implemented as follows:

1. A string of loci, numbered consecutively, each of which represents one particular location. For example, if L locations need to be passed through, then the locations may be listed alphabetically and assigned the values 1, 2, ..., L.
2. Five different means of creating initial strings were used. Each of them involved the concept of next-node probabilities. These probabilities are an array of probabilities, normalized to sum to 1.0, which combined give the statistical likelihood of selecting each of the remaining nodes. These probabilities are recomputed each time a node is selected (using an RNG that is mapped into the probabilities). (a) The first methods used include naïve, in which all of the next-node probabilities are identical,

$$p_{i \in S(\text{unassigned})} = \frac{1}{N_{i \in S(\text{unassigned})}}.$$

The probabilities are equal for every unassigned node in the particular path being initialized. The set of unassigned nodes are in the set S(unassigned). (b) The second method assigns the next-node probabilities by relative distance,

$$P_{i \in S(\text{unassigned})} = \frac{D(\text{node}_i - \text{node}_{\text{current}})}{\sum\limits_{k \in S(\text{unassigned})} D(\text{node}_k - \text{node}_{\text{current}})},$$

where $D(*)$ is the Euclidean distance. (c) The third method assigns the next-node probabilities by cluster. Effectively, this reduces the problem space to a search within a cluster until all of the locations in a cluster are visited, and then the next cluster is appended. This is in effect a hybrid approach, and depends on an appropriate clustering algorithm—such as k-means clustering—to be performed first. The constraint can be relaxed such that the weighting within the clusters is simply relatively higher than the weighting outside of the clusters in comparison to what would otherwise be assigned by methods (a), (b), (d), or (e). Speaking of which, method (d) creates the next-node probabilities based on the direction of travel. If the current direction of the travel is clockwise (or counterclockwise) around the centroid of all the locations, then the next nodes in the clockwise (or counterclockwise) direction being traversed can be weighted more highly relative to its otherwise-assigned weight. The final method (e) weights the next-node probability based on the inverse of the distance; that is,

$$P_{i \in S(\text{unassigned})} \propto \frac{1}{D(\text{node}_i - \text{node}_{\text{current}})}.$$

Taking into account all remaining unassigned nodes,

$$P_{i \in S(\text{unassigned})} = \frac{\sum\limits_{k \in S(\text{unassigned})} \dfrac{1}{D(\text{node}_k - \text{node}_{\text{current}})}}{D(\text{node}_i - \text{node}_{\text{current}})}.$$

3. The fitness measurement for each path is trivial to compute: it is the sum of the distances from the starting node through all the other nodes and back to the starting node. The results were not particularly surprising. As reported in Simske and Matthews (2004), when applying the inverse distance approach to both initialization and crossover (discussed below), excellent fitness and rapid convergence were observed.
4. The means to select a next iteration of search strings involved crossover, which involves simple reversal of a length of nodes: for example, if nine cities are visited in order from $\{4,2,5,8,3,9,1,7,6\}$ and crossover is dictated to occur from nodes 3 to 7, then the order of cities visited becomes $\{4,2,1,9,3,8,5,7,6\}$. This effectively swaps part of one gene with a previous version of itself, and leads to generally much more change than a point mutation. Mutation in such a situation includes substitution, which for the example results in the swapping of nodes 3 and 7, producing the order $\{4,2,1,8,3,9,5,7,6\}$.
5. The algorithm described in points 2–4 is terminated after a certain number of iterations in which no overall improvement in fitness is observed.

The results of applying these steps to the TSP were immediate. In particular, the inverse distance approach resulted in much faster convergence and much less residual error after convergence than the other methods. However, in the earlier work (Simske and Matthews, 2004), I did not have the opportunity to experiment on all five of the different means of creating initial strings. With the type of crossover used, it turns out that method (b) sometimes works, especially with asymmetric systems, since it helps create genes that are exploring new local optima. This overcomes one of the most common problems in genetic algorithms; namely, the difficulty of creating new searches that escape from the current local optimum. The mutation type implemented also helps prevent convergence on what is only a local optimum. Also, for both symmetric and asymmetric TSPs, the clustering method (c) works quite well when the clustering is attached to an F-score metric such as described later in this chapter.

The TSP solutions investigated here are a simplification of the real complexities involved in directing traffic. In real-world applications, the node–node transition costs are based on, among other things, both the distance and the traffic congestion between nodes. Other pragmatic considerations include the intelligent clustering of nodes based on similarity in the navigator's intents at each destination (e.g., cities to visit may be clustered by language in Europe), maximum distance preferred between nodes (so the traveler will not have too much driving for a day), and other pragmatic considerations.

Like neural networks described in the previous section, genetic algorithms are used to search a very wide space in an efficient manner. Genetic algorithms are not expected to find the global optima efficiently, but are expected to find something close to the overall optimum rather quickly. Thus, certain genetic algorithm approaches can provide the types of machine intelligence plasticity proffered by ANNs. One easy way to see a relationship is to consider how the initial conditions on the connections in a neural network are set. Suppose we choose to create a population of MLPs as introduced in Section 1.7.2. We may then initialize the MLPs using an RNG and score each MLP after just a small number of iterations using a fitness metric. The genetic "crossover" can then be used on the initial weightings to initialize a new MLP that can be deployed next, and this will presumably lead to a more optimal MLP more quickly than running a single MLP for many iterations.

Genetic algorithms have long been a favorite adaptive stochastic optimization algorithm, with the analogies to biological evolution often touted for robustness and convergence advantages. Another evolutionary analogy has often been overlooked: that of punctuated equilibrium (Gould and Eldredge, 1977). Punctuated equilibria comprise static periods in the evolution of a sequence of species that are interrupted by relatively short periods of rapid change. Genetic algorithms run the risk of remaining in stasis unless there is an impetus sufficient to effect change. As illustrated herein, this impetus can stem from changes in initialization, crossover, and/or mutations. It can also, importantly, stem from hybridization of a genetic algorithmic approach with that of clustering or other techniques.

1.7.3   Markov Models

Markov models are an important form of reinforcement learning (Marsland, 2009). This type of learning is based on knowing the right answer only, but not the correct means to arrive at it. This is analogous to many genetic algorithmic and ANN approaches, as described in the previous two sections. Markov models, however, have a very simple architecture. At each of the possible states in a Markov process, the conditional probability distribution for all possible next states is only dependent on the current state.

If the system state is fully observable, a Markov chain is used. If the system state is only partially observable, then a hidden Markov model (HMM) is used. The general Markov model is described by a joint probability model in which the probability of a sequence of S events is based on the conditional probabilities leading to each of the states $\{o_1, o_2, \ldots, o_S\}$. Thus, the probability of obtaining S specific outputs in this order is given by

$$p(o_1, o_2, \ldots, o_S) = \prod_{k=2}^{S} p(o_k \mid o_1, o_2, \ldots, o_{k-1}).$$

In most cases, the output set is too large, the relevancy of the many-termed conditional distributions is too low, and/or the training set is too small for all of these probabilities to be calculated. In most applications, a first-, second-, or third-order Markov model suffices. For a first-order model, the probability for the state sequence $\{o_1, o_2, \ldots, o_S\}$ is, therefore,

$$p(o_1, o_2, \ldots, o_S) = p(o_1) \prod_{k=2}^{S} p(o_k \mid o_{k-1}).$$

A first-order Markov model can be applied readily to classifier output. Examples of different states include characters in text recognition problems, phonemes in speech recognition problems, and different classes in any general classification problem. As an illustration, in Figure 1.13 the transition diagram associated with a three-state system is given. The conditional probabilities are first-order; that is, the probability of the next state is conditional only on the current state.

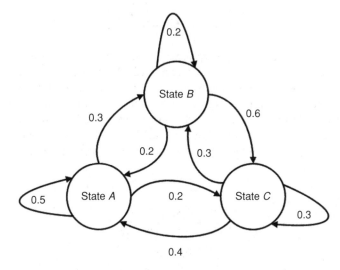

Figure 1.13 Transition diagram for a three-state Markov model. The model is first-order as all of the state transition probabilities depend only on the current state. The associated probabilities are indicated by the numerical values associated with the arrowed arcs. These probabilities are used to create the transition matrix, **T**, as described in the text

Extracting the transition probabilities from Figure 1.13, we obtain the transition matrix, **T**, given here:

$$
\begin{array}{c}
\text{State} \\
n-1
\end{array}
\begin{array}{cc}
 & \begin{array}{ccc} & \text{State} & n \\ A & B & C \end{array} \\
\begin{array}{c} A \\ B \\ C \end{array}
&
\left[
\begin{array}{ccc}
0.5 & 0.3 & 0.2 \\
0.2 & 0.2 & 0.6 \\
0.4 & 0.3 & 0.3
\end{array}
\right]
\end{array}.
$$

This is a very convenient matrix to use for state probabilities. For the probability equation above, suppose we wish to compute probability of starting at state A and three transitions later ending at state C. Then

$$
p(A, X_1, X_2, C) = \sum_{X_1 \in \{A,B,C\}} \sum_{X_2 \in \{A,B,C\}} [p(C \mid X_2)p(X_2 \mid X_1)p(X_1 \mid A)].
$$

The first of these is $p(C \mid A)p(A \mid A)p(A \mid A)$, or $(0.2)(0.5)(0.5) = 0.050$. Summing all nine of these probabilities, we obtain

$$
\begin{aligned}
p(A, X_1, X_2, C) = {} & 0.050 + 0.090 + 0.030 + 0.012 + 0.036 \\
& + 0.054 + 0.016 + 0.036 + 0.018 = 0.342.
\end{aligned}
$$

Similarly, we can obtain the following probabilities:

$$
\begin{aligned}
p(B, X_1, X_2, C) &= 0.346, \\
p(C, X_1, X_2, C) &= 0.343.
\end{aligned}
$$

Perhaps not surprisingly, these values are all very similar, meaning that the state three cycles before is more or less unrecoverable for such a first-order Markov process. One would, based on these results, conclude that the system is in state C roughly 34.4% of the time.

I wrote software to perform millions of iterations on the states in this system, using an RNG to decide the state transitions. In fact, after 3×10^7 iterations, I found that the system spent 38.4% of its time in state A, 27.3% of its time in state B, and 34.3% of its time in state C. This result makes sense when we sum the columns in the transition matrix as shown:

$$
\begin{array}{c}
\text{State} \\
n-1
\end{array}
\begin{array}{cc}
 & \begin{array}{ccc} & \text{State} & n \\ A & B & C \end{array} \\
\begin{array}{c} A \\ B \\ C \end{array}
&
\left[
\begin{array}{ccc}
0.5 & 0.3 & 0.2 \\
0.2 & 0.2 & 0.6 \\
0.4 & 0.3 & 0.3
\end{array}
\right] \\
& \begin{array}{ccc} \hline 1.1 & 0.8 & 1.1 \end{array}
\end{array}.
$$

Based on these sums alone, we expect the system to spend 36.3% of its time in state A, 26.3% of its time in state B, and 36.3% of its time in state C. Further eyeballing, we can see that state A is more likely to stay in state A (50%) compared to state C (30%), and thus the overall 38.4–34.3% results observed, favoring state A over state C, also make sense.

The state sequences described herein are often captured in a lattice or trellis diagram. These diagrams are especially helpful, in my experience, for understanding the optimal pathways, such as the max-sum optimum as identified by the Viterbi algorithm (Viterbi, 1967).

Markov models with large output sets, farsighted conditional distributions, and insufficient training sets are unlikely to have reliable conditional probability values to use. The same caution should be noted for Bayesian networks, introduced in Section 1.6.3.

1.8 Data Mining/Knowledge Discovery

Data mining, often referred to as "knowledge discovery," brings together some of the algorithms and system approaches outlined in earlier parts of this chapter; in particular, probability, statistical NLP, and artificial intelligence. This is an important field to introduce here since it is so highly dependent not just on the algorithms used, but on how the algorithms interact with the hardware used to house the data. Data mining is tightly coupled to the type of data to be analyzed as well as the database upon which it performs search and analysis. For example, column-oriented storage organization significantly accelerates sequential record access within a database. However, this approach does not help for so-called object-oriented database (OODB) operations: these include commonplace transactional operations including refresh/update, delete, insert, and single-record search and retrieval. For OODB-like data mining, linked lists and other modern addressing and indexing approaches are more relevant.

In the end, data mining is the extraction of data from a data store (usually a database) for the purpose of upgrading its content: this can mean the creation of meta-data, the creation of new data structures, or raw analytics (data statistics). After this new data is created, it too should be represented in such a way as to optimize its downstream search and data mining on the given database, using the given data management system, and using the salient access protocols and structures. Thus, an important aspect of data mining is not just the updated content, but how the updated content will be represented and accessed thereafter.

Whereas genetic algorithms were shown to be associated with search—finding existing information from a large set of data—data mining is associated with discovery; that is, the finding of new content. Data mining is therefore analysis of large data sets in order to discover one or more of the following:

1. Patterns within and between large or multiple data sets. This *analysis of content* is designed to uncover relationships within data sets, including temporal relationships, occurrence frequencies, and other statistics of interest. The patterns can be based on model fitting (templated data discovery) or can be open-ended (statistical data discovery).
2. Unusual data, data sets, or data structures. This *analysis of abnormal content* is termed anomaly detection, and can include analysis of context. Data outside the range normally associated with a particular event will be tagged as anomalous, and will trigger an associated response (notification, corrective response, etc.).
3. Data dependencies, wherein terms co-occurring are identified and used to define association rules. These rules are used extensively in suggestive selling, advertising, promotional

pricing, and web usage mining. Combined, these rules are used to profile a person, group, or other user pool, and comprise *analysis of context* and/or *analysis of usage*.

Data mining is usually not the end of the analysis of the data. In many case, the analysis of content, context, and/or usage is the input to secondary, more specific analysis. For example, the data dependencies can be used to comprise a predictive model for later behavior. This predictive model may be used in synchrony with other input—for example, business policies, operational rules, personal knowledge, and business intelligence—to help define a decision support system that is used in the appropriate business context to more effectively solve problems and make better-informed decisions.

Knowledge extraction (KE), associated with data mining, is concerned with the repurposing of the data extracted for use in downstream processes. As such, KE is concerned with mapping mined data to a particular schema or structure that can be used in other contexts. Both structured and unstructured data are remapped into an ontology or taxonomy that can be more effectively incorporated into other decisioning processes. A key point here is that data mining and KE are more effective when they are part of a bigger, hybrid system comprising two or more machine intelligence algorithms, systems, and engines.

1.9 Classification

In many ways, classification is the central subject in intelligent systems, including the most sophisticated one: the human brain. Classification is the systematic placement of objects—concrete or abstract—into categories. Classification can be performed on struc-tured data (in which the categories are known beforehand) or on unstructured data that has previously undergone cluster analysis (in which case *clustering* is performed first and later these clusters are associated with categories or otherwise assigned/combined).

SVMs are powerful classifiers, and they are overviewed in Section 1.6.2. SVMs are boundary-based approaches, which by definition focus on the *support vector*, or the set of samples that define the border zone between multiple classes. An example of a boundary-based classifier is given in Figures 1.14, 1.15, 1.16, and 1.17. In Figure 1.14, the 2D data is shown, the two dimensions are the two features measured for each sample. There is no clear linear boundary between the data.

Figure 1.15 shows the results of applying a boundary-based classification on the data in Figure 1.14. Here the boundary is loosely fit (e.g., a cubic spline fit).

Figure 1.16 shows a tightly fit decision boundary (TFDB) for the same data set. Here the boundary is able to avoid misclassification of any of the training data. This can be achieved by several of the methods described earlier in the chapter, including the SVM with the kernel trick and ANNs.

Figure 1.17 illustrates the samples that are part of the support vector for the boundary in Figure 1.16. More than half of the samples in the set are used for the definition of the vector space. As a consequence, the decision boundary in Figure 1.17 is likely fit too closely with the training data. The decision boundary created in Figure 1.16 will likely perform as well or better on test data.

Other classification approaches that are focused on decision boundaries are linear classifiers—which distinguish classes based on linear combination of the features (Simske,

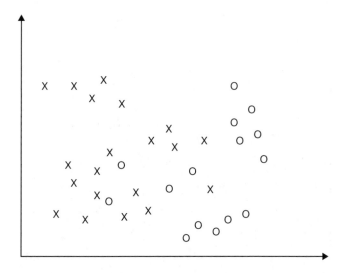

Figure 1.14   Simple 2D sample set ($n_X = 20$, $n_O = 15$) to be classified

Li, and Aronoff, 2005)—and quadratic classifiers, which separate classes based on a quadric surface.

All classifiers benefit from the availability of training data. For example, in Figures 1.14, 1.15, 1.16, and 1.17, if the amount of training data were greatly increased, we would have a better means of comparing the deployment accuracy of the two decision boundaries, shown in Figures 1.15 and 1.16. Figure 1.18 provides such a comparison through the addition of 55 more samples, which are considered a *validation set*. After adding the validation set data samples, the loosely fit decision boundary (LFDB) of Figure 1.15 has 87.8% accuracy on

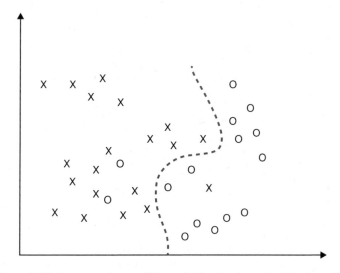

Figure 1.15   Data sample set of Figure 1.14 with a loose decision boundary

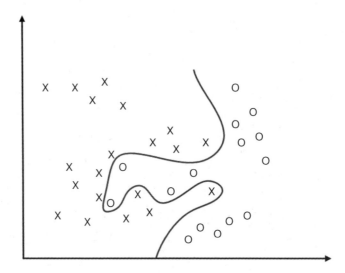

Figure 1.16 Data sample of Figure 1.14 with a tightly fit (overfit?) decision boundary

the combined training + validation data, slightly but statistically nonsignificantly greater than the 86.7% accuracy on the combined training + validation data observed for the TFDB of Figure 1.16. This is in spite of 100.0% accuracy when using the TFDB on the training set compared to the lower 91.4% accuracy when using the LFDB. Thus, the LFDB performed far better (85.5% accuracy) than the TFDB (78.2% accuracy) on the validation data, and is therefore the boundary of choice for deployment.

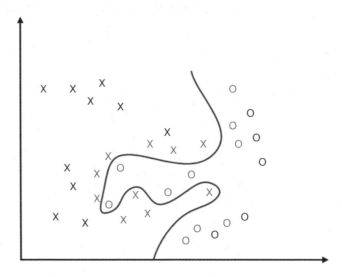

Figure 1.17 Data sample set of Figure 1.14 with the TFDB of Figure 1.16, with the support vector-relevant samples in light gray. In general, overfit decision boundaries will involve a higher than expected number of samples in the support vector

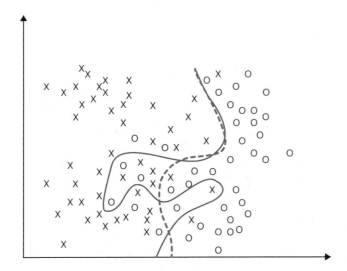

Figure 1.18 Data sample set of Figure 1.14 augmented with 30 more X and 25 more O samples ($n_X = 50$, $n_O = 40$). Here it becomes clear that the decision boundary of Figure 1.15 is most likely as good an approximation of the optimal decision boundary as is that of Figure 1.16. See text for details

Training data is the basis for the weightings used in boosting and the probabilities used in Bayesian classification and HMM classification. Both Bayesian- and HMM-based classifiers select as the assigned class the output with the highest summed probability.

Boosting classifiers such as AdaBoost (Freund and Schapire, 1996; Schapire, 1999) are ensemble classifiers that effectively used the consensus decisions of a (usually large) set of relatively inaccurate classifiers to provide a highly accurate decision. The random forest classifier (Breiman, 2001) is also an ensemble classifier, and it consists of many decision trees. After a large number of decision trees are formed, the most common output is accepted as the classification.

Clustering approaches are also used to create "bottom-up" classifiers; that is, classifiers that are based on aggregating rather than defining boundaries. Figure 1.19 shows a GMM clustering for the data of Figure 1.14, where five clusters are used to represent the "X" class, and four clusters are used to represent the "O" class.

The k-NN classifier is also an aggregation-based classification method. This classifier compares the test sample to the k nearest training samples and assigns the class as the modal neighbor class. Typically, k is in the range $\{1, \ldots, 5\}$, and is generally higher when the number of training samples is greater.

A helpful analytical measurement for clustering and for aggregation-based ("bottom-up") classifiers such as GMMs or simple linear combinations of Gaussian classifiers (Simske, Li, and Aronoff, 2005) is the F-score, which I use as the shorthand for the test statistic of an F-test such as is used in analysis of variance (ANOVA) and other statistical comparisons. The F-score is defined as

$$F = \frac{\text{MSE}_b}{\text{MSE}_w},$$

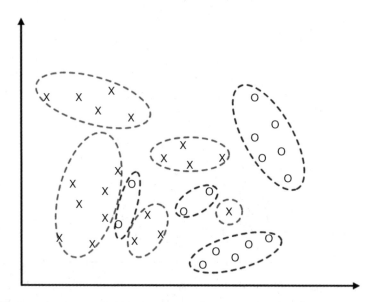

Figure 1.19 Data sample set of Figure 1.14 using a Gaussian mixture model for the classes. Here the number of Gaussians in the class are overfit, five for the X samples and four for the O samples

where MSE_b is the mean-squared error between the clusters and MSE_w is the mean-squared error within the clusters. Since the mean-squared errors are defined by the sum squared errors (SSE) and the degrees of freedom in the data sets, we rewrite the F-score as

$$F = \frac{SSE_b/df_b}{SSE_w/df_w},$$

where df are the degrees of freedom, generally 1 less than the members in a group. Next, we consider the values, V, of the data set. The mean value of cluster c, designated μ_c, is given by

$$\mu_c = \frac{\sum_{s=1}^{n(c)} V_{s,c}}{n(c)}.$$

Here, $V_{s,c}$ is sample s in cluster c. The number of samples in cluster c is $n(c)$ and the total number of clusters is n_c. From this, the value of MSE_w is given by the following:

$$MSE_w = \frac{\sum_{c=1}^{n_c} \sum_{s=1}^{n(c)} (V_{s,c} - \mu_c)^2}{\sum_{c=1}^{n_c} n(c) - n_c}.$$

The value of MSE_b is readily derived as follows:

$$MSE_b = \frac{\displaystyle\sum_{i=1}^{n_c} \sum_{j=i+1}^{n_c} (\mu_i - \mu_j)^2}{n_c(n_c - 1)/2}.$$

Using the fact that

$$\sum_{s=1}^{n(c)} (V_{s,c} - \mu_c)^2 = \sum_{s=1}^{n(c)} (V_{s,c}^2 - 2V_{s,c}\mu_c + \mu_c^2) = \sum_{s=1}^{n(c)} V_{s,c}^2 - 2\mu_c^2 n(c) + \sum_{s=1}^{n(c)} \mu_c^2$$

$$= \sum_{s=1}^{n(c)} V_{s,c}^2 - 2\mu_c^2 n(c) + n(c)\mu_c^2,$$

the value of MSE_b is more conveniently written as

$$MSE_b = \frac{\displaystyle\sum_{c=1}^{n_c} (\mu_c - \mu_\mu)^2}{n_c - 1}.$$

Here, μ_μ is the mean of the means, which is the mean of all the samples if all of the clusters have the same number of samples, but usually not so if they have different numbers of samples. In terms of reducing computational overhead, then, MSE_b is

$$MSE_b = \frac{\displaystyle\sum_{c=1}^{n_c} \mu_c^2 - n_c\mu_\mu}{n_c - 1}.$$

Similarly, the computations for MSE_w can be reduced to

$$MSE_w = \frac{\displaystyle\sum_{c=1}^{n_c} \sum_{s=1}^{n(c)} V_{s,c}^2 - \sum_{c=1}^{n_c} n(c)\mu_c^2}{\displaystyle\sum_{c=1}^{n_c} n(c) - n_c}.$$

We now consider how the ratio of MSE_b/MSE_w—that is, the F-score—can be used to optimize clustering using a simple example. Consider the simple data set in Figure 1.20, which have two possible clusterings. The dashed lines indicate the three clusters:

$$A = \{(1, 4), (3, 4), (2, 3)\},$$
$$B = \{(3, 1), (4, 2), (5, 1)\},$$
$$C = \{(8, 4), (9, 5), (10, 4)\}.$$

The solid lines indicate two clusters, $D = A \cup B$, and C.

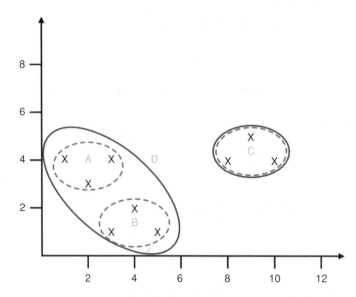

Figure 1.20   Data set for the F-score clustering example (see text). The dashed lines indicate a three-cluster representation, A, B, and C. The solid lines indicate a two-cluster representation, D = A ∪ B, and C

It is left to the reader to show that $F_{\{A,B,C\}} = 11.62$ and $F_{\{D,C\}} = 7.06$. From this, we determine that the three-cluster representation is optimal.

This example, while relatively trivial, illustrates the value of a simple metric—the F-score—to provide very important output—the clusters associated with a data set of any size. The F-score is of huge importance in parallelism by meta-algorithmics, as each of the clusters of data may be analyzed differently based on their different characteristics.

1.10   Recognition

Pattern recognition (Tveter, 1998) plays an important role in many intelligent systems, particularly signal processing and image understanding. The brief overview of recognition is placed here since it builds on many of the technologies introduced in earlier sections of this chapter.

Recognition can be considered in two distinct—both important—ways. The first is *absolute recognition*—wherein an item is categorized as being an object of a certain type, or class. This *identification* can be performed even if the object has not been observed before; for example, it can be identified based on rules or an intelligent grammar of rules (expert system). The second is an awareness that something observed has been observed before. This *relative recognition* is often achieved using *correlation*. Correlation in the broad sense is any measure of dependency of one set of data on another set(s): this means the k-NN classification method is a correlative method, since the sample to be classified best correlates, in a positional sense, with the modal class of its k nearest training samples.

Examples of relative recognition abound in NLP, in which lemmatization, part-of-speech tagging, summarization, keyword generation, and semantic analysis can be used to describe a portion of text. Different text elements can be matched relatively, then, by having similar keywords or other salient meta-data. Words with matching lemmas are related conceptually to each other (absolute recognition), whereas words with matching stems may not be (relative recognition).

In imaging applications, correlation between an image and a distortion-free reference image is used to provide absolute recognition. Correlation between an image and a nonreference image is used to provide relative recognition. The differences in correlation can be used for quality assurance, surveillance alerts, tracking, and other object recognition and image understanding applications. Not surprisingly, relative differences in correlation can be used as the input data for clustering.

1.11   System-Based Analysis

This chapter has overviewed many of the core technologies used in machine intelligence systems. As systems get larger—cloud computing scale and beyond—it will be increasingly important to be able to treat the individual knowledge generators as black boxes. Though their internal workings vary significantly, their inputs, outputs, and transfer functions should be well understood by the would-be system architect, as this will allow them to be combined in a variety of parallel approaches.

Systems analysis is the study of interacting components that create a system, with the intent of providing a recommendation, usually related to improvement of the deployed system architecture. In this book, *system-based analysis* is used to further optimize not only the architecture of the system (interaction of components) but also the system components themselves. This is achieved by using three different forms of parallelism—the subject of the next chapter—to optimize the system components. For intelligent systems, these components range from algorithms to large intelligence engines such as automatic speech recognition and image understanding systems.

System-based analysis would be overwhelming without a set of tools to help perform the task. Patterns are essential to direct and even constrain the ways in which to optimize the intelligence components of the system. Much of this book will, in fact, be focused on these patterns.

1.12   Summary

This chapter outlined some of the most important technologies used in intelligent systems. These smart systems include machine intelligence, artificial intelligence, data mining, classification, and recognition systems. This chapter provided enough background to highlight the commonalities and differences among the key machine learning approaches; from SVMs to ensemble methods and from Bayesian to regression methods. Adaptive systems, such as genetic algorithms and neural networks, were shown to be robust to different input to output maps. Others, such as Bayesian systems, were shown to improve as more training data and/or successfully completed tasks were available. The wide diversity of machine learning approaches should be viewed as an opportunity, not a worry. The goal of this book is to show how this great diversity can be used to our advantage in designing systems.

This book is aimed at allowing one with a moderate understanding of statistics, calculus, logic, linear systems, and design patterns to be able to architect and deploy more intelligent systems than even the best individual intelligent algorithm designer can achieve. The so-empowered meta-architect must understand both the relative advantages and disadvantages, and both the flexibility and the limitations, of each of the component systems. The meta-architect need not understand these individual technologies well enough to improve them as they are; rather, she must understand how to make them more valuable within a larger system involving two or more intelligent components. Let us now consider the types of systems that will require these new system-based analysis skills.

References

Bailey, C.H. and Kandel, E.R. (2008) Synaptic remodeling, synaptic growth and the storage of long-term memory in Aplysia. *Progress in Brain Research*, **169**, 179–198.

Berk, R.A. (2004) *An Introduction to Ensemble Methods for Data Analysis*, Department of Statistics, UCLA, 34 pp., July 25.

Belkin, M. and Niyogi, P. (2003) Using manifold structure for partially labelled classification, in *Advances in NIPS*, vol. 15, MIT Press, p. 8.

Bishop, C.M. (2006) *Pattern Recognition and Machine Learning*, Springer Science + Business Media LLC, New York, 738 pp.

Breiman, L. (2001) Random forests. *Machine Learning*, **45**, 5–32.

Breiman, L., Friedman, J.H., Olshen, R.A., and Stone, C.J. (1984) *Classification and Regression Trees*, Wadsworth & Brooks/Cole Advanced Books & Software, Monterey, CA.

Cortes, C. and Vapnik, V. (1995) Support-vector networks. *Machine Learning*, **20**, 273–297.

Cox, T.F. and Cox, M.A.A. (1994) *Multidimensional Scaling*, 2nd edn, Chapman & Hall, London, 328 pp.

Dempster, A.P., Laird, N.M., and Rubin, D.B. (1977) Maximum likelihood estimation from incomplete data via the EM algorithm (with discussion). *Journal of the Royal Statistical Society: Series B*, **39**, 1–38.

Dreyfus, P. (1962) *L'informatique*. Gestion, Paris, pp. 240–241.

Fogel, D.B. (1995) *Evolutionary Computation: Toward a New Philosophy of Machine Intelligence*, IEEE Press, Piscataway, NJ, 272 pp.

Freund, Y. and Schapire, E. (1996) *Experiments with a New Boosting Algorithm*. Proceedings of the 13th International Conference on Machine Learning, pp. 148–156.

Goldberg, D.E. (1989) *Genetic Algorithms in Search, Optimization & Machine Learning*, Addison-Wesley, 412 pp.

Gould, S.J. and Eldredge, N. (1977) Punctuated equilibria: the tempo and mode of evolution reconsidered. *Paleobiology*, **3** (2), 115–151.

Grimes, C. and Donoho, D.L. (2005) Image manifolds isometric to Euclidean space. *Journal of Mathematical Imaging and Vision*, **23**, 5–24.

Hastie, T., Tibshirani, R., and Friedman, J. (2009) *The Elements of Statistical Learning*, 2nd edn, Springer Science + Business Media LLC, New York, 745 pp.

Ho, T.K. (1998) The random subspace method for constructing decision forests. *The IEEE Transactions on Pattern Analysis and Machine Intelligence*, **20** (8), 832–844.

Hothorn, T. (2003) *Bundling Predictors in R*. Proceedings of the 3rd International Workshop on Distributed Statistical Computing, vol. 3, March 20–22, 10 pp.

Jacobs, R.A., Jordan, M.I., Nowlan, S.J., and Hinton, G.E. (1991) Adaptive mixtures of local experts. *Neural Computation*, **3** (1), 79–87.

Jain, A.K., Duin, R.P.W., and Mao, J. (2000) Statistical pattern recognition: a review. *The IEEE Transactions on Pattern Analysis and Machine Intelligence*, **22** (1), 4–37.

Jolliffe, I.T. (1986) *Principal Component Analysis*, 2nd edn, Springer, New York, 487 pp.

Kandel, E.R., Schwartz, J.H, and Jessell, T.M. (2000) *Principles of Neural Science*, 4th edn, McGraw-Hill, New York, 1414 pp.

Kohonen, T. (1982) Self-organized formation of topologically correct feature maps. *Biological Cybernetics*, **43** (1), 59–69.

Leondes, C.T. (ed.) (1998) *Image Processing and Pattern Recognition*, Academic Press, San Diego, CA, 386 pp.

Levenberg, K. (1944) A method for the solution of certain non-linear problems in least squares. *Quarterly of Applied Mathematics*, **2**, 164–168.

Liu, C. and Rubin, R.B. (1998) Maximum likelihood estimation of factor analysis using the ECME algorithm with complete and incomplete data. *Statistica Sinica*, **8**, 729–747.

Marsland, S. (2009) *Machine Learning: An Algorithmic Perspective*, CRC Press, Boca Raton, FL, 390 pp.

Roweis, S. and Saul, L. (2000) Nonlinear dimensionality reduction for locally linear embedding. *Science*, **290** (5500), 2323–2326.

Rumelhart, D.E., Hinton, G.E., and Williams, R.J. (1986) Learning internal representations by back-propagating errors. *Nature*, **323** (99), 533–536.

Schapire, R.E. (1999) *A Brief Introduction to Boosting*. Proceedings of the 19th International Joint Conference on Artificial Intelligence, vol. 2, pp. 1401–1406.

Sewell, M. (2007) *Ensemble Learning*. Department of Computer Science, University College London, April, revised August 2008, 16 pp.

Simske S. and Matthews, D. (2004) *Navigation Using Inverting Genetic Algorithms: Initial Conditions and Node-Node Transitions*. GECCO 2004, 12 pp.

Simske, S.J., Li, D., and Aronoff, J.S. (2005) A Statistical Method for Binary Classification of Images. *ACM Symposium on Document Engineering*, pp. 127–129.

Shawe-Taylor, J. and Cristianini, N. (2004) *Kernel Methods for Pattern Analysis*, Cambridge University Press, 476 pp.

Tveter, D.R. (1998) *The Pattern Recognition Basis of Artificial Intelligence*, IEEE Computer Society Press, 369 pp.

Vapnik, V. (1995) *The Nature of Statistical Learning Theory*, Springer, Berlin, 334 pp.

Viterbi, A.J. (1967) Error bounds for convolutional codes and an asymptotically optimum decoding algorithm. *IEEE Transactions on Information Theory*, **IT-13**, 260–267.

Wang, J. and Chang, C.-I. (2006) Independent component analysis-based dimensionality reduction with applications in hyperspectral image analysis. IEEE Transactions on Geoscience and Remote Sensing, **44** (6), 1586–1600.

Wolpert, D.H. (1992) Stacked generalization. *Neural Networks*, **5** (2), 241–259.

2

Parallel Forms of Parallelism

Si Dieu n'existait pas, il faudrait l'inventer.

—Voltaire (1768)

2.1 Introduction

If Voltaire were alive today, he might note that "if parallelism did not exist, it would be necessary to invent it." Necessary because so many problems worth solving naturally—perhaps inevitably—lead to parallel approaches. For example, parallelism naturally arises when there are multiple paths to search. It is obvious that covering two paths at once will, in general, halve the time necessary to find a lost item—the frustrated cry, "I've looked there already!" perhaps stems from the perceived wasting of this parallelism.

In the twenty-first century, parallelism has become the norm because of the harmony between the development of advanced graphical processing units (GPUs), multi-core processors, advanced caching architecture and approaches, and the surge in solid-state devices. This hardware enabling of parallelism partners with software such as hybrid multi-core parallel programming (HMPP) and concurrent programming languages to ensure system parallelism occurs. However, the overwhelming majority of applications and services are based on C, C++, Java, and related, not explicitly parallel, software languages, necessitating the use of threading.

Complementing these parallel computing approaches are distributed file system software frameworks such as Apache Hadoop, based on Java, which provides parallelism of task through distribution of task. In some ways, this is analogous to threading. Combined, these different parallel approaches emphasize the natural progression from monolithic, serial processing and storage to distributed, parallel processing and storage. These trends are not just exciting—they are also necessary for computer architecture to match the capacity and capabilities of the brain. In the brain, parallelism is ensured by the mean 10^4 synaptic connections each of the 10^{11} neurons makes with other neurons. The brain is an analog computer—a synapse is not just on or off—so that analogies with digital computers are by definition forced. However, it is safe to say that this high degree of connectivity is currently unmatched by parallel computing software.

Meta-algorithmics: Patterns for Robust, Low-Cost, High-Quality Systems, First Edition. Steven J. Simske.
© 2013 John Wiley & Sons, Ltd. Published 2013 by John Wiley & Sons, Ltd.

In this chapter, I introduce a different type of parallelism. While this parallelism ultimately benefits from the myriad hardware and software parallelisms described above, it is also agnostic to the details of how these hardware and software parallelisms are achieved. *System parallelism* is the focus of this book, and system architecture can and should be reconsidered in light of the new reality of parallelism. To date, much of the focus of cloud computing has been on distributed data storage. And rightly so. The advantages of having ubiquitous access to the same data at any time, any place, by any party with sufficient access rights, is obvious. But this form of cloud computing is more distributed than parallel, more server–client than peer-to-peer.

System parallelism, instead, is focused on explicitly architecting intelligence-creating systems that innately benefit from separating tasks, components, and/or algorithms into distinct threads for downstream processing (e.g., interpretation and execution). In this chapter, I consider three different types of parallelism. It must be noted that there are other, quite likely equally valid, cladograms for the overall set of parallel processing approaches. I chose this taxonomy for two reasons: (1) it provides excellent functional distinction between the three forms of parallelism, and (2) it aligns each form nicely with specific, highly salient topics that can benefit directly from the science of these forms of parallelism.

In the case of parallelism by task, the system design strategies I discuss herein can be applied to "function parallelism" or "control parallelism" tasks in which execution processes are performed in parallel. These execution processes need not be related. Two patterns that illustrate fundamental differences in parallelism by task are what I term *queue-based* and *variable-sequence-based* task parallelism, as described in Section 2.2.1. Parallel-separable operations (PSOs) are classified as latent, unexploited, and emergent, and some implications of the PSOs for data mining, assembly lines, and scheduling are discussed.

In the case of parallelism by component, system design strategies naturally have relevance to the broad field of software development. Parallelism by component is effectively the same as data parallelism or "loop-level" parallelism, and has obvious extensions to data mining, search, and other map-reduce parallel processing approaches.

Finally, in the case of parallelism by meta-algorithmics, there is ready application of the parallel processing strategies—herein defined as first-, second-, and third-order meta-algorithmic patterns—to a wide host of machine intelligence approaches. While these patterns are the main focus of this book, the nuclei of these patterns will be recognizable in the older parallel processing and hybrid machine intelligence fields.

2.2   Parallelism by Task

2.2.1   *Definition*

With parallelism by task, a larger process, sequential (series) or series-parallel in nature, is reconstructed so that one or more of the sequential set of steps is restructured as one or more parallel processes. The first pattern is the simplest: the *queue-based* pattern shown in Figure 2.1 simply breaks up a given process into a set of necessarily sequential tasks—in the example of Figure 2.1 there are four tasks {A, B, C, D} that must be completed in order. These tasks, therefore, have built-in dependencies: the input of Task B requires the output of Task A, the input of Task C requires the output of Task B, and so on. Such a dependency is often referred to as a pipeline, and the sequence is shorthanded as A | B | C | D. The pipeline A | C | B | D will

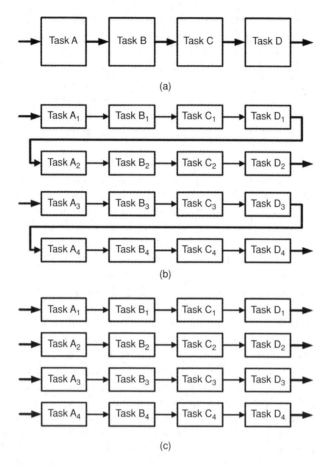

Figure 2.1 Queue-based parallelism by task example. In (a), the four different tasks to be completed for a single pipeline are labeled A, B, C, and D. These form a pipeline, and so must occur in the given order. In (b), two parallel processors are available and multiple pipelines are assigned, sequentially, to them. In (c), four parallel processors are available, and pipelines are assigned to each of them simultaneously. More complicated sequential architectures are possible, for example, where each processor handles only a specific task. These scheduling architectures are discussed in Section 2.2.3

produce different, and presumably unusable, results. Pipelines are commonplace in machine understanding tasks such as image segmentation and optical character recognition. In the latter, the pipeline may consist of the following steps: binarization (thresholding) | connected component formation | character tagging | word tagging and correction. If binarization does not occur before connected component formation, many nonwhite and/or nonuniform backgrounds will form large, solid connected components, completely obfuscating any associated text. Tagging likewise has to be performed on connected components. Finally, word formation without character tagging is much less accurate, since (1) the individual variances associated with each character add to the overall word variance, and (2) there are many more words than characters.

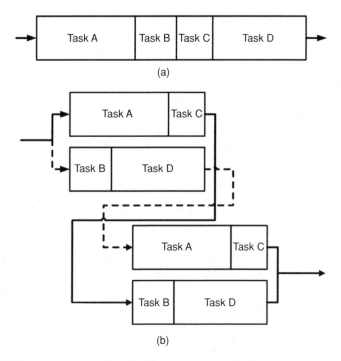

Figure 2.2 Variable-sequence-based parallelism by task example. In (a) the length of the boxes Task A, Task B, Task C, and Task D are proportional to the processing time required to complete each task. In (b), a four-processor design is shown. Two of the processors are used to perform Tasks A and C, which happen to take a similar time to process as the time the other two processors require to perform Tasks B and D. In this example, our sequential variability is such that Task C requires output from Task A, and Task D requires output from Task B, with no other dependencies. See text for details

The queue-based parallelism by task shown in Figure 2.1 is very simple. Each processor— one, two, and four are shown in Figures 2.1a, b, and c, respectively—can perform every task in the pipeline. This requires either replication of code or multi-processor access to the same code, and in general is not a poor design choice. However, in some cases, depending on the overall system architecture, a more complicated processor design, where individual processors have access to different sets of codes and thus perform different operations, is desirable. This type of scheduling is discussed in more detail in Section 2.2.3.

A second simple pattern for parallelism by task is the *variable-sequence-based* pattern as shown in Figure 2.2. There is sequence variability whenever a larger pipeline can be split into two or more smaller pipelines—even as individual tasks—that can proceed independently of each other. An example may be an automatic speech recognition (ASR) system in which A is a determination of amplitude (loudness), B is an assessment of spectrum peaks in the voice, C is a determination of mood (e.g., anger), and D is an assessment of gender/nationality/regionality. A simple ASR engine may determine C based on A but not B, and determine D based on B but not A. Such a simplified ASR pipeline matches the example of Figure 2.2.

Just because you can do something does not mean you should. Why would we select the architecture in Figure 2.2 over the architecture in Figure 2.1, even with the reduced number of dependencies? Mainly because such a design provides more robust behavior to "job hang," the situation when a particular task takes much longer than anticipated. Suppose, for example, one specific Task A effectively freezes a processor. With the design in Figure 2.2, we can route the corresponding Tasks B and D to a now freed-up processor, completing those tasks before the job hang is resolved. With the architecture of Figure 2.1, we would need to wait for the job to un-hang before performing the corresponding Tasks B and D, compounding the delay. Such robustness to job hang is an important consideration in all parallelism, and a reason more complicated scheduling approaches than the relatively simple architectures of Figures 2.1 and 2.2 are generally selected for complex machine knowledge tasks.

2.2.2   Application to Algorithms and Architectures

In addition to the rather straightforward assignment of multiple tasks to multiple parallel paths, parallelism by task can be used *introspectively*; that is, to look inside specific algorithms, systems, or services and identify separable operations. Such PSOs include (1) latent, (2) unexploited, and (3) emergent operations. All three types of introspective separable operations are opportunities for parallelism. I will next provide examples of how these operations can be used.

A *latent introspective parallel-separable operation* (LI-PSO) is an operation amenable to parallelism that is not evident in the original operation—for example, algorithm or system—but can be made evident when the original operation is deconstructed for the purposes of making it parallel-ready. An LI-PSO can, therefore, be thought of as a PSO that is born of redesign or rearchitecting. A simple example is the determination of the mean value of an $A \times B$ dimension matrix **M**. The original code (shown here without the normal error-catching, method variables, etc., for simplicity) may look like the following:

```
double mean = 0.0;
for( int j=0; j<B; j++ ) {
   for( int i=0; i<A; i++ ) {
     mean += M[j*A+i]; } }
   mean /= (A+B);
```

This is an $O(AB)$ or $O(N^2)$ operation. It is obvious that this code can be rewritten to provide B parallel operations, each of $O(A)$ as shown here:

```
mean[j] = 0.0;
for( int i=0; i<A; i++ ) {
   mean[j] += M[j*A+i]; }
mean[j] /= A;
```

The overall mean is then determined in an $O(B)$ operation, resulting overall in the transformation of an $O(N^2)$ operation into an $O(2N)$ operation, assuming the parallel processing hardware is available. This offers the possibility of an up to $(N/2)$ improvement in throughput speed. While this is in some ways a trivial example (and is actually an example of parallelism by component—see Section 2.3), note that the same method can be applied directly to more complicated imaging operations like computing histograms, entropy calculations, and projection profiles—not to mention less trivial mathematical operations like decryption and decompression.

The second type of PSO considered is the *unexploited introspective parallel-separable operation* (UI-PSO), which can be thought of as a PSO that is born of deployment. A UI-PSO is an algorithm already known to be parallel that is currently used in serial form for any of a number of reasons, including hardware limitations, fear of multi-threading, or simply the lack of motivation to change legacy code. An example of a UI-PSO may be a *factory method pattern* (Gamma *et al.*, 1994) that simultaneously produces many different objects, each with an interface suitable for subclasses to instantiate the later fully defined object. Since each of these factory objects is independent, the factory is innately parallel-friendly.

Another example of a UI-PSO is in data mining. When a database is fully data-mined, both structured and unstructured data sets are analyzed, and usually largely different types of content are examined. Several levels of parallelism are thus obvious—structured versus unstructured; text versus image versus video versus audio versus other media; and even distinct passes or "looks" at the same data set (e.g., extracting key words and indexed words from a text field). Where possible, each of these independent analyses should be performed using parallelism by task.

The third type of PSO is the *emergent introspective parallel-separable operation* (EI-PSO). Unlike a UI-PSO, the EI-PSO is not obvious in the original system; unlike an LI-PSO, the EI-PSO does not become obvious in the deconstruction of the original system. Thus, the EI-PSO can be thought of as a PSO that is born of a change in the system language. Usually, this is thought of in terms of moving from compiled or interpreted coding techniques to parallel coding techniques, but I will use a hardware rather than software example for illustration here.

Many traditional manufacturing centers have been designed with a specific set of inputs and outputs used to define the overall architecture. There are significant advantages to such a black box approach, which is then recursively applied to different stations within the factory. Such systems are modular, scalable, and thus very efficient at producing large numbers of the same, *noncustomized* products. Production lines are serial, with components added at different locations, or individualized workstations. This results in *standardization* of the stations, with the associated throughput advantages. However, this type of design approach is associated with a number of assumptions, some of them quite subtle. The first assumption, implicit in the very name, is that production itself is paramount. The economic implications of this approach are many: that overproduction and later waste of unsold product is allowable; that differences in supply and demand are inevitable; and that the degree of customization is relatively small compared to the commonality among all the products in the overall production run. However, these implications are not always valid in the digital age. In printing and publishing, for example, there are strong indications that print-on-demand and customized printing—in which the customer sets the printing process and increasingly the printing content in motion—will be the norm for physical and not just electronic publishing. Similarly, rudimentary production-on-demand has been implemented in industries as complex as the computer and automobile ones. With the complementary trends in digital mobility, cloud-based services, and location awareness, there is no reason to believe that other forms of production will not follow this example, extend it, and move entire industries to a more local, more just-in-time, and thence more sustainable practice.

The second assumption inherent in the use of assembly-line-inspired sequential, focused-task station-based production lines that were the signature of twentieth century manufacturing, is more subtle. The widespread adoption of assembly lines is indicative of an assumption that there are advantages to production inherent in serial architectures. But is this necessarily the case? Most systems of value—be they automotive, electronics, clothing, even food—involve multiple, efficiently parallelizable systems that argue instead for a more complicated architecture. In the automotive and electronics industries—just as is the case for the software industry—it has become obvious that *modularity of interface* for components creates tremendous advantages. The automobile itself can be considered a system, with recursively smaller systems—electrical, combustion, suspension, passenger comfort, safety, and so on—themselves comprised even smaller systems—ignition, dashboard, lighting, and so on—until individual components are the "subsystems." At this level—the "parts" level—there is now an entirely different, in some sense orthogonal, ecosystem, with many parts capable of being used in a wide variety of the largest systems. Nuts, bolts, clamps, and belts are perhaps the most obvious of these, but gaskets, cams, and even shocks fit into this separately woven ecosystem. The automotive industry, being large enough, can simultaneously support a production and parts ecosystem, and so maintain a serial manufacturing architecture for each. But the modularity of interface, allowing for the same parts to be used in multiple sequential assembly processes, bespeaks of the underlying—if latent—parallelism.

Modularity of interface therefore enables a parallelism of task more subtle than the queue-based, variable-sequence-based, and PSO forms described above. It allows a parallelism by task through intersection of two or more sequential processes. In such large ecosystems like the automotive industry, the overlap of these two sequential processes—vehicular and parts manufacturing—is sufficient, and the profits ample enough, that each process can act in a somewhat sequential manner. However, this brings us to a third assumption innate in the widespread adoption of assembly lines.

The third assumption is that assembly lines are more efficient than alternative approaches to product development. While this may be true in terms of raw throughput—for example, maximum peak production rate—in most cases, such a measure of efficiency is overly simplistic and generally shortsighted. Looking at production from a different perspective, the efficiency is proportional to maximum peak production rate, to be sure, but is also dependent on the percentage of produced goods that are actually used, the value of the use of all of these goods, the costs of warehousing and preserving goods produced that may be sold later, the cost of decommissioning or recycling goods never sold, and the cost of recycling goods once they have reached the end of their life cycles.

From the perspective of PSOs, it is clear that in many industries, the additional considerations for costs as outlined above provide at least two opportunities for the application of parallelism by task. The first is the discovery of LI-PSOs, the acknowledgement of UI-PSOs, and the creation of EI-PSOs by restructuring the assembly lines to provide better parallelism by task for the different procurement, assembly, shipment, logistics, and recycling steps in the product life cycle. The second is to apply parallelism by task to the *modeling tasks* associated with the defining, monitoring, evaluating, and redefining of an ecosystem's return on investment (ROI) model. It may well turn out that the use of parallelism results not only in more robust systems but also in more robust system models.

One of the key goals of this book is to provide a set of approaches, usually in the form of patterns, to effectively implement parallelism. Queue-based and variable-sequence-based approaches are simple patterns for parallelism by task. A third, more complex pattern is *recursively scalable task parallelism*, which is based on the ability to create and optimize a cost function related to the choice between processing a given Task A and a subset of tasks $\{A_1, A_2, \ldots, A_N\}$ that together comprise A. This cost function is therefore dependent on the

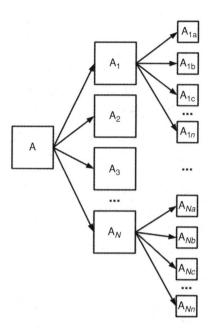

Figure 2.3 Recursively scalable task parallelism. The Task A (far left) itself may have been extricated from a larger task. Tasks $\{A_1, A_2, \ldots, A_N\}$ are extricated from Task A. Tasks $\{A_{1a}, A_{1b}, \ldots, A_{1n}\}$ are extricated from Task A_1, and so on, down to Tasks $\{A_{Na}, A_{Nb}, \ldots, A_{Nn}\}$ being extricated from Task A_N. Note than N is not necessarily equal to n

additional cost, or overhead, of unwrapping Task A into subtasks $\{A_1, A_2, \ldots, A_N\}$ and then rewrapping $\{A_1, A_2, \ldots, A_N\}$ into A. Further subdivision of subtasks into smaller subtasks, recursively, can continue as possible/appropriate (see Figure 2.3). The overhead cost in turn depends on the cost of analyzing and deconstructing the code, recoding as necessary—including translating from compiler code to GPU code, along with associated differences in cost of hardware—processors, cache, storage, and so on. These comprise fixed system costs, which are of less concern in this book—I assume that there is an adequate parallel processing "cloud" available.

The run-time costs are typically bandwidth, processing time, accuracy, and robustness. In recursively scalable task parallelism, we assume that system accuracy and robustness are equivalent for all deployment architectures, since the same algorithms are performed. This simplifies the choice of whether to continue the recursion to a trade-off between bandwidth and processing time. Bandwidth is associated with the repackaging of information for each subtask, and so is an added cost every time a task is broken into subtasks. Processing time is dependent on multiple factors: constructing the task-related data from the subtask-related data, performing the subtask processing, and performing the task processing. However, task-specific processing—the assigning of imaging-related tasks to GPUs and random-access tasks to CPUs, for example—will usually result in significant processing time savings, as illustrated in Figure 2.4.

The general concept of recursively scalable task parallelism is not new. MapReduce (Dean and Ghemawat, 2004) can be viewed as a simplified form of recursively scalable task

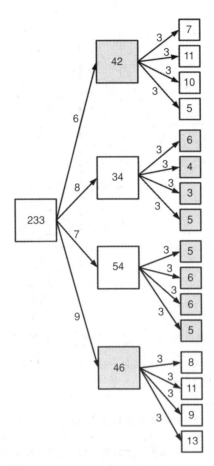

Figure 2.4 Recursively scalable task parallelism example of Figure 2.3 with cost of unwrapping/rewrapping (numbers on arrows) and processing (numbers in boxes) indicated on each path and node. The selected nodes are shaded. The total cost is [6+42] + [8+3+3+3+3+6+4+3+5] + [7+3+3+3+3+5+6+6+5] + [9+46] = 48 + 38 + 41 + 55 = 182, or a reduction of 21.9% over the largest Task A (223) or a reduction of 11.6% over the use of the first layer of subtasks A_1, A_2, A_3, A_4, which is [6+42] + [8+34] + [7+54] + [9+46] = 206

parallelism that effectively maps a single, highly parallel representation of a task to many like subtasks. The greater relative complexity of the pattern shown in Figures 2.3 and 2.4 is provided by the recursion capabilities. The recursive nature of the pattern extends to the manner in which each task associates with each of its subtasks. The association includes (1) a list of subtasks; (2) the means to unwrap the data and operations in the invoking task to create input for each of the subtasks, which includes in some cases the replication of data; (3) the means to perform the subtasks; and (4) the means to rewrap the output of each subtask to create the full task output.

These associations are directly related to the costs shown in Figure 2.4. Note that the shaded boxes in Figure 2.4 are the levels of parallelism by task chosen for the overall deployment. Turned sideways, it can be seen that Figure 2.4 is actually a tree structure and so the optimal costs can be determined using simple tree evaluation methods. The evaluation problem is also greatly simplified in Figure 2.4 because of the constraints involved. At Stage S, the choice to unwrap involves a binary decision between choosing S versus choosing the set $\{S_1, S_2, \ldots, S_N\}$. Keeping a partial set unwrapped, for example, as $\{S_1, S_2, \ldots, S_M\}$ where $M < N$ and then unwrapping for parallelism by task the set $\{S_{M+1}, S_{M+2}, \ldots, S_N\}$—along with other partial unwrappings of this kind—very quickly allows a more exhaustive search that is nevertheless a reasonable approach in some cases. However, in practice it is likely that in re-examining the functionality in S, the task owner will create a reasonably atomic set $\{S_1, S_2, \ldots, S_N\}$ such that no further partial unwrappings require evaluation.

In larger serial systems, however—especially those wherein the original system developer is no longer available—a complete overhaul of the task structure may be required, resulting in a set of subtasks $\{S_1, S_2, \ldots, S_N\}$ that may or may not be the optimal set. In this case, it is possible that another application of parallelism—for example, parallel application of path evaluation methods such as Dijkstra's method (Dijkstra, 1959)—may be used to determine optimal subtask definition. Alternatively, representing the tree chart of Figure 2.4 as a flow diagram with the costs for unwrapping and rewrapping explicitly designed as flows, and operations as nodes, may provide amenability to flow network optimization techniques such as min-cut, max-flow, and others. This is left for future research, but nicely bridges us to the next section on scheduling.

2.2.3 Application to Scheduling

It is certain that sometimes we will need more complicated designs than shown in Figures 2.1 and 2.2; for example, when we have mixed GPU/CPU architecture, special processors such as application-specific integrated circuits (ASICs), limited cloud resources, and so on. In most cases, the decision of how far to go in task parallelization depends on how well the particular processing resources match the task at hand. Since the overhead for scheduling can be substantial, the throughput gain must be able to offset the inefficiencies of scheduling. For example, if scheduling creates a 25% reduction in throughput, then the advantage of the multiple-processor-type parallel processing must provide a 33.3% improvement (since $0.75 \times 1.333 = 1.0$) in processing time. This is because the scheduling costs are upfront. It is analogous to needing to drive 100 km in an hour, but being slowed to 60 km/h for the first 50 km due to road construction. Thereafter, one must proceed at 300 km/h to reach the destination in time (not 140 km/h). Caution must therefore be taken to ensure that the cost of unwrapping is not too high. Structured data mining, such as performed by MapReduce (Dean and Ghemawat, 2004), and column-oriented database (Monash, 2007) approaches are very powerful—and widely deployed—in part because of the low overhead required to "unwrap" the data in preparation for analysis, or "mining." Column-oriented storage increases the throughput of sequential record access through elimination of common transactional operations, and so is well suited to a number of machine intelligence tasks, so long as these tasks involve a similar type of processing for large sets of data.

However, in many important areas of machine intelligence, domain expertise is needed. Domain expertise is certainly needed in order to deconstruct any sizeable chunk of code into parallelizable parts. Domain expertise is also required in order to determine what method of parallelization to deploy. For example, should an algorithm be reconstructed for deployment on a CPU, a GPU, or wrapped up as services for distribution to a cloud with the cloud service provider deciding how to handle the data analysis task? The answer to this question depends on the type of data, the type of task, the type of multi-threading in the code, and the degree of storage, processing, and transaction coding used.

Scheduling is a broad science concerned with deciding how to commit resources when there is a plurality of tasks to complete within a given time period. In the example of Figure 2.2, I argue that a more robust scheduling algorithm takes advantage of parallel paths to break up scheduled sets of (associated) tasks, since the effect of system "hangs" (long-term, nonterminating jobs) is lessened. The main concerns of any scheduler are (1) throughput of processes, (2) response time latency, or the time between task submission and system response (usually a scheduled time to begin and/or complete the task), (3) turnaround latency, or the time between task submission and completion, and (4) real-time adaptation to provide fairness when unanticipated strains on resources—from system "hangs" to a large influx of new tasks—occurs. There are many scheduling algorithms, of which the most familiar may well be the FIFO (first in, first out, or "queue"), multi-level queue, round robin, fixed priority pre-emptive, and the shortest remaining time (or "shortest job first") algorithms. A tremendous amount has been written on how to select one scheduler versus another based on concerns such as the four listed earlier in this paragraph. However, it may well be that using multiple scheduling algorithms in parallel and intelligently combining them may result in a more robust, more effective scheduling approach. We will revisit this topic later after having considered parallel processing patterns in more depth in the following chapters.

The discussion above, to be continued—perhaps less explicitly—throughout this book, is based on the perspective that system design and system architecture are processes, or means to an end—and not ends in and of themselves. Adaptable, customizable, modular, and thereby robust architectures must be reconfigurable on the fly in order to merit the upfront investment in these systems. Parallel architectures, with the increased flexibility in maintaining, evolving, and deprecating components, offer distinct advantages over more rigid, serial/sequential systems. This section on parallelism by task highlights some of these advantages, and hopefully has helped justify the assertion that system design and system architecture are verbs, not nouns—for the large, highly efficient, and lithe ecosystems (systems of systems) needed in the digital age, the fields of *system designing* and *system architecting* are paramount. I use the present progressive tense to indicate both the need for continual upkeep of the design and architecture, and also to indicate the improvement, or progress. The three major branches of parallelism—by task, by component, and by meta-algorithm—will help make these ecosystems a reality. I now turn to the second of these—parallelism by component.

2.3 Parallelism by Component

2.3.1 *Definition and Extension to Parallel-Conditional Processing*

Parallelism by component involves the reconstruction of a large data set so that two or more partitions of the data can be processed in parallel. Often, the same processing is used on each component, making this form of parallelism by component even more amenable to MapReduce (Dean and Ghemawat, 2004) or column-based (Monash, 2007) analytics than is parallelization by task. Parallelism by component is effectively the same as data parallelism or "loop-level" parallelism. Repetitive tasks usually addressed in program "loops" are instead assigned to

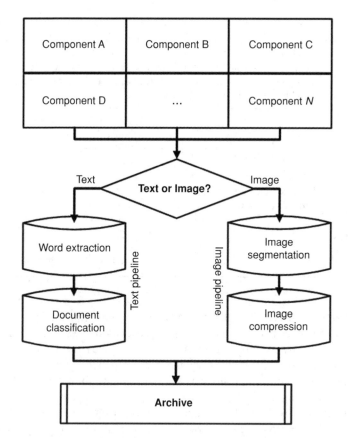

Figure 2.5 Parallel-conditional processing example. In this simplified example, individual components are zones, or regions, from mixed documents. Each component is processed in parallel to determine its content type. A conditional decision is then made. Zones classified as "Text" are assigned to a text pipeline, while zones classified as "Image" are assigned to an image pipeline. All zones are archived in parallel after one or another of the two parallel pipelines has been performed on the zone

parallel branching codes, as shown in the discussion under LI-PSOs above. These can be considered uniform split-and-merge processes. The same process or pipeline of processes is performed on each component, simultaneously in a separate parallel processing apparatus or thread.

In addition to these uniform split-and-merge processes, parallelism by component is also amenable to more complicated parallel-conditional processing as shown in Figure 2.5.

Parallel-conditional processing can be viewed as a means of recursively defining opportunities for parallelism by component. In Tables 2.1 and 2.2, I will show the impact of domain expertise for defining the parallelism through a form of sensitivity analysis. In this example, we have a fixed number of processors, each of which is allowed to process either text or image data, but not both. We assume that the licensing costs and the inefficiencies of dynamic re-allocation of licensed software is not an option in order to illustrate the parallel processing approach.

Table 2.1 Sensitivity analysis of parallel-conditional processing when the expected processing time of each of two conditional tasks closely matches. In this example, the expected value of text and image processing time are equal at 3 h. With 10 processors available, 5 can be assigned to each of the two conditional pathways and the separate parallel branches are each expected to complete in 0.6 h. When an unexpectedly high (95%) proportion of text or image processing becomes necessary, the overall processing time is increased by as much as 137.5% (1.425 h)

Factor	Expected Values	High-Text Scenario	High-Image Scenario
% Text content	60	95	5
% Image content	40	5	95
Text processing at 100% usage of a single processor	5 h	5 h	5 h
Image processing at 100% usage of a single processor	7.5 h	7.5 h	7.5 h
# Processors text	5	5	5
# Processors image	5	5	5
Text processing time	0.6 h	0.95 h	0.05 h
Image processing time	0.6 h	0.075 h	1.425 h

In Tables 2.1 and 2.2, there are two different parallel processing applications. The example in Table 2.1 is a well-balanced parallelism by component situation in which the conditional leads to two branches with equivalent numbers of processors assigned to each branch. The expected workload is 60% text, 40% image, but since the image pipeline requires 50% more relative processing, the two conditional branches can each be assigned half of the available processors. This is in contrast to Table 2.2, which presents an example of an expected unbalanced workflow. The expected workload is 86% text, 14% image, for which the optimal assignment of processors is eight processors for text processing and two for image processing. The expected processing

Table 2.2 Sensitivity analysis of parallel-conditional processing when the expected processing time of each of two conditional tasks is not closely matched. In this example, the expected value of text and image processing hours after parallelization are nearly equal at 0.5375 h and 0.525 h—with 8 of the 10 processors available assigned to text processing and 2 to image processing. When an unexpectedly high (95%) proportion of text or image processing becomes necessary, the overall processing time is increased by as much as 562.8% (to 3.5625 h from 0.5375 h). This is much more than the 137.5% increase observed for the same scenario in Table 2.2

Factor	Estimated Values	High-Text Scenario	High-Image Scenario
% Text content	86	95	5
% Image content	14	5	95
Text processing at 100% usage of a single processor	5 h	5 h	5 h
Image processing at 100% usage of a single processor	7.5 h	7.5 h	7.5 h
# Processors text	8	8	8
# Processors image	2	2	2
Text processing time	0.5375 h	0.5938 h	0.0313 h
Image processing time	0.525 h	0.1875 h	3.5625 h

time is just over half an hour, 0.5375 h, for the text. Each table also reports on a simple form of sensitivity analysis. If the workload suddenly moves far from the predicted balance, what is the impact on the overall processing time? In Table 2.1, the worst results are obtained when 95% of the content is image, in which case the image processing parallel path requires nearly $2\frac{1}{2}$ times as long to complete its work (1.425 h instead of 0.6 h). However, in Table 2.2, the results are far more drastic. With the same unanticipated "high-image scenario," the image processing time now jumps to 3.5625 h, a 560% increase.

The sensitivity analysis data from Tables 2.1 and 2.2 shows that balanced conditional pathways in a parallelization by component architecture make the design much less sensitive to unexpected deviations from the predicted workloads. The more effort placed upfront to design a system to have matching number of processors in each conditional branch, the more robust the system will be to unexpected changes in workload composition. In the example of Table 2.2, one approach toward balancing the load would be to include more of the class *text* in the *image* category. If, for example, color text, headline text, and so on, can be accommodated by the image pipeline, the balance can be tipped in favor of the image pipeline. More realistically, however, we may not be able to do anything to balance the tasks. In this case, we may prebalance the conditional pipelines to minimize the downstream effects of extremes in the workload. If we change, for example, the number of processors deployed to seven for the text pipeline and three for the imaging pipeline, then under normal conditions the text processing time will be 0.6143 h (compared to only 0.35 h for image processing time). This is a 14% decrease in throughput at the expense of greater resilience to unexpected workflows. Under these new conditions, a sudden change in workflow to a 95% image processing load balance increases the image processing time to 2.375 h (a 342% increase in time to completion, but much less than the 560% increase associated with Table 2.2).

These results demonstrate that an optimal parallel-conditional processing architecture must strike a balance between two concerns. The first is balanced deployment of processing capacity for the expected workload, which provides optimum throughput under normal/expected conditions. This is the default deployment in Tables 2.1 and 2.2. The second concern is system robustness to extremes in the composition of the workflow. The balance between these depends on the variance in the workflow composition as well as the imbalance in the expected workflow itself. A very balanced *expected workflow* (Table 2.1) is much more resilient to variability in the *actual workflow*.

2.3.2 Application to Data Mining, Search, and Other Algorithms

Data mining and search are tasks well suited to parallelism by component. Text mining algorithms, for example, begin with word counts. Word-forming rules are used to decide, for example, whether to stem or lemmatize. Lemmatization is the grouping together of the inflected forms of a root word for analysis as a single item—for example, the words "running," "runs," and "ran" are all assigned to the same grouping, or *token*. Stemming is a special form of lemmatization in which the root of the various forms of the word is extracted—for example, "running," "runs," and "ran" all map to "run." Tokenization of the words simplifies the parallel pathway to simply counting the occurrences of each token. Tokenization allows language independence and optimal scheduling across all of the parallel processors. This is accomplished primarily by converting a *language-dependent* operation into a *language-independent* operation. Once the words are tokenized, the parallel processing used to perform word counts—or the complementary operation of searching for specific words—handles the tokens no differently than it handles any other set of binary patterns. Once the appropriate preprocessing has been performed, therefore, word count is the same as image object count, and word search is the same as image feature search. The only difference is in how the data has been processed prior to the counting or searching.

This is an important point, which follows the parallelization tenet that *uniformity of parallel process* allows the most balanced architecture. The implications: this balance provides insensitivity to change in workload composition (along the lines of Table 2.1), but an upfront cost in preprocessing the data for the parallel processes. However, the preprocessing has an additional advantage the exploitation of which will be a key concern of this book. Performing preprocessing to allow multiple domain-specific tasks (image search, word search, audio search, etc.) to be performed with the same (exact) set of parallel processing code constitutes the identification and elaboration of a specific, reusable pattern. Several parallel processing patterns—relatively straightforward in conceptualization and presentation—have been described already in this chapter. But, as the book builds to more and more complex operations, engines (for analysis, machine intelligence, etc.), systems, and services, I intend to show the value of advanced patterns for parallel processing, particularly in the field of meta-algorithmics.

Let us return to scheduling here to see further value in specifying parallel processing patterns. Real-world problem-addressing systems, even when concerned with massive parallelization by component workflows, gain much from a transactional approach to scheduling. I use the term *transactional* here to contrast with the term *pipeline* as used above. In a transactional system, such pipelines (1) are not readily identified, (2) are not efficiently assigned to parallel pathways, or (3) simply do not exist. In Figure 2.5, the pipelines are readily identified, so we need only be concerned with the second of these three considerations when deconstructing the architecture. One criticism of the parallel-conditional diagram of Figure 2.5 might be: "Why include word extraction and document classification in the same parallel branch, and why include image segmentation and image compression in the same parallel branch? Breaking them up into atomic parallel pathways should give more scheduling flexibility. Are these really efficiently assigned to pipelines?" In this case, the answer is yes, since, for example, if there is an operation hang on any of the pipeline components, no additional amount of parallelization will speed up throughput since the maximum number of parallel pipelines are still available. So, for the relatively simple set of operations performed by the system in Figure 2.5, it is convincing to say that the parallelization is optimized.

Other systems, though, do not have such obvious—and static—pipelines. Consider, for example, an image processing pipeline in which there are a number of potential branches, or conditionals, and it is difficult to predict a priori what processing needs to occur downstream. This often happens in image processing, since differences in image capture quality and characteristics—illumination, blur/focus, angle between camera and image surface, albedo variability based on these other variables, and so on—affect the processing that must take place downstream. In addition, the content of the image may—or may not—trigger downstream processing. For example, if a document is captured, it may or may not need to be deskewed. Depending on the illumination and blur characteristics, an image may or may not need to have illumination compensation and/or deblurring algorithms applied. Often, these operations may require significantly more processing resources than the intended processing associated with the imaging pipeline.

This type of complex branching makes the application of Amdahl's law to the parallel processing system architecture less straightforward. Much traditional parallel processing system design trade-off is based on Amdahl's law (Amdahl, 1967) and the definition of efficiency so implied. If we identify, for example, that P percent (where $0.0 \leq P \leq 1.0$) of the image processing can be made parallel, then using Amdahl's law the minimum possible time to complete the overall task, $t_{overall}$, is proportional to the time to complete the task in a fully serial fashion, t_{serial}, by the following relationship:

$$t_{overall} = \left(1 - P + \frac{P}{N}\right) t_{serial},$$

where N is the number of parallel processors available.

Branching, however, is among the factors preventing Amdahl's law from being achieved. The other factors, as discussed above, include the overhead of scheduling and the overhead of structurally reframing the components for parallel processing. This introduces, then, the following relevant modifications to Amdahl's law:

$$t_{overall} = k_{BO}k_{SchO}k_{SRfO} \left(1 - P + \frac{P}{N}\right) t_{serial},$$

where k_{BO} is the branching overhead factor, k_{SchO} is the scheduling overhead factor, and k_{SRfO} is the structural reframing overhead factor. For each of these coefficients k,

$$k \geq 1.0.$$

Note that when the task is truly—and efficiently—made parallel, then $k_{BO} \approx 1.0$. When the scheduling overhead is relatively negligible compared to the processing time for the specific task and component, then $k_{SchO} \approx 1.0$. Finally, when the structural reframing overhead is relatively negligible compared to the processing time for the specific task and component, then $k_{SRfO} \approx 1.0$.

Let us now consider an example applied to breaking up a large crowd surveillance image into multiple parts and optimizing the number of parts. For this process, $P = 0.9$, which means that 90% of the processing time can be made parallel. In this case, the coefficients k_{BO}, k_{SchO}, and k_{SRfO} are dependent on the number of images, N_I, into which the original image is subdivided. Suppose we obtain the following relationships for $2 \leq N_I \leq 16$:

1. $k_{BO} = 1.0 + 0.02 \times (N_I - 1)$
2. $k_{SchO} = 1.0 + 0.001 \times (N_I - 1)$
3. $k_{SRfO} = 1.0 + 0.05 \times (N_I - 1)$.

This makes the equation for Amdahl's law become

$$t_{overall} = (1.0 + 0.02 \times (N_I - 1))(1.0 + 0.001 \times (N_I - 1))(1.0 + 0.05 \times (N_I - 1))$$
$$\times \left(1 - P + \frac{P}{N_I}\right) t_{serial},$$

Table 2.3 Values of the salient factors in the modified version of Amdahl's law. The optimum parallel processing approach is the one that minimizes the ratio of $t_{overall}/t_{serial}$. Of the four cases shown, $N_I = 8$ gives the optimal ratio

N_I	k_{BO}	k_{SchO}	k_{SrfO}	$1- P + (P/N_I)$	$t_{overall}/t_{serial}$
2	1.02	1.001	1.05	0.55	0.589 6
4	1.06	1.003	1.15	0.325	0.397 4
8	1.14	1.007	1.35	0.212 5	0.329 3
16	1.30	1.015	1.75	0.156 25	0.360 8

which simplifies, should we prefer such a form over the alternate forms above, to

$$t_{overall}/t_{serial} = (0.98 + 0.02 \times N_I)(0.999 + 0.001 \times N_I)(0.95 + 0.05 \times N_I)\left(1 - P + \frac{P}{N_I}\right),$$

Assuming the image is divided into 2, 4, 8, or 16 subimages, the values for $t_{overall}$ as a proportion of t_{serial} are given in Table 2.3. For this configuration and for these relationships between the coefficients and the number of subimages, the optimum configuration uses $N_I = 8$, as shown in Table 2.3. The importance of Table 2.3 is not in the absolute values we obtain for this somewhat simplified model of the overall system behavior per se; rather, the importance is in the fact that in moving from a system architecture implementing $N_I = 8$ to one implementing $N_I = 16$, we see a deleterious effect on system performance. Within this context, we can therefore be assured that we have the ability to optimize the system within the range specified. In fact, if we confine the allowed values of N_I to even values, the optimum value for $t_{overall}/t_{serial}$, of 0.3280, is obtained for $N_I = 10$. It is left as an exercise for the reader to show that $t_{overall}/t_{serial} = 0.3280, 0.3346$, and 0.3460, respectively, for $N_I = 10, 12$, and 14, respectively.

However, there is another factor in predicting the relative processing time for moving a task from serial to parallel form that Amdahl's law as stated above does not address. This is the fact that, since smaller data sets are often processed much more efficiently than larger data sets, it is commonplace for component processing to require disproportionately less processing time than for a larger component. For example, consider the surveillance task above in which M passes through the image are required. The processing time is thus proportional to $M(HW)^T$, where H = the height of the image and W = the width of the image, and T is often $\gg 1.0$. Breaking the image into N^2 subcomponents (smaller images), the processing time is then proportional to $MN^2(HW/N^2)^T$, or $M(HW)^T/N^{2T-2}$. The ratio of processing time is, therefore, proportional to $1/N^{2T-2}$. If $N = 4$ and $T = 2$, then the proportion is 1/16, a huge impact on processing. This will be discussed in more detail in Chapter 5, but should be captured here by another factor in the updated version of Amdahl's law as applied to parallel processing system architecture optimization, as given here:

$$t_{overall} = \frac{k_{BO}k_{SchO}k_{SRFO}(1 - P + \frac{P}{N})}{N^{2T-2}} t_{serial},$$

where T is the processing time factor as described above. Note that this effect is herein considered separately from the impact of structural reframing per se, although it would not

Table 2.4   Values of the ratio of $t_{\text{overall}}/t_{\text{serial}}$ for the full parallel processing system architecture optimization equation, where T has a mean value of approximately 1.5 (the values were determined experimentally for a single large image for purposes of illustration here). Of the four cases shown, $N_I = 12$ gives the optimal ratio

N_I	k_{BO}	k_{SchO}	k_{SRfO}	$1 - P + (P/N_I)$	T	$1/N^{2T-2}$	$t_{\text{overall}}/t_{\text{serial}}$
2	1.02	1.001	1.05	0.55	1.58	0.4475	0.2639
4	1.06	1.003	1.15	0.325	1.51	0.2432	0.0966
6	1.10	1.005	1.25	0.25	1.48	0.1791	0.0619
8	1.14	1.007	1.35	0.2125	1.45	0.1539	0.0507
10	1.18	1.009	1.45	0.19	1.42	0.1445	0.0474
12	1.22	1.011	1.55	0.175	1.40	0.1370	0.0458
14	1.26	1.013	1.65	0.1643	1.37	0.1419	0.0491
16	1.30	1.015	1.75	0.15625	1.35	0.1436	0.0518

be inappropriate to consider this part of structural reframing. In the determination of the coefficient k_{SRfO} above, only the data access timing improvements incumbent to breaking up the images is considered.

With the factor T taken into account, the optimum configuration for the surveillance system described in Table 2.3 changes. The new data, taking into account the factor T, is presented in Table 2.4. Here, the optimum value of $t_{\text{overall}}/t_{\text{serial}}$, of 0.0458, is obtained for $N_I = 12$ (slightly more than the $N_I = 10$ result above). This implies that no further parallelism—such as, for example, setting $N_I > 16$—need be considered to further improve the ratio $t_{\text{overall}}/t_{\text{serial}}$.

This section, therefore, provides a more accurate predictor for the effect of parallelism on processing time than does the simple application of Amdahl's law. As demonstrated here, there are a considerable number of factors to be considered when moving a serial processing architecture to a parallel one. To summarize, these factors include the following:

1. Percentage of the serial task that can be made parallel (P).
2. Number and range of parallel processing pipelines allowable (N).
3. Amount of branching necessary in the programming (k_{BO}).
4. Complexity and overhead of scheduling necessary for the processing (k_{SchO}).
5. Structural reframing (k_{SRfO}, T).

Having explored parallel processing optimization for this imaging task, we turn to another topic amenable to—and necessary for the implementation of—parallelization by component: software development itself.

2.3.3   Application to Software Development

Parallelism by component readily extends to the world of component formalism; that is, the world of software development itself. Component-based development (CBD), or component-based software engineering (CBSE), is an important branch within software engineering that is concerned with the separation of functionality within a larger software system into software modules, packages, or services containing a related set of functions. For the purposes of parallelism, I consider CBD to be focused on an optimized component definition; that is, a component definition that maximizes the F-score as defined in Section 1.9. As discussed there, the F-score is a relevant measure of the ratio of the variability within each component to the variability between components. For example, while design patterns such as the Flyweight and Composite (Gamma *et al.*, 1994) are efficient at separating related functionality from nonrelated functionality, other design patterns—such as Strategy, Observer, and Visitor (Gamma *et al.*, 1994)—are less obviously tightly related.

Beyond software patterns such as described in Gamma *et al.* (1994), there are many modern systems in which parallelism by component is an important consideration in the design phase. Mobile applications are a good exemplar. Suppose you wish to develop mobile applications for the five primary mobile platforms, requiring development in different languages; for example, different combinations of Objective C, C, C++, Java, JavaScript, and HTML 5.0. We would like the codebase to be as small as possible and the choice of coding languages to be best suited to the task. Finally, we would like to make the decision of where to place the processing such that it minimizes the use of bandwidth and, simultaneously, processing needs on the mobile device.

In order to optimize the code architecture, then, we need to provide an optimized plan for processing, packaging, transmitting, and parallelizing the operations to be performed. Code architecture optimization before the ubiquity of cloud computing and mobile computing devices largely focused on the individual device (laptop, workstation, etc.) and so GPU/CPU and multi-core/multi-thread trade-offs were two of the main design considerations. However, the combined impact of cloud-based computing and mobile device on-ramp has led to a new dominant consideration—the trade-off between bandwidth and local (mobile device) processing. To address this consideration, there are several possible design options to take for code optimization, each involving different levels and types of parallelism. One power set of possibilities is the following:

1. Perform all processing on the mobile device.
2. Transmit all data from the mobile device to a server, distributed system, cloud, or other back end computing.
3. Perform some of the processing on the mobile device and the rest on the back end.
4. Perform no processing. This option is included to complete the power set, but it is a realistic option if the amount of data to be transmitted exceeds the bandwidth × time limitations of the system and the amount of processing to be performed in a specific time exceeds the limitations of the mobile device.

Using this power set of processing options, the design constraints are rather obvious. We assume that bandwidth is limited and that processing capabilities are effectively infinitely greater on the back end than the mobile device. Thus, those portions of the processing that would otherwise be performed on the mobile device but have a high processing to size-after-compression ratio—that is, require a lot of processing but do not require a lot of bandwidth to send from the mobile device to the cloud—are more favorably processed off the device. Good examples of this are the plethora of mobile barcode reading applications and voice recognition applications that have been adopted by mobile platforms over the years 2008–2012. One approach—taken by the developers of applications intended, for example, to provide analysis of a wide array of bar coding symbologies or to perform potentially processing-intensive image restoration on the images captured—is to have the image sent over the internet (e.g., by http or MMS protocols). This approach is bandwidth-intensive. A second approach—taken by developers wishing to allow off-line functionality, feeling the need to support only a limited set of barcodes (e.g., usually QR and Data Matrix), or only interested in supporting the most powerful mobile platforms—is to have all of the image processing done on the device. Obviously, both solutions have their advantages. However, the former can benefit significantly from parallel processing, while the latter benefits more from multi-threading.

The processing considerations above are very important for determining the code optimization strategy. Bandwidth versus on-device processing as the main consideration directly impacts the design strategy for each of the following factors:

1. Code repetition/consolidation
2. Device-specific settings
3. Adaptability (real-time configuration)

We now consider each of these three factors in some depth. *Code repetition* is minimized and code consolidation is supported by moving much of the processing off the mobile device and instead performing it in the cloud or other back end processing system. Here, the amount of code—that is, in terms of software development metrics like difficulty of code writing, expense of code support, expense of code testing, and raw lines of code—becomes a third factor (the other two being processing resources required and bandwidth for transmission between mobile device and the cloud, as described above) to consider in the design. The system costs, however, now diverge. The overall cost of creating the code comprises (1) development costs, (2) testing costs, and (3) support costs. Of these, (1) and (2) are—or should be—almost wholly associated with the upfront development of the system, while (3) is an ongoing cost associated with issues such as coding error correction (bug fixes), rolling out new features, and supporting new services, plug-ins, and add-ons. The costs of (1) and (2) are therefore relatively fixed, as opposed to the costs of bandwidth—which, if reasonable, are usually passed on to the mobile device owner. For the software producer, then, if all other factors are equal, the code development preference is to have a single codebase. This means either a single (or multiple, compatible, e.g., all Linux-based) mobile platform is supported (e.g., the most prevalent platform), or else the entire codebase, as possible, is written for the back end.

Since that is not possible, we next address *device-specific settings*. If these are absolutely minimized—thus, maximizing the amount of code residing in the back end that can be used in common by all platforms—then the mobile devices effectively become thin clients; that is, conduits of information gathered by the mobile device. The processing on the mobile devices is therein limited to data capture—for example, taking a picture or collecting audio—and preparing the data for transmission. This latter cost should not, however, be underestimated. First, the data must be compressed. Lossless compression, while relatively straightforward, still requires development of, for example, a variable-length code table for Huffman coding, or the numbers and subintervals required for arithmetic coding. Next, the information may need to be encrypted. Since a compressed signal should have roughly the same signal entropy as an encrypted signal—unless the compression is incomplete or the encryption is not cryptographically secure—compression should always occur before encryption for two reasons:

1. Encryption is an expensive operation from a processing standpoint, so the amount of data to be encrypted should be minimized.
2. Once data is encrypted, it should not be possible to compress it, unless the encryption is not secure. This is the worst possible decision from a bandwidth perspective.

If the amount of information to be transmitted is now acceptably small (meaning it can be reliably transmitted in an amount of time acceptable to the user of the software), then no additional code is required on the mobile device. If, however, the time to compress, possibly encrypt, and send is too high, more processing (and relevant code) must now reside on the mobile device. If all of the costs are perfectly determinable before the programming starts, then the right trade-off between coding costs, downstream system bandwidth, and processing costs can be struck. However, this level of foreknowledge is not typical, and so there is always some estimation involved. It should be noted that there are some provider costs associated with cloud computing, too. Nevertheless, under most conditions, these costs are far less than the other costs of bandwidth, consumer time if excessive processing occurs on the mobile device, and of course the costs of software development, testing, and support throughout the lifetime of the software application or service.

Regardless, some of the uncertainty in designing the codebase and the system architectures for mobile imaging applications can be removed when *adaptability* is built into the design from the ground up. Adaptability can be provided by a system that has real-time configuration options. There are two primary types of such options considered here. The first, and less important, is task-estimation-related adaptability. Here, the processing, bandwidth, and time for the task to complete are estimated using a subset—for example, on a subimage of a larger image. This subtask can be used, for example, to determine if the fourth option of the power set above—namely, *perform no processing*—should be chosen based on the fact that the amount of data to be transmitted exceeds the product of the bandwidth and time constraints of the task, and the amount of processing required exceeds the capabilities of the mobile device over the same allowed time.

The second, and more important, form of adaptability is real-time system optimization adaptability. The system architecture should be designed from the bottom up to allow the deployment architecture settings to adapt as the relative resources change. The primary relative resources are bandwidth, mobile processing, and code development expense. An adaptability design can simultaneously address each of these in the following fashion. Firstly, the design should allow the data (e.g., image, audio clip, etc.) that is to be processed to be variably partitioned—for example, into a variable number of subimages as discussed for Tables 2.3 and 2.4. Secondly, the design should allow, after the image partitioning, the option of processing on the device and/or on the back end, depending on the real-time analysis of local processing capabilities, bandwidth availability, and compressibility of the data to be transmitted to the back end. This design approach provides a scalable, readily deployed ability to process one or more partitions of the image on the device, responsive to the current network capabilities and the estimated time for completion of the task on the device compared to processing only on the back end. This avoids a one-size-fits-all strategy, and means that whatever parallelism must occur will be a combination of processing on the device and on the back end. Currently, most mobile devices have a single processor, and the back end can be considered to have a nearly limitless number of processors. This approach is also flexible to changing processing resources on the mobile device. For example, consider a processing task on a mobile device with the following design constraints and decisions:

1. The processing needs to be completed within 5 s or it will have unacceptable performance.
2. The available bandwidth allows the transmission of 15% of the data each second, which includes the compression and encryption.
3. Packaging data up for transmission reduces available processing time by 5%.
4. The back end is able to perform its processing within 0.1 s and return the results, including a modest amount of data transmission, back to the device in 0.2 s.
5. The mobile device processor is able to analyze 8% of the data per second.

Based on these data, data can be sent to the back end for 4.7 s (in order to receive the information back before 5.0 s elapse). In 4.7 s, 70.5% of the data can therefore be transmitted and analyzed. In general, in T seconds, $15\% \times (T - 0.3 \text{ s})$ of the processing can occur on the back end. Meanwhile, on the mobile device, in T seconds, $8\% \times T$ of the processing can occur. Thus, in 5 s, 40% of the processing can occur on the device. The relative ratio is thus $70.5\%/40\% = 1.7625$. A close approximation to this ratio can be obtained by dividing the data in 14 partitions and assigning 9 of these (64.3%) to the back end and 5 of these (35.7%) to the mobile device itself. The back end processing will complete when $64.3\% = 15\% \times (T - 0.3 \text{ s})$; that is, after $T = 4.59$ s. The mobile device, meanwhile, will complete its processing when $8\% \times T = 35.7\%$; that is, after $T = 4.46$ s. The safety margin is just over 0.4 s, but the task can be completed within the specified time.

Interestingly, the *effective parallelism* of the task described above is 2.8 mobile processors. That is, 1.8 times as much data is processed on the back end as on the mobile device. This is a relatively low value considering the back end is capable of nearly limitless parallel processing. In fact, if the back end can process 4.7 s worth of data in just 0.1 s, the comparative processing ratio is $1.8 \times 4.46/0.1 = 80.3$ times as high for the back end. Thus, the bandwidth significantly reduces the effective parallelism of the back end.

An even lower effective parallelism of the back end results if the processing capability of the mobile device is increased. Suppose that the mobile device is now updated to have two processors, each capable of processing 8% of the data in 1 s. The governing equations are now $15\% \times (T - 0.3 \text{ s})$ of the processing on the back end, as before, and $16\% \times T$ of the processing on the mobile device. Suppose we again divide the data into 14 partitions. Now we assign eight partitions to the mobile device and six to the back end. Using the approach as above, the mobile device completes it task in 3.57 s, while the back end completes its task in 3.16 s. The effective parallelism of the back end is reduced to only 0.75.

Now, suppose, for the original mobile device, bandwidth capacity is doubled. Here, the governing equation for the back end becomes $30\% \times (T - 0.3 \text{ s})$, and if we divide the image into 14 partitions again, we might assign 3 to the mobile device and 11 to the back end. The mobile device thus completes its 21.4% of the processing in 2.68 s, and the back end completes its 78.6% of the task in 2.92 s. The effective parallelism of the back end is 3.67 (a more than 100% improvement over the original 1.8), but the task completes in 63.6% of the time (a 36.4% improvement).

These examples illustrate the advantages of a flexible architecture that allows parallelism involving both the mobile device processor and the back end. Several factors were omitted for conciseness—for example, the cost of the back end—but the key design factor was assumed to be minimizing the time to task completion. Importantly, the design overviewed allows flexibility in optimization (e.g., to cost, bandwidth use, etc.) beyond simply that of time to completion.

Having overviewed some considerations of parallelism by task and by component, the next section introduces parallelism by meta-algorithmics. This will be the main theme of this book, and will also be shown to incorporate many of the findings of these other two major forms of parallelism.

2.4   Parallelism by Meta-algorithm

If Voltaire were alive today, well, he would be even older looking than in the famous Dali painting ("The Slave Market with the Disappearing Bust of Voltaire"). But, more importantly, if he were both alive and computer literate, he might say that if meta-algorithmics did not exist, it would be necessary to invent them. There are a number of reasons that we can say this. First off, the cloud has made—for a wide and growing number of applications—the old assumptions about processing and storage scarcity irrelevant. Secondly, increased availability, form factors, and power of parallel processing have made hitherto unthinkable approaches not only thinkable but also *de rigueur*. As a consequence, even when bandwidth, cost, and/or availability limits access to cloud computing, many stand-alone systems—from workstations to laptops to touch pads—have significant parallel processing capabilities that obviate this lack of access. Thirdly, years, and in some cases decades, of work on intelligence engines for content digitization—especially document, text, speech, and image understanding ones—have resulted, in many cases, in highly accurate but relatively inflexible systems. The amount of increased accuracy that can be derived from further optimization of a single algorithm, engine, or system is rather small. Fourthly, it has become clear that when multiple "intelligence generators"—or algorithms, engines, or systems that perform useful digitization, analysis, interpretation, classification, and so on—are evaluated, the individual generators tend to make mistakes on different types of content. These differences are a legacy of the differences in how those generators were created, tested, deployed, changed, and upgraded over the years. These differences are often used by their owners to highlight the advantages of one generator over its competitors, which from a financial standpoint is both expected and rational. However, from a parallel processing standpoint, these differences are an opportunity. In fact, they are a huge opportunity. In this book, I argue that being able to process with multiple intelligence generators in parallel is the most significant opportunity offered by the combination of cloud, parallel processing, and intelligence generator maturity.

This section puts forth the argument for parallelism by meta-algorithm and introduces some of the meta-algorithmic patterns that will be a core focus of this book. As described in Chapter 1, meta-algorithmic approaches and their more formal patterns provide a toolbox of potential solutions for intelligent system design and deployment architectural choices. Meta-algorithms are related to and extend data fusion, classifier fusion, ensemble methods, and other hybridizing or combining approaches. Importantly, meta-algorithmics are intuitive, relatively easy to apply, and can be readily adapted to domain-specific nuances. That is, while meta-algorithmic approaches like weighted voting, predictive selection, and confusion-matrix-based classification are easily understandable as generic patterns, they are also readily made specific for tasks as different as image segmentation and biometric voice identification.

In this book, meta-algorithmic patterns will be described in Chapter 6. The applications of meta-algorithmics will focus on the *first-order meta-algorithmic patterns* in Chapter 7 and elaborate to the application of more complex—but not necessarily more effective—second- and third-order meta-algorithmic patterns of Chapters 8 and 9.

First-order meta-algorithmics are characterized by their relative simplicity. Meta-algorithmics are designed to provide the means of combining two or more sources of knowledge generation—that is, algorithms, engines, or systems—even when, or especially when, the combined generators are known only at the level of black box (input and output only). Five primary first-order patterns will be described. These include the (1) Sequential Try, in which knowledge generators are applied in a specific order until a sufficient accuracy or other specification is obtained. The second first-order meta-algorithmic pattern is the (2) Constrained Substitute pattern, which allows the choice of a suitable reduced-expense (in terms of cost, processing time, bandwidth, a combination thereof, or other metric) algorithm, engine, or system to—effectively—replace a higher-expense approach. The third of these types of patterns is the (3) Voting pattern, including its often more powerful variant, the (3a) Weighted Voting pattern. This pattern is the first to include the output of multiple algorithms, services, or systems in the final output, rather than simply selecting the best knowledge generator. The fourth pattern, (4) Predictive Selection, is quite powerful, and usually involves choosing the information generator that has the highest precision in a specific predictor test. The last of these patterns, (5) Tessellation and Recombination, is shown to be especially useful for creating correct results even when none of the individual generators produces a correct result—a process called emergence. It should be noted that voting and weighted voting can also result in emergence.

The first-order meta-algorithmics provide a relatively broad set of basic patterns that can be deployed, with some domain expertise, to a wide array of systems of intelligence. More complicated patterns comprise the *second-order meta-algorithms*. A new set of analysis tools—namely, output space transformation, confusion matrices, and expert decisioners—are required for these second-order meta-algorithmic patterns. In addition, second-order meta-algorithmics incorporate system thresholding. If a specific degree of certainty (the *threshold*) is not reached when using a simple meta-algorithmic pattern, then a second decision approach is taken. In order to proceed to these combinatorial meta-algorithmic patterns, the confusion matrix—a powerful tool for temporal and series-parallel design of meta-algorithmic systems—is explored in full first. The first of the second-order meta-algorithmic patterns, therefore, is the Confusion Matrix pattern—which is useful in, for example, Predictive Selection (a first-order pattern). Its variant, the Weighted Confusion Matrix pattern, is more generally applicable to the combinatorial second-order patterns. The next second-order meta-algorithmic pattern is in fact such a combinatorial pattern; namely the Confusion Matrix with Output Space Transformation pattern. With this pattern, the multiple intelligence generators can not only work on the same input data and create compatible output data but also produce well-behaved output data. This means that the output probability curves have similar behavior across the input set for all of the generators. This pattern is usually deployed in the case where one or more of the engines report probabilities as part of their output. Thus, this pattern can also be termed Confusion Matrix with Probability Space Transformation.

Continuing on second-order meta-algorithmic patterns, other combinatorial patterns are then considered. Tessellation and Recombination with Expert Decisioner, Predictive Selection with Secondary Engines, and Single Engine with Required Precision are three such patterns. All three are built on first-order meta-algorithmics: Tessellation and Recombination, Predictive Selection, and Sequential Try. Two variations on the first-order Voting pattern are then described: Majority Voting or Weighted Confusion Matrix, and Majority Voting or Best Engine. These relatively simple second-order patterns are especially useful when a certain level of confidence in the output is required. Another such pattern is the Best Engine with Differential Confidence or Second Best Engine pattern, which performs a minimized Sequential Try if and only if the engine with the highest precision provides an output with too low of a value of confidence to accept. Similarly, the Best Engine with Absolute Confidence or Weighted Confusion Matrix provides another confidence-dependent combinatorial pattern. While the names of these patterns may seem a bit arcane at first read, they are relatively simple in implementation, as they build simply from the pattern building blocks, commonly called subpatterns.

The third-order meta-algorithmic patterns—generally focused on feedback from the output to input—add a further level of complexity. They also add a higher degree of flexibility and tunability, since they provide multiple subpatterns joined together. The first third-order meta-algorithmic pattern is, in fact, the simple Feedback pattern. This is in some ways closely related to the next third-order meta-algorithmic pattern, the Proof by Task Completion pattern, which dynamically changes the weighting of the individual knowledge-generating algorithms, systems, or engines. The confusion matrix repertoire of approaches is used in the next pattern, the Confusion Matrix for Feedback pattern. Similar, but reliant on rules and learned constraints, is the Expert Feedback pattern. The fifth third-order meta-algorithmic pattern is termed the Sensitivity Analysis pattern, which is focused on identifying stable points in the solution space. This includes stable areas within the confusion matrix for intelligent engine combinations, and stable areas within the correlation matrix for algorithmic combination. The next pattern is concerned with what could be considered "introspective meta-algorithmics," in which individual engines are tuned for subclasses of the overall task. This means that different meta-algorithmic combinations may be used for subsets of the data—akin to the Predictive Selection pattern—and in addition the intelligence-generating algorithms, systems, and engines themselves may be configured differently for each subclass of the overall problem space. This pattern is termed the Regional Optimization pattern, but could also be termed the Extended Predictive Selection pattern. The seventh and final third-order meta-algorithmic pattern is termed the Generalized Hybridization pattern. This pattern is concerned with optimizing the combination and sequence of first- and second-order meta-algorithmic patterns for a given—generally large—problem space.

2.4.1   Meta-algorithmics and Algorithms

Using a loose definition, an algorithm is a set of tasks that transform data of one type, termed "input," into data of another type, termed "output." Mathematically, then, an algorithm is a *function*. In meta-algorithmics, we are concerned with multiple functions—each output is a transformation of the input, and the function (the mathematical description of the transformation) is generally different for each algorithm. This is advantageous, however, because there are two fundamental principles of algorithms as applied to meta-algorithms:

1. No single algorithm encapsulates the complexity of most highly challenging artificial intelligence tasks, including machine learning, machine vision, and biometrics. A plurality of algorithms is more likely to be able to provide a correct answer, from at least one of the algorithms, than a single algorithm is.
2. Given a plurality of algorithms, it makes more sense for a system designer to optimize each of the algorithms for a partition of the input range, and leave the remainder of the input range to the other algorithm(s) in the plurality.

With these two principles in mind, algorithm development proceeds differently for meta-algorithmics than for traditional intelligent systems. A new algorithm may be developed, not for providing high accuracy across the input domain, but instead for providing high accuracy where the other algorithms fail. This type of "targeted algorithm," by standard measurements of such important metrics as precision, recall, and accuracy, is ineffectual. But, within the context of cooperating with the other algorithms in the meta-algorithmic set, a targeted algorithm can be very effective—so long as it is only used—or at least predominantly used—where it actually is accurate.

2.4.2 Meta-algorithmics and Systems

Meta-algorithmic systems are unique not just in terms of how algorithms are developed; they are unique in how they can be comprised. The targeted algorithms that comprise the meta-algorithmic system need not be broadly useful—they simply need to be useful for, at minimum, the targeted partitions of the input range for which they are brought into the system. Thus, commercial off-the-shelf (COTS), open source, and custom designed algorithms can be combined together, as appropriate. If cost is an important factor in the overall system, then the designer may wish to choose more open source and custom designed systems. Since most COTS algorithms, being for sale and therefore of differential value, outperform the various forms of freeware, it is often the case that a meta-algorithmic system without any COTS algorithms will require more total algorithms for a given level of accuracy (or other measurable system performance metric). While this system approach may require more storage, more processing time, and more code maintenance, it can also provide significantly more flexibility. In general, the more algorithms added to the system, the greater the overall system behavior. This is because of the following two complementary, yet simultaneous, behaviors:

1. As more algorithms are added, the aggregate behavior is more likely to be highly accurate by the *central limit theorem*. As less and less accurate algorithms are added to the repertoire deployed for the meta-algorithmic system, the overall accuracy of the system continues to increase so long as there is not a systematic bias away from the correct interpretation.
2. The variance of the system tends to increase, since, in general, variances add, assuming there is no systematic correlation among the distinct algorithms. Thus, in addition to mean behavior improving, the robustness of the system—in terms of having greater coverage of the input domain wherein at least one of the algorithms can provide a correct response—improves.

2.4.3   Meta-algorithmics and Parallel Processing

A meta-algorithmic system, at the highest level of its architecture, is certainly parallel. This is because each of the meta-algorithmic system components—whether they are simple algorithms, complicated systems, or feature-rich, highly integrated engines for information generation—can be performed in parallel upfront. As such, the connection to traditional parallel processing is innate. If each meta-algorithm is considered to be performing the same task—namely, the transformation from a specific input to a specific output type—then meta-algorithmics are analogous to parallelism by task. If, however, each individual meta-algorithm is considered to be operating on a different partition or set of partitions of the input domain, then meta-algorithmics are analogous to parallelism by component.

These analogies are more than just empty comparisons. It should be clear even from the introductory discussion of meta-algorithmic patterns in this chapter that meta-algorithmic approaches always involve at least some parallelism. Given sufficient processing and storage resources, then, meta-algorithmic performance is limited by that of the slowest meta-algorithm.

At another level of complexity, meta-algorithmic parallelism comprises the simultaneous consideration of two or more meta-algorithmic patterns in parallel. This poses, simultaneously, the following two types of parallelism:

1. Parallelism of the meta-algorithmic patterns
2. Parallelism within the meta-algorithmic patterns

This may seem obvious, but the consequences of it are trickier than the words above convey. Suppose, for example, that meta-algorithmic pattern 1 involves using meta-algorithmic engines A, B, and C, while meta-algorithmic pattern 2 involves using meta-algorithmic engines A, B, and D. Now suppose that different input and output operations are required for the two distinct meta-algorithmic patterns. Using identification of the input, processing, and output operations as small alphabetic letters, then we have the following operations for meta-algorithmic pattern 1:

1. Aa, Ab, Ac
2. Ba, Bb, Bc
3. Ca, Cb, Cc

Aa, Ba, and Ca are the input operations; Ab, Bb, and Cb are the processing operations; and Ac, Bc, and Cc are the output operations. Next, for meta-algorithmic pattern 2, let us suppose the simplest case in which the input and output operations change but the processing operations remain the same. In this case, we have the following operations for meta-algorithmic pattern 2:

1. Ad, Ab, Ae
2. Bd, Bb, Be
3. Da, Db, Dc

Ad, Bd, and Da are the input operations; Ab, Bb, and Db are the processing operations; and Ae, Be, and Dc are the output operations. From the above, we have the following set of operations and in parentheses the number of times they are invoked (if more than once):

1. Aa, Ab (2), Ac, Ad, Ae
2. Ba, Bb (2), Bc, Bd, Be
3. Ca, Cb, Cc
4. Da, Db, Dc

From this, it is more clear that these could be run (1) in two parallel paths, meta-algorithmic pattern 1 and 2; (2) in four parallel paths, {Aa, Ab (2), Ac, Ad, Ae}, {Ba, Bb (2), Bc, Bd, Be}, {Ca, Cb, Cc}, and {Da, Db, Dc}; and (3) in six parallel paths: {Aa, Ab, Ac}, {Ba, Bb, Bc}, {Ad, Ab, Ae}, {Bd, Bb, Be}, {Ca, Cb, Cc}, and {Da, Db, Dc}, among other combinations. Comparing the options here to those in the sections on parallelism by task and component (Sections 2.2 and 2.3), it is clear that meta-algorithms offer at least as wide a range of serial/parallel design options as those two types of parallelism.

2.4.4 Meta-algorithmics and Data Collection

Any approach that truly provides a new motif for data analysis must not only affect the way in which data is processed but must also affect the way in which data is gathered. Meta-algorithms, indeed, will be shown to significantly impact not just the quantitative analysis of data but also the qualitative manner in how data is created. In fact, using, for example, summarization (Section 10.2), I will show that if the data collection model is designed with later meta-algorithmics in mind, it can lead simultaneously to more scalable, more efficient, and more valuable data gathering.

Because meta-algorithmics are concerned with differential algorithm, system, or other intelligence-generating engine response to the same set of input data, it is clear that training data sufficient to describe the range and variability of the input is necessary. With more training data come the following advantages:

1. *Better statistical behavior of the classes*: As more elements in each class are collected, by the central limit theorem, the behavior of each class will become more Gaussian. This allows better identification of classes, and thus a better estimate of the number of classes. This also provides the dimensions for the meta-algorithmic confusion matrix used in many second- and third-order meta-algorithmic patterns.
2. *Better identification of domain/range of input*: With extensive training data, areas in the domain space that contain no data are much more likely to be actual nonrelevant domain sections. Such gaps in the domain of one or more intelligence-generating systems significantly aid in the selection of individual systems for the application of a meta-algorithmic pattern.
3. *Improved data layering*: With more data comes better association among the data. That is, more nuanced data sets afford greater possibilities for multiple meta-algorithmic patterns to be evaluated on the same data set in order to elucidate an optimum meta-algorithmic

approach downstream. As more data is collected, higher-confidence relationships among primitive data elements can be attained, creating composite data elements. This may allow meta-algorithmics to be performed both on the primitive and the composite data.

4. *Repurposability*: Data should be collected in light of the fact that the data collector does not know a priori which meta-algorithmic pattern will be most effective for the task at hand. Thus, the ground-truthed data should be as broadly useful as possible. In general, this is achieved by making the data as atomic as possible. That is, tagging should be made as atomic as is possible without unnecessarily burdening the person providing the tagging. In text, for example, labeling the key sentences may be of sufficient atomicity to avoid asking for the keywords to be labeled, since the keywords can be determined using these key sentences. Labeling only the key paragraphs, however, may be insufficiently atomic to have future utility for keyword and semantic tagging.

5. *Scalability*: The data should be collected such that when new algorithms, engines, and/or systems are built, borrowed, bought, or otherwise brought into the system, the previously collected training data is still relevant. Thus, training data comparing one algorithm with another is not scalable, but training data based on ranking the elements within the data is a scalable approach: it requires no additional training irrespective of the number of meta-algorithmic approaches—algorithms, engines, systems, or patterns—added after the data has been created.

6. *Nonprovinciality*: The training data should not be too closely tied to the specifics of one of the engines. Another way of saying this is that the engine should be as generic as possible. One means of helping assure this is to have the training plan architect be a different person than the system and meta-algorithmic design architect(s).

7. *Ground truth is extremely expensive, and so where possible, the collection of ground truthing data should be "compressively sampled"*. This is easier to achieve than it might seem. A set of potential training cases can be prefiltered, even automatically, so that very obviously similar samples are not manually ground truthed (assuming that it will not be important to keep all of them). Some care must be taken here, however, not to—in performing such compressive sampling—violate the rule (6) about nonprovinciality. For example, such an approach may tend to favor boundary-based classification approaches (such as support vector-based methods) over cluster statistics-based methods (such as Gaussian mixture models and expectation maximization approaches).

Aside from these concerns, somewhat specific to meta-algorithmics, the normal concerns with data collection are fully applicable. The samples collected should be representative, updated through time to reflect the currently relevant types of data samples to be analyzed, and occasionally re-analyzed to make sure that the interpretation of the training data has not drifted over time.

2.4.5   Meta-algorithmics and Software Development

In concluding this introduction to meta-algorithmics, the heritage of meta-algorithmics should be acknowledged. While significantly different—and more diverse—than the boosting techniques described in Section 1.5, meta-algorithmics nevertheless share in the common the means to provide system and algorithmic adaptability to changes in input.

One recent mention of meta-algorithmics is in the field of software development. While this use of meta-algorithmics is different in nature than how they are used in this book, it is reviewed here to provide comparison. Programming by optimization, or PbO, reviewed in Hoos (2012), is an interesting form of parallelism in which software developers architect a large design space of programs to complete a specific task. Optimized programs for accomplishing the task under different context are automatically selected through a given cost function—for example, speed. Many of the important knowledge generation problems discussed in this book—machine learning, classification, informatics, artificial intelligence, and so on—are NP-hard problems, and as such traditional programming methods can result in systems lacking flexibility and robustness to changes in the context in which the programs operate. A good example of such a system is a global climate model. Increasing rate of change in climate is resulting in increasing complexity of climate prediction, making linear models nonrobust (Benestad and Schmidt, 2009). Similarly, there is increasing complexity in surveillance, web use tracking, and biometric identification due to increased global on-line consumers. In each of these cases, more robust systems are almost guaranteed to come from hybrid, or parallel systems, which can adapt to changes in input without losing relevance.

In Hoos (2012), the term meta-algorithm refers to the optimization procedure. One example is the stochastic-optimization procedure (Spall, 2003). In Hoos (2012), three classes of meta-algorithmic methods are identified: (1) racing procedures, which focus on parameter optimization; (2) model-free searching procedures including stochastic local search, which include perturbations that are in some ways analogous to mutations in genetic algorithm-inspired approaches; and (3) sequential model-based optimization (SMBO), which uses information gained from parameter configuration—such as garnered from racing procedures—to determine promising overall configurations for the system. These optimizations, clearly, have more in common with boosting than they do with first-, second-, and third-order meta-algorithmic approaches that are the primary focus of this book.

2.5 Summary

This chapter overviewed and provided insights into how the three main forms of parallelism provided the ability to improve the design of intelligence systems. Parallelism by task defines the structural blocks of the architecture to be individual processes that are performed on the data. Parallel designs using tasks allow the definition of parallel pipelines. Parallelism by component, in contrast, defines the structural blocks of the architecture to be subsets of the overall data set. Parallel designs using components allow the partitioning of data to closely map to the processing capabilities throughout the distributed system. Finally, parallelism by meta-algorithmics was introduced. Meta-algorithmics are the means by which parallel processing strategies are brought to multiple intelligence generators, each of which acts on the same data and provides output of the same type. First-, second-, and third-order meta-algorithmic patterns are introduced, corresponding to increasing complexity (more variable factors in the design) and in some cases increased tunability. Meta-algorithmics are used for a plethora of algorithm, system, or other intelligent engine optimizations, and examples in a wide array of fields will be illustrated. These patterns and their applications, in fact, will be the main focus of this book, and the nuclei of these patterns will be recognizable in the older parallel processing and hybrid machine intelligence fields.

References

Amdahl, G. (1967) *Validity of the Single Processor Approach to Achieving Large-Scale Computing Capabilities.* AFIPS Conference Proceedings, vol. 30, pp. 483–485. AFIPS Press, Reston, VA.

Benestad, R.E. and Schmidt, G.A. (2009) Solar trends and global warming. *Journal of Geophysical Research-Space Physics*, **114** (D14101), 18 pp.

Dean, J. and Ghemawat, S. (2004) *MapReduce: Simplified Data Processing on Large Clusters.* Proceedings OSDI '04, pp. 137–150.

Dijkstra, E.W. (1959) A note on two problems in connexion with graphs. *Numerische Mathematik*, **1**, 269–271.

Gamma, E., Helm, R., Johnson, R., and Vlissides, J. (1994) *Design Patterns: Elements of Reusable Object-Oriented Software*, Addison-Wesley, 416 pp.

Hoos, H.H. (2012) Programming by optimization. *Communications of the ACM*, **55** (2), 70–80.

Monash, C. (2007) Are row-oriented RDBMS obsolete? DBMS2, January 22, 2007, http://www.dbms2.com/2007/01/22/are-row-oriented-rdbms-obsolete (accessed January 11, 2013).

Spall, J. (2003) *Introduction to Stochastic Search and Optimization*, John Wiley & Sons, Inc., New York, 618 pp.

3

Domain Areas: Where Are These Relevant?

All generous minds have a horror of what are commonly called 'Facts'. They are the brute beasts of the intellectual domain.

—Thomas Hobbes

Never question the relevance of truth, but always question the truth of relevance.

—Craig Bruce

3.1 Introduction

The primary focus of this book is on meta-algorithmics, the "third form" of parallelism. Parallelism by task and parallelism by component, as overviewed in the previous chapter, are hugely important in systems demanding optimum performance. However, for systems requiring optimal accuracy, robustness to changing input, or flexibility in terms of system architecture, meta-algorithmics are the most promising form of parallelism.

In this chapter, the breadth of the application space for meta-algorithmics as elaborated in the rest of the book is addressed. Turning Hobbes sideways, these "intellectual domains" will be the brute beasts for carrying forth the facts regarding each of the forms of parallelism: by task, by component, and by meta-algorithmics. There are four primary domains, which will be illustrated in each of the chapters on parallelism; that is, Chapters 4–9. In addition, there are eight secondary domains that will be used more sparingly to illustrate either more subtle points about parallelism by task and component, or to illustrate one or more of the 21 different patterns for meta-algorithmics introduced in Chapter 6. Combined, these domains may seem unrelated, and unnecessarily broad for a single book. However, this book aims to show that the types of parallelism described are broadly applicable and perhaps more importantly *readily* applicable to the architect of any/all intelligent systems. There is no better way to show this than by example.

Meta-algorithmics: Patterns for Robust, Low-Cost, High-Quality Systems, First Edition. Steven J. Simske.
© 2013 John Wiley & Sons, Ltd. Published 2013 by John Wiley & Sons, Ltd.

3.2   Overview of the Domains

The primary domains encompass both text and image content, both one-dimensional (1D) and two-dimensional (2D) signal processing, both natural and human-crafted content, and various forms of security. In addition, the primary domains reflect subject areas in which I have some expertise. Real expertise: as in patents, publications, and product development. In order to achieve these goals simultaneously, I selected the following four domains (for the reasons in parenthesis):

1. Document understanding (2D signal processing, human-crafted content).
2. Image understanding (2D signal processing, natural content).
3. Biometrics (1D, 2D, and three-dimensional (3D) signal processing, natural content, authentication security).
4. Security printing (1D and 2D signal processing, natural and human-crafted content, various forms of security).

From the above, security printing is noticeable for the breadth of analysis challenges it presents. Indeed, security printing has been a favorite of mine over the years for just that reason. Combining multiple forms of security, imaging, printing, and system architecture technologies into a single ecosystem is certainly hard to resist. Importantly, security printing requires overdesign. If a system can provide variable data printing (VDP), security, and imaging-based data recovery all in one printed mark, then the associated technologies can readily be "backed off" to provide other printed data, secure printing, and content decoding solutions.

Unfortunately, in spite of its great breadth, security printing is the domain least covered by existing literature. Many of the approaches to security printing that have been used in recent years are not covered by books or review papers. Document understanding, image understanding, and biometrics literature is far more mature. For that reason, I will cover security printing in more depth than the other topics in this chapter.

The fields of document and image understanding are also quite broad, with perhaps greater depth in addition. Biometrics is also a personal favorite, as I cannot see its importance ever diminishing with the increasingly difficult CAPTCHA (completely automated public Turing test to tell computers and humans apart) and other Turing test approaches necessary to differentiate man and machine. Even when biometrics for telling two humans apart is no longer a challenge, biometrics for disambiguating a real human from a computer-based representation of that human will be important. Any intelligent system designer in the 2010s and beyond will have to be able to incorporate biometrics into their architecture, just as any artificial intelligence student of the 1990s had to know how to build an artificial neural network.

The secondary domains, while less broad than the four primary domains, are used to elaborate further the parallel processing and parallel analysis approaches comprising the following eight chapters. Image segmentation provides a different, less structured, set of image analysis techniques than document segmentation. Speech recognition, focused on 1D signal processing, borders the technology of biometrics on one side and the broader field of audio analysis on the other side. Medical signal processing also requires advanced 1D signal processing in the case of biopotentials such as electromyograms (EMGs), electroencephalograms (EEGs), and electrocardiograms (ECGs), while readily bridging to 2D medical imaging in the case of vectorcardiograms (VCGs). Other medical imaging applications cover an interesting domain between that of surveillance and image segmentation/processing. Natural language processing (NLP), on the other hand, focuses on text-based processing, and as such can be entirely independent of signal processing or image processing (though it can be performed on the output of optical character recognition (OCR)). Surveillance combines 2D signal processing (image processing) with frame-to-frame analysis, effectively requiring 3D image processing. OCR is a specialized technology within the broader field of document image processing, used to convert images of characters into their electronic (e.g., ASCII or Unicode) representations. The last of the secondary domains is that of security analytics. These analytics are concerned with providing visibility into the activities, accesses, applications, and agents involved in a given network. What are the agents, or "personas," accessing, what are they doing with this access, and what applications are they using after gaining access? These analytics share some domain space with NLP, but are more directly concerned with identifying specific patterns of behavior, which is somewhat analogous to some aspects of surveillance as well.

3.3   Primary Domains

The four primary domains are (1) document understanding, (2) image understanding, (3) biometrics, and (4) security printing. Roughly, the first two are concerned with understanding the content other people have created. Biometrics is concerned with understanding content nature has created. Security printing is concerned with understanding both of those types of content, along with recovering content created intentionally for downstream decoding and authentication.

3.3.1   Document Understanding

Document understanding systems are focused on understanding, with equal facility, scanned documents and electronic documents. This broad field of research includes the subdomains of NLP (Section 3.4.5) and OCR (Section 3.4.7). Document understanding is primarily concerned with adding meta-data to document data. This upgrading of content can take many forms, including the following four important classes of content:

1. File information, or content, meta-data
2. File context meta-data
3. File use meta-data
4. File analytics meta-data

File content meta-data includes structural (syntactic) and semantic (meaning-based) content. Structural meta-data is a catalog of the content in the document, and includes much of the important traditional format, presentation, and content information associated with creating and rendering a document. Format information includes structural relationship between other sets of data; for example, the relationship between header, footnotes, titles, and paragraphs in an article or the relationship between legend and image in a figure. Presentation information includes the number of columns into which a page is flowed, which does not affect the format information in, say, an article or figure, but it certainly affects the way it looks in the document form. Article formats are perhaps unchanged when viewing web content on a laptop or tablet compared to when viewing content on a mobile device, but the presentation is greatly different. The structural content itself is the traditional image, text, graphics, tables, links, and so on, that comprise the "information" in the document.

Semantic, or meaning-based, content is information that changes—it is actually enriched—when the document participates in an ecosystem of other documents. Semantic content tagging is related to NLP, and includes diagnosis of the relationships between different words. Absolute semantic tagging is derived from—or used to generate—*taxonomies*. The set of all semantic tags for the entire set of potentially related documents is an *ontology*. Relative semantic information, however, may prove to be more relevant in the long term, as changes in meaning associated with specialized usage patterns (slang, trade jargon, etc.) and differences in semantic content among related articles are highly significant for establishing authorship, for uncovering plagiarism, for identifying related documents, and for a host of other linguistic tasks.

Document understanding is also used to produce file context meta-data. In some ways, this is similar to relative semantic content as described above. However, contextual meta-data is also concerned with situational awareness: what type of content is especially used in a given document in comparison to any other document? What class of document is this? Which of a set of related—or seemingly related, anyway—documents is the most typical, or most representative, document? These are difficult questions to answer, and often relate to or directly rely on the analytics meta-data to be described shortly, but are very important for context. A key element in contextual meta-data is keeping the referent document set clear. For example, consider a document about Tallahassee in the context of a large set of documents about the state of Florida. Then, consider the same document in the context of a large set of documents about different US state capitals. In the first case, a differentiating set of terms might be {capital, government, legislature, Florida A&M, Florida State University}, while in the second case the differentiating terms might be {Florida, Leon County, Florida A&M, Florida State University}. Clearly, there are some overlapping terms—Florida A&M and Florida State University—and some terms specific to the particular referent. Terms that are shared irrespective of the referent may be deemed *absolutely differentiating* terms, while terms unique to a given referent may be consider *relatively differentiating* terms.

The third type of document understanding meta-data is that for file usage. File use meta-data is focused on the events—time, user, device, version, and so on—associated with the stage in the workflow in which the document is acted upon. This type of meta-data includes the unique tag to identify the document, tags to associate the document with other documents, timestamps associated with versioning of the document, salient user authentication information (usually digitally signed), and identifiers (IDs) associating the document with other, related documents. Examples of related documents are documents associated with the same workflow, documents created by or worked on by the same author or team, and documents that are discovered to contain similar content, context, or use history as the document. For the latter, we require document analytics, which happens to be our next topic.

File analytics meta-data includes absolute and relative analytics about the document content (data) and/or meta-data as described in the previous paragraphs. Analytics, for example, can be used to generate the absolutely and relatively differentiating terms, as described above for context meta-data. Analytics, however, go beyond the tabulating of term occurrences and their relative frequencies. Analytics may include document fingerprinting, in which expressions, phrases, word co-occurrences, and other style-related metrics are computed. This fingerprint can be used to identify original author, to uncover plagiarism, and to provide suggestions for greater readability—for example, by suggesting phrase variation—if so desired/appropriate.

Considering the breadth and depth of meta-data that can be added during document understanding processing, it is obvious that document understanding is the broad field of extracting and defining the content in a document—either during digitization or during analysis of an already electronic document. The technologies salient primarily to digitization—that is, scanner or camera capture of a physical document, label, sign, and so on—are reviewed first:

1. Background versus foreground and zoning analysis. Here, the salient foreground content is separated from the background, so that the text over a colored background, for example, can be readily recognized using OCR (see Section 3.4.7).
2. Text versus nontext. Next, text areas, which are important for document indexing and classification, are extracted and converted into electronic form (ASCII or Unicode characters) using OCR, handwriting recognition, barcode reading software, and/or template matching. The nontext regions are treated the same as electronic nontext regions (see point 4).
   Next, both the now-digitized and the native-electronic documents are processed and their meta-data upgraded using one or more of the following techniques:
3. Shape recognition and special item recognition is used to identify both general (e.g., barcodes) and specific (e.g., slides, negatives, custom logos, etc.) regions for downstream analysis.
4. Nontext "image" data is further classified as being a photo, line art, graphic/business graphic, large text (e.g., colored text and headline text), drawing, logo, or map region. Tabular regions can also have their text fields extracted for table analysis (e.g., to internalize a table into a spreadsheet).
5. With the set of regions segmented and classified, the document can now be matched against a set of possible document templates.
6. Compound text regions, such a bibliographical entries, table and figure legends, headers and footers, and so on, can then be split into their salient fields, such as "author," "journal title," "volume," "number," "pages," and "year."
7. Different analytical algorithms, such as NLP, text and image entropy calculation, and correlation with other documents, are then performed.

Note that documents of all types—digitized or native electronic—can have their meta-data upgraded at any time in their lifecycle; for example, after new analytics technologies become available. Thus, a wide variety of algorithms can be performed on document content, in parallel, at different times, on different portions of the document content, and for different downstream purposes.

3.3.2 Image Understanding

In the previous section, document understanding was discussed in broad terms. One of the tasks of document understanding was shown to be the extraction of images as separate zones from the text. Images were then classified as specific classes of content; for example, photo, line art, graphic, business graphic, colored text, headline text, drawing, logo, map, or specialized images such as slides or negatives.

Image understanding starts where this classification leaves off. Typical image understanding applications include scene recognition, face detection and identification, shape matching and recognition, product inventory, location detection, object extraction, and intelligent image processing—for example, to automatically improve the image quality. Inspection systems are also image understanding applications in which a desired, predefined metric or set of metrics is automatically generated and compared to a desired range of values, for example, for pass/fail of the printing quality, captured image quality, or scene readability. Several subdomains of image understanding are considered separately in later sections: image segmentation (Section 3.4.1), medical imaging (Section 3.4.4), and surveillance (Section 3.4.6).

Technologies used for image understanding are quite broad, and include thresholding and segmentation approaches not dissimilar from those used for document processing. Object recognition is performed using pattern matching, which is often based on cross-correlation between the intended and actual images. Image processing techniques such as image contrast, exposure, and color balance detection are used to decide on the image restoration approach for images of unacceptably low quality.

Another important consideration for image understanding is image compression. Functional image compression is desired, whereby the compression does not prevent effective downstream use of the document, for example, for document classification, object tracking, or inspection pass/fail. It is very important during image understanding to consider what the later uses of the image will be.

In the context of this book, image segmentation is an excellent candidate for parallel operations. Different images in a video stream and different partitions of a larger image can be processed in parallel. In addition, parallel image processing approaches can be used to find the modal image in a series of images and to create 3D images from a set of 2D images.

Since image processing tasks have a high error rate, being able to use multiple parallel image processing *algorithms* has great potential for improving accuracy and robustness. Accuracy can be improved through the intelligent combination of distinct algorithms, while robustness can be improved by taking advantage of the often large differences between data (input) observed in training data as opposed to actual deployment data through feedback approaches. Different algorithms tuned to different subclasses of the possible input, properly cooperating, provide an optimal analysis for each of the distinguishable subclasses and thus can improve, often greatly, the overall image understanding. As will be shown in the chapters that follow, image understanding is especially amenable to performance improvement through traditional forms of parallelism, with simultaneous improvement in accuracy and/or robustness through meta-algorithmic parallelism.

3.3.3   Biometrics

A biometric is the measure of an attribute of a living organism for the purpose of uniquely identifying that organism compared to all other members of its class. Biometrics include 1D signal processing (e.g., audio for speaker identification) and 2D image processing (e.g., for fingerprint identification). While largely dissimilar in spectral composition, both 1D and 2D biometrics can benefit from the same signal normalization processes (Wallace *et al.*, 2012). Normalization is important since biometrics represent a broad class of problems, involving a mixture of pattern matching (identification) technologies as well as classification (one vs. many) technologies.

Table 3.1 Important physical biometrics and their measured attributes

Physical Biometric	Sample Measured Attributes
Face	Facial feature location, shape, size, and inter-feature distances
Fingerprint	Whorls, points of interest, pores
Hand	Size, shape, perimeter, lines
Iris	Distribution of high interest features
Retina	Vasculature distribution
Vein	Location map of earlobe or hand

Biometrics include physical biometrics, behavioral or "continuous" biometrics, and innate or chemical biometrics (Simske, 2009). These are captured in Tables 3.1, 3.2, and 3.3, respectively. While the physical and behavioral biometrics use image and signal processing techniques, the innate biometrics are based on biochemical analysis techniques, including bioinformatics for sequencing.

Biometrics are highly amenable to parallel processing approaches for at least three reasons. The first is that different biometrics, generally, require different techniques for optimal identification accuracy. Support vector machines (SVMs) are quite useful, as expected, for image-based biometrics, while bioinformatics are used for genetic innate biometrics. The second reason is that multiple sample windows (data streams) may be used to identify a person; for example, multiple phrases in a conversation, multiple images, multiple fingerprint images captured, and so on. The third reason is that multiple biometrics can be analyzed in parallel to provide higher overall identity (authentication) confidence. For this combination of biometrics, we wish to use the highest accuracy approach on each of the individual biometrics—and we also wish to use the highest accuracy *pattern for combining these biometrics*. This is precisely where meta-algorithmics come in.

3.3.4 Security Printing

Security printing is a broad set of technologies used to add, and later recover, identifying information to a physical object. As mentioned above, security printing as a domain of

Table 3.2 Important behavioral or "continuous" biometrics and their measured attributes

Behavioral Biometric	Sample Measured Attributes
Arm sweep	Location and velocity patterns
Fingerwriting	Location and velocity patterns, pressure kinetics
Gesture	Location, velocity, size, and shape of hand
Handwriting	Location and velocity pattern
Heartbeat	Electrocardiogram (ECG), vectorcardiogram (VCG), pressure, sound
Keystroke	Latencies, pressure kinetics
Voice	Cepstrals, formants, accent, timing in idiomatic expressions
Walking	Gait analysis (location and velocity patterns)

Table 3.3   Important innate or chemical biometrics and their measured attributes

Innate Biometric	Sample Measured Attributes
Genetic	DNA, RNA, mDNA, HLA
Tissue assay	Protein composition, protein expression
Mass spectroscopy	Chemical composition

machine intelligence is less well described in the literature than the other domains addressed in this book, and so it is reviewed in more depth here.

Broadly speaking, security printing consists of (a) determining the manner in which data will be added to the physical object; (b) associating the data with the object, a process called encoding; and (c) recovering the data from the object, usually through decoding. As the name implies, this information is usually associated with printing. However, the technologies to provide forensic authentication have improved to the point that printing is not always required. Data creation techniques useful for printing have also proven to be useful in other, nonprinted physical and electronic-native objects such as RFID and near-field devices, and file-associated meta-data used for secure access or secure authentication of electronic information.

There are a number of different factors to consider during the planning of a security printing deployment. In this section, six main factors are considered:

1. Nature of encoded information
2. Level of image analysis
3. Role of person performing the authenticating
4. Utility of the encoded information
5. Extent of print variability
6. Complexity of the encoding

These factors are now considered sequentially.

By the *nature of encoded information*, we mean whether the information is encoded in overt, covert, steganographic, or forensic functionality. Overt objects are both visible and understood to contain readable information by the average person. Barcodes are perhaps the most obvious example of this, and they have blossomed as the number of mobile devices capable of capturing images of sufficient quality to be decoded has increased. However, there is no innate security in reading barcodes. The barcode may redirect your mobile device to a website that downloads a virus to your device. The barcode may otherwise direct your device to a honeypot website, which lures you into downloading rogue software. All the data on your mobile device is therefore at risk. This type of risk is not unique to barcodes: any other overt mark—serial number, graphical alphanumeric, color data mark, and so on—can be copied or readily replicated by a fraudulent agent. However, the task of fooling consumers and retailers is made easier when no variable data is used, and the overt mark is simply a—supposedly—difficult to reproduce mark such as a hologram or guilloche. Even when these cannot readily be duplicated, they can be obtained from insiders. Having the overt feature contain data allows it to be connected to on-line verification, which offers a more reasonable degree of protection: the same code cannot be used multiple times and codes that are nonvalid will be so identified by the on-line service.

Covert printed objects contain information that is hidden in plain sight, and the means of decoding—perhaps even the very existence—of the content is not conveyed to at least some of the people having access to the physical item. Digital watermarks are often covert (in addition to being steganographic) and their location and the secure access rights to decode them are only provided to certain persons. Other covert marks include copy detection patterns, ultraviolet (UV) and infrared (IR) inks, MICR inks, and color combinations that depend on specific ink palettes.

Steganographic security printing objects are marks that contain hidden information, even if the marks themselves are overt and well understood to contain decodable information. When a digital watermark is known to be present in an image, the manner in which it contains information is hidden, or steganographic. However, the mark itself may visibly impact the image (strong watermark) or be difficult to notice (weak watermark). Strong watermarks are expected to survive copying, and thus are used for copyrighting material. Weak watermarks are not expected to survive copying, and are used therefore for authentication. Steganographic information can be contained by subtle text or logo manipulations, within the halftoning patterns themselves, or by the tacit addition of hue, intensity, and/or saturation variation to printed areas.

Forensic security printed information is used to identify with a certain degree of statistical confidence that a document, label, package, or other surface is authentic. If properly protected and/or difficult to reproduce, forensic patterns will be reused by fraudulent agents, rather than reproduced. In this case, the forensic mark must be reuse or tamper evident, as discussed below. Forensic materials—such as security substrates, security ink or substrate additives, and secure finishing coatings, laminates, or procedures—must be protected from theft, and so constitute controlled substances/procedures. Many different forensic materials exist. However, high-resolution imaging obviates the need for "special" substrates, additives, or finishing by allowing the item (printed or otherwise) itself to provide forensic authentication. High-resolution imaging hardware can be used for the forensic identification of a printed document or label using only a single printed character (Simske and Adams, 2010; Simske, Pollard, and Adams, 2010). This capability has been extended to other printed content, including small logos containing steganographic content, as well as significantly improving the statistical confidence in the forensics. No special forensics are needed; the printing itself *is* the forensic. Current work focuses on extending this to a portion of any surface, allowing any object to be forensically authenticated.

The related factor, *level of image analysis*, is important in security printing as it defines the type of applications that can be initiated by the security printed object. Forensic image analysis is the most difficult to reproduce, with other forms of image authentication providing acceptable accuracy at the item level and forensic confidence at the cluster level (Simske *et al.*, 2009). At the low end of statistical confidence for authentication is image inspection, in which images are graded for quality. Image similarity is often used for quality measurements—the lower the image variability within a set of legitimate samples, the more likely counterfeit samples will be correctly classified (Simske *et al.*, 2009).

The level of image analysis required for authentication is therefore dependent on the density of information contained in the security printing object, the entropy in the information, and the inter-cluster specificity of the measurement. The greater the density of information, all other factors being equal, the greater the statistical confidence and the greater the chance a printed object can be used for true forensic (e.g., less than 1 in 10^9 chance of a false positive identification) confidence. The greater the entropy of the information extracted, the greater the separability (e.g., Hamming distance) between any two randomly selected samples. However, low entropy is preferred for samples that should be decoded identically. In summary, then, for security printing objects we prefer (a) high bit density, (b) high entropy for sets of unrelated samples, and (c) low entropy for sets of related samples.

The *role of person authenticating* the security printed object is another means of defining the downstream applications and services that are initiated by decoding the object. Security printing is associated with a supply chain, value chain, or other logistics-managed move-ment of content between different actors. The seven primary actors, generalized across most domains, are the manufacturer, the warehouse, the distributor, the retailer, the consumer, the inspector, and the forensic analyst. The manufacturer and forensic analyst, one on each end of the overall object lifecycle, are generally interested in the highest level (forensic) of object authentication. It behooves the manufacturer, then, to provide forensic-level inspec-tion on the packaging, labels, or other product-associated materials. In between, different motivations—inspection, auditing, track and trace, individual product authentication, supply chain integrity validation—drive the different actors. Hybrid, VDP allows a single printed region to be used for many purposes simultaneously, allowing each actor to achieve her goals without compromising other actors. The fact that multiple actors can use the same printed regions—or even a single region—for multiple aims shows the high value of parallel process-ing in security printing.

The *utility of the encoded information* is an important factor when a hybridized security printing design is used. This is preferable when using VDP because the amount of effort to craft multiple variable data regions is only marginally more than the effort required to implement one VDP feature. For this modest upfront effort, a wide variety of downstream advantages are garnered. In order to produce a truly secure printed feature, all three of the following utilities must be provided: (1) unique ID, (2) copy prevention, and (3) tamper-evidence.

A unique ID is readily produced using VDP. The simplest unique ID is a serial number. Usually, the order of serial numbers is randomized using a random number generator, encryp-tion, or digital signing. The end result is that each printed item has a unique ID suitable for look up in a (cloud-accessed) database. The unique ID can be used as an entry field for the other printed information—overt, covert, steganographic, and/or forensic—associated with the same object. For example, the descriptor for the forensics of a specific printed character can be stored in the database and then compared to the descriptor of the same character on the object tagged with the unique ID. A nonmatch indicates the object has been copied, reprinted, or otherwise counterfeited.

Copy prevention is important for a minimum of one decodable object. Otherwise, a direct copy of the image can be made and falsely "authenticated." Clearly, the forensic character descriptors associated with the use of high-resolution imagers cannot be copied. However, it is also difficult to achieve forensic confidence for individual images or clusters of images, if the images have sufficient complexity (Simske *et al.*, 2009).

The third requirement for a security printing object is tamper-evidence. If an object has a unique ID and cannot be copied, it can still be reused. Tamper-evident objects, however, are compromised when they are authenticated, making them single use. Associating a secu-rity feature with the tear strip means that it will be bisected when the package is opened.

Scratch-off surfaces that must be rubbed away to access the unique ID underneath are another rational form of tamper-evidence.

The *extent of print variability* is used to craft the analytics for a security printing campaign. The analytics are, of course, the collection and digestion of information associated with the use of the security printing objects. Static printing, which is generally cheaper, does not provide a unique ID, and so, mass serialization is usually provided by a low-cost thermal or other in-line printer. This unique ID is typically of low quality, and thus provides no copy protection and usually no tamper-evidence. However, static printing can certainly provide cluster-level image forensics (Simske *et al.*, 2009), making even relatively small-scale counterfeiting stand out in the analytics: they cluster together and are distinct from the legitimate samples.

As argued above, VDP opens the door for hybridization; that is, using different types of variable regions for different tasks. These include inspection, point of sale, authentication, unique ID/mass serialization, forensic authentication, and URL embedding. From a security standpoint, VDP is very powerful: hybridization means that the *relationship* between multiple VDP objects can be varied from one security printing campaign to the next without requiring a change in the VDP objects used. This is an excellent way to make the counterfeiters spend more in reverse engineering the system.

The logical extension of VDP is full customization. In full customization, everything printed can be made variable from one object to the next. This includes the layout, the relative spacing between text characters, and the amount of steganographic information added, among others. While full customization requires extensive processing overhead, modern printers boast massive—and massively parallel—processing capabilities, making this approach far less daunting than in years past.

The final factor considered is the *complexity of the encoding*. Here, the changing nature of printing plays a huge role. With 3D printing promising to replace many manufacturing processes in the years to come, it is obvious that many aspects of printing—both on the substrate and finishing ends—may eventually become absorbed into the printing process itself. From lamination of the substrate to textured finishing, 3D printing technologies stand on the brink of reducing the length and the complexity of the printing line. The simplest printing will continue to be "flat"; that is, a matter of printing one type of ink onto one type of surface. The security of this approach depends entirely on the degree of VDP implemented.

In the case of liquid electrophotography, ink layers can be peeled off one another in a process called "sandwich printing." More complex encoding can occur when multiple layers of ink—for example, combinations of visible, UV, IR, conductive, and other inks—are used together (Simske *et al.*, 2008).

Further complexity is accommodated by security printing when a thorough understanding of the printing and downstream imaging processes is applied to optimizing the information originally printed (Simske *et al.*, 2008). In this process, the printing is "precompensated" for the expected downstream effects. The two most effective forms of precompensation are structural (Simske *et al.*, 2008; Vans, Simske, and Aronoff, 2009) and spectral (Simske *et al.*, 2008; Simske, Sturgill, and Aronoff, 2009). Structural precompensation is largely concerned with anticipating the manner in which ink will spread on a given substrate. For example, inkjetted inks will generally spread out more on a plastic or coated substrate than they will on a porous, cellulose-based substrate unless rapid drying or curing is implemented. Figure 3.1 shows the benefits of structural precompensation on barcode reading. Here, structural precompensation was implemented by making the dark (inked) modules in the barcode narrower. When printed,

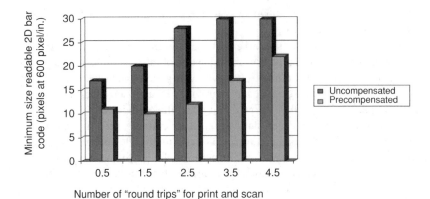

Figure 3.1 Minimum size of the individual module, or tile, in a 2D barcode with readability rate of 90% or higher (*y*-axis) versus the number of print/scan round trips (*x*-axis) for regular 2D Data Matrix barcodes (uncompensated values) and 2D Data Matrix barcodes with structural precompensation in the first printing

the ink spread resulted in black and white modules in the barcodes being equal in size. Without such structural precompensation, the black modules bled into the surrounding white tiles, resulting in failed readability at a much larger tile size. Structural precompensation was found to have as profound an effect as three successive copies of the barcode—in the opposite direction. To understand this, note that in Figure 3.1, one copy (print/scan cycle) of the barcode represents one "round trip," and so the original print is 0.5 round trips. After making a copy of a copy of this original print, the *x*-axis is at 2.5 round trips. In Figure 3.1, the precompensated barcodes read as well after 3.5 round trips as the uncompensated barcodes read after original printing (0.5 round trips). This means that structural precompensation removes three rounds of copying from the would-be counterfeiters.

In addition to structural precompensation, spectral precompensation is used to significantly improve the payload bit density. Spectral precompensation was introduced in Simske *et al.* (2008) and consists of reversing the hue-changing effects of printing and then scanning. Those effects are illustrated in Figure 3.2, in which all 13 dry electrophotographic ("LaserJet") and thermal inkjet ("DeskJet") printer/paper combinations are shown to make the magenta more "red," the blue more "cyan," and the cyan more "blue." Spectral precompensation for these printer/paper combinations therefore consists of printing the magenta with a slight blue shift, the blue with a slight magenta shift, and the cyan with a slight green shift. For the Indigo printer, spectral precompensation encompasses printing the magenta with a slight blue shift, the blue with a slight cyan shift, and the green with a slight cyan shift. The actual shifts in each case are readily determined with a single test sheet containing patches with specific hues, as described in Simske *et al.* (2008). Spectral precompensation was shown (Simske, Sturgill, and Aronoff, 2009) to double payload bit density, and its effect on color barcodes was equal to the impact of a single copying cycle.

The ability to hybridize security deterrents, as mentioned above, enables many downstream applications and services. VDP security printing adds conditional patterns to the system architecture, since the designer has the opportunity to choose (a) what each VDP mark is used

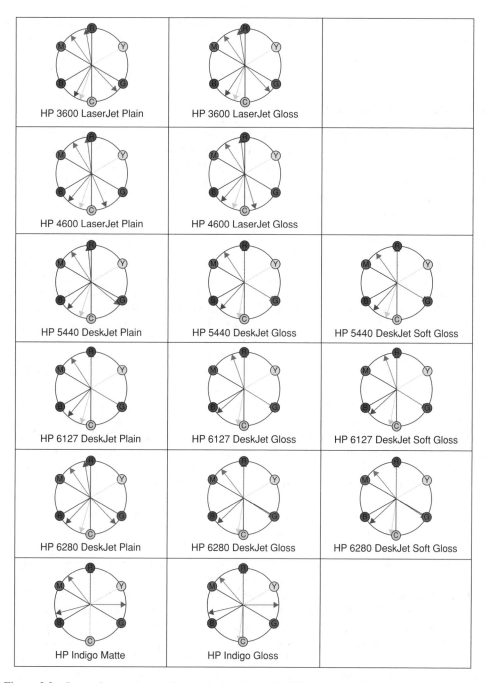

Figure 3.2 Spectral precompensation as deployed on 15 different printer/paper combinations for a six-color (red (R), yellow (Y), green (G), cyan (C), blue (B), and magenta (M), in clockwise order starting with red at the top) 2D barcode. The targeted color tile hues are given by the filled circles ("hues-as-printed"). The actual hues obtained after the printing and scanning cycle are indicated by the filled triangles and measured hues after printing and scanning ("hues-as-read"). The results are given for 15 different printers: all were scanned using the same desktop scanner (Simske *et al.*, 2008)

for, and (b) how the data in the different security marks relate to one another. The implications for parallel processing are obvious. If multiple VDP marks need to be analyzed, each can be analyzed in its own thread.

3.4 Secondary Domains

The secondary domains of interest covered in this book include signal processing (speech recognition, medical signal processing), image processing (image segmentation, medical imaging, surveillance, OCR), text analysis (NLP), and security (security analytics) applications. As with the primary domains, there is a strong emphasis on image processing applications; however, there is enough variety in these domains to further illustrate the broad utility of parallel processing approaches.

3.4.1 Image Segmentation

Image segmentation is a very broad field, and these short paragraphs will not do it justice. Most of the other domains of interest (document understanding, image understanding, biometrics, security printing, medical imaging, surveillance, OCR) in this book depend on image segmentation for at least part of their machine intelligence, and so a relatively in-depth overview will be provided here.

There are many useful approaches to image segmentation. I will, with an admittedly Procrustean flair, attempt to describe these as belonging to either *bottom-up* or *top-down* approaches. Bottom-up approaches are concerned with local, or regional, pixel (or voxel—I will continue to use the term pixel for both 2D and 3D imaging) behavior that is used to decide whether to combine neighboring pixels or leave them separate. This is a *constructive approach*, and it may result, as is the case for document understanding, in unassigned pixels. Top-down approaches, on the other hand, impose a condition on the entire image at once, which then allows the segmentation to proceed based on this condition. Originally, all the pixels belong to a single segment, but the top-down model is *deconstructive* inasmuch as it peels away pixels to create new segments. In overviewing six types of bottom-up and four types of top-down approaches, this section will provide a flavor of the complexity and diversity of image segmentation technologies.

The bottom-up approaches I will overview are (a) thresholding, (b) clustering, (c) edge-based, (d) watershed, (e) histogram-based, and (f) region growing. *Thresholding* is usually performed as binarization, resulting in a foreground and background set of pixels. Multi-level thresholding can be used when there are several backgrounds, or when objects with different textures are to be identified, but as that approach is similar to the histogram approach, I will leave off discussion of that for the nonce. Figure 3.3 shows the results of binarization using two excellent algorithms, those of Kittler and Illingworth (1986) and Otsu (1979). I call these excellent algorithms because the design of the algorithms is "impedance matched" to the task: for the Kittler and Illingworth (1986) algorithm, the minimum error from two Gaussian distributions is minimized, and for the Otsu (1979) algorithm, the two-class intra-class variance is minimized. Both, therefore, attempt to optimize binarization by optimizing the definition of two classes. It should be noted that these are both global thresholds: there are many excellent

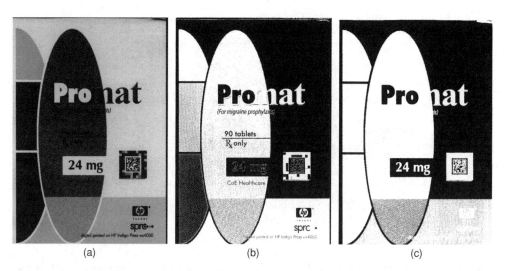

(a)   (b)   (c)

Figure 3.3 Thresholding-based segmentation examples. (a) Original image with multiple background colors and intensities. (b) Image after thresholding using a modified version of the algorithm in Kittler and Illingworth (1986). (c) Image after thresholding using the method of Otsu (1979)

adaptive thresholding approaches, as well. These base their binarization on local variability, and thus provide better robustness to variable backgrounds and to nonuniform illumination.

Figure 3.3 illustrates the benefits provided by using *both* of the binarizations. The Kittler and Illingworth (1986) algorithm, with output shown in Figure 3.3b, identifies several areas of different texture, specifically in the lower left quadrant of the image. The Otsu (1979) algorithm, with output shown in Figure 3.3c, does not identify as many textured areas, but does an excellent job of extracting the small, 2D Data Matrix barcode in the right center of the image. This is readily identified as a specific "region" in the image, and thereafter assigned to a barcode decoding pipeline.

The bottom-up *clustering approach* is also quite straightforward. Connected components, or regions, are created by joining neighboring pixels that are similar enough—in intensity, hue, saturation, edge direction, texture, and so on—to be considered alike. There are many variables that can be adjusted for this approach, including the threshold for similarity, the definition of neighbor (is it the four pixels to the left, right, top, and bottom, or can the four corner diagonal pixels or an even more distant set of pixels be considered neighbors?), and the number of passes through the image (larger numbers of passes tend to create larger clusters). Also, do all pixels have to be assigned to a cluster? The latter is usually true for general image segmentation, but not so for document images. Usually, the clustering approach continues aggregating clusters until a predefined or desired number of clusters exist. This benefits tracking systems when the number of objects of interest is known, but can lead to strange aggregations when the number of clusters selected is inappropriate.

Edge-based approaches provide segmentation by defining edge boundaries, edge directional fields, or both. The maps of these edge data are then used to guide the creation and aggregation of connected components. I consider this approach to be bottom-up on the basis of the implementation details. The two primary approaches I have used are (a) forming connected

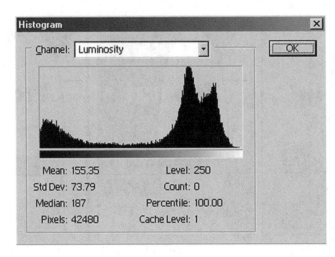

Figure 3.4 Sample image histogram illustrating three obvious peaks in the luminance channel. These three peaks are used to create clusters of pixels belonging to only one of the peaks, and thus provides excellent segmentation by differences in luminosity

components from the nonedge pixel map of the image, and (b) allowing clustering to continue only when a plurality or majority of neighboring pixels have similar edge directionality properties. In either of these two cases, connected components are formed from aggregating neighboring pixels with like properties.

The *watershed approach* is also a bottom-up algorithm, in which segments are defined based on where shared water would flow into the topographical representation of the image. The isobars can be in various spaces—chroma, luminance, and so on—but the principal is the same for each space. Wherever the rainfall over an area of the image pools with rainfall elsewhere, these pixels are joined into a single segment. This approach is excellent for identifying optima in the image, making it highly suitable for tracking and object extraction, but less suitable for more structured images (like documents).

The *histogram-based approach* is functionally equivalent to multi-thresholding. The histogram of the entire image, as shown in Figure 3.4, is computed, and the peaks in the histogram are used to define the target values for different clusters in the image. If there are N_P peaks in the histogram, the histogram-based approach proceeds identically to having $N_P - 1$ thresholds.

The last of the bottom-up approaches is the *region growing* approach. This approach shares much in common with the clustering method described above. Neighboring pixels are joined based on some criteria. Unlike the clustering approach, however, region growing approaches are not tied to a specific number of region types. Generally, region growing approaches benefit from intelligent postprocessing; for example, similar large regions may be interrupted by ectopic, narrow boundary regions. Such an error can be overcome through a simple dilation + erosion postprocessing pipeline.

Turning now to the *top-down approaches*, we explore how a simple, functional metric can be used to drive segmentation. The premise of the compression-based approach is that the optimal segmentation will result in the minimum compressed file size. This is a reasonable assumption since compression ratios are higher when like pixels are grouped. Interestingly, this approach explicitly encourages parallelism, since more than one segmentation candidate must be provided in order to ascertain an optimal compression. Thus, this method could, for example, attempt segmentation using any combination of the bottom-up approaches outlined above—along with any combination of the top-down approaches still to be introduced—and select the best segmentation based on compressed file size. As a quick illustration, suppose we decided to vary the global threshold (see Figure 3.3) over the entire reasonable range; for example, from the 5% to the 95% cumulative points on the (usually luminance) histogram used to create the binarized image. It was found that the Otsu (1979) threshold led to the creation of a smaller (lossless) compressed image than did the Kittler and Illingworth (1986) threshold.

A second top-down image segmentation approach is the *model-based approach.* As the name implies, this method is driven by a model of one or more objects to be segmented. Segmentation—or multiple candidate segmentations—proceeds and each connected component is evaluated in a probabilistic manner for its degree of matching to the model or models. Shape, palette, and various distribution methods (e.g., for face or other composite shapes) can be used to employ this method. Another feature of this method is that it is very suitable to training. For example, as extracted objects are validated to be correct matches—even automatically, for example, in tracking problems—then the model itself may be updated to accommodate the range of, or change in, objects matching the model.

Another top-down method of interest is the split and merge. The "split" part of this algorithm is the division of the image into four quadrants if the image is not homogeneous. This implies that there is a threshold for homogeneity, which can be based on a wide variety of image features such as luminance, chroma, texture, and so on. After being split into quadrants, these quadrants—by themselves, in any combination, or in any combination of one or more of them with their neighboring quadrants—can be merged into a single connected component. This "merge phase" provides the means to form more complex connected component boundaries. The rectangular splitting process is highly advantageous for downstream compression if the original (presplit) regions are aligned with the block boundaries.

The final top-down image segmentation approach to be overviewed in this section is the partial differential equation (PDE)-based approach. PDEs are typically used to define arcs for connected component boundaries through the definition of contours. Under some conditions, this can perform like a watershed approach. The PDE approach usually operates on texture or contrast, and so focuses on the definition of boundaries rather than building regions from the bottom up. This results, typically, in less region clean-up after the boundary definition than is required after the watershed approach. Tying this in with the discussion in Chapter 1, the PDE-based approaches are roughly analogous to the formation of a manifold or support vector in classification problems, whereas the watershed approach is more analogous to a Gaussian mixture classification approach.

Is there a single best technique from this dizzying array? That is unlikely, even for a specialty domain such as, for example, face detection or shape extraction. One reason is that the technique that will perform best for a given problem is less dependent on the domain to which it is applied than it is on the specific attributes of the images (quality, contrast, exposure, camera used, lighting conditions, blur, etc.). Another reason is that each domain has its own nuances that may favor a given image segmentation approach over another; for example, bottom-up approaches tend to work better for document segmentation while top-down approaches tend to work better on shape recognition and object extraction.

3.4.2   Speech Recognition

Voice recognition, as mentioned above, is a form of biometric. Automatic speech recognition (ASR), however, is deployed in many circumstances where the voice cannot be recognized with sufficient confidence for biometrics; for example, for entering commands into a computer or for selecting options on an automatic voice menu. Speech recognition is extremely important based on the high value of spoken language in communication, a value that has only risen as we have moved from a text-based to a multimedia communications society. Speech recognition is based on 1D signal processing, and it is easy to ground truth for native speakers.

Speech processing is concerned with the relative magnitude of the spectral coefficients. This includes the mel-frequency cepstral coefficients (MFCCs, described in Vaseghi, Yan, and Ghorshi (2009) and elsewhere), which represent an audio spectrum with a set of coefficients that are uniformly distributed over a log scale of the frequency. In other words, the cepstrum is a sampling of the audio spectrum that provides a "fingerprint" or "signature" of the spectrum (useful for classifying or distinguishing different speakers, etc.). Additional frequency representations include perceptually motivated MFCCs (Davis and Mermelstein, 1980; Krishnamurthy and Childers, 1986), which open the pathway to task-specific cepstral coefficients (TSCCs). TSCCs can be crafted to more adequately cover the expected range of response when completing a specific auditory task. As an example, if an emotive response is measured (or triggered), a TSCC that better represents maximum and mean of the first derivative of the pitch contour (Yacoub *et al.*, 2003) will provide better emotion recognition. TSCCs can also be crafted using transformations of the MFCC, such as through cepstral mean subtraction, to provide more accurate speaker identification (Rosenberg, Lee, and Soong, 1994).

Other work on emotion recognition has shown that auditory data streams can be used for simultaneous establishment of identity and emotion determination (Cowie *et al.*, 2001). Emotion detection is important to allow emotional state to be compensated for during identity determination. In this case, emotion can be used as another factor in identification of the speaker. This supports the use of hybridized classifiers (Chaudhuri, Ghosh, and Oja, 2009; Mohamad, Likforman-Sulem, and Mokbel, 2009), directly in line with the emphasis on parallel algorithms throughout this book. There is growing recognition that the combination of cognition-based, machine intelligence, and NLP approaches are necessary to move voice and speech recognition forward (Baker *et al.*, 2009). These approaches can be processed using parallelism by task or component. The output of these approaches may be combined using parallelism by meta-algorithmics.

3.4.3   Medical Signal Processing

With medical signal processing, we are concerned with 1D signals, usually biorecordings such as the ECG or EMG, biomechanical recordings (e.g., stress-strain recordings), and a wide variety of other biosensing recordings such as biochemical, biomagnetic, bioacoustic, and bioimpedance recordings (Bronzino, 1995; Bankman, 2000; Enderle, Blanchard, and Bronzino, 2000). These recordings are made to enable downstream diagnoses of tissues, organs, and organ systems. Diagnoses are, simply, analyses: they enable us to cluster biorecordings by types, perform classification of the biorecording events, and combine multiple biorecordings to make a more accurate and systematic diagnosis.

Table 3.4 Types of medical signal processing and imaging biorecordings

Type of Signal/Image	Characteristics of the Signal/Image	Examples
Internal passive	Electrode or sensor is within the body	Electrochemical DNA probes, physicochemical property sensors, fiber optic sensors, catheter and stent-associated sensors, and so on.
Internal evoked	A specific response is elicited and recorded	Internal bioacoustic, biomagnetic, bioimpedance sensors, used for gait analysis, limb motion analysis, internal heart monitors, and so on.
External passive	Electrode or sensor is on the surface of the body, or not abutting the body at all	ECGs, EMGs, EEGs, optical sensors, sphygmomanometers, stethoscopes, and so on.
External evoked	External sensors record a specific task-based elicited biosignal	Patellar reflex test, lie detector test, papillary light reflex test, Glasgow Coma Scale test, and so on.

There are many ways to categorize the biorecordings that underpin medical signal processing, including deterministic versus stochastic (Bronzino, 1995), continuous versus discrete (Enderle, Blanchard, and Bronzino, 2000), and by frequency range of the biosignal (Bankman, 2000). Another way to categorize these biosignals is by how they are recorded. I find this a useful way of organizing both medical signal processing and medical imaging (discussed in Section 3.4.4). The two primary axes are internal versus external biorecording, and passive versus evoked biorecording, as collected in Table 3.4. Internal biorecordings include *in vivo* and *in utero* measurements, the use of catheters, and internally placed bioamplifiers and other sensors. Sensors include chemical, pH, thermal, mechanical and impedance, and other detectors. Bioamplifiers are used to ensure that the output signal is in the appropriate range for either transduction into an electrical signal or for direct use as an electrical signal. Sampling and filtering of the signals can be performed before or after transduction/amplification.

Generally, internal biorecordings are more accurate, and often more noise-free. However, they are invasive, and so must meet biocompatibility requirements. They are also more localized, and so may represent a local effect that does not represent a more systemic measurement that is actually not as clinically relevant. Passive internal biorecordings record physicochemical processes as they are; for example, blood glucose concentration, local milieu temperature, or radial/lateral impedance in a myocyte. Evoked, or elicited, internal biorecordings collect the signal resulting after a particular event or forcing function has been applied to all or part of the measured system. Mechanical measurements (e.g., localized strain measurements using implanted strain gages) and electrical measurements (using implanted electrodes, including those associated with larger implants such as pacemakers) can be used to measure the response to volitional movements for gait analysis, muscle health assessment, muscular dystrophy assessment, and other diagnoses.

Table 3.4 also lists examples of external passive and evoked signals. External passive biorecordings include the three most significant surface recordings of body electrical activity: the ECG, the EMG, and the EEG. All such surface recordings are, in reality, measurements of coordinated extracellular charge flow. This averaging, along with the shunting of current flows caused by the high capacitance and impedance of subcutaneous fat, results in some lowpass filtering of the actual electrical activity. Diagnosis of the surface potentials, however, is highly clinically relevant, and the relative ease, low cost, and biological relevance of surface electrical recording—in comparison to invasive or 2D/3D imaging, for example—has made these measurements very popular for patient monitoring in addition to diagnostic purposes.

The signal processing of these surface potentials consists of both time-series and spectral analysis. For the ECG, time-series analysis includes determining the times at which the key waveforms—P, QRS, T, and U—of the cardiac cycle occur. Spectral analysis includes determining the frequency content of the QRS wave, as its slew rate is indicative of the propagation velocity of the depolarization event through the ventricles. Importantly, multiple electrodes are used for the ECG (and for the EEG and many EMGs, including the electrooculogram), which allows the ready computation of a 2D or 3D image such as the VCG. This, in turn, opens up opportunities for improving the 1D signal. For example, the VCG is generally smooth, meaning that simple moving average filtering in the VCG space can be used to remove correlated noise (such as breathing and movement artifacts and 60 Hz noise) and even uncorrelated noise (due to shunting, electrode deterioration, etc.) from the signal before the VCG is converted back into the ECG (Simske and Blakley, 2012).

In addition to surface potential recordings, there are an increasing number of optical sensors that can be used for external biomedical recordings (Bronzino, 1995; Baldini *et al.*, 2008; Soria *et al.*, 2011), particularly for the partial pressure of gases such as oxygen and carbon dioxide. However, the "Holy Grail" of transcutaneous optical sensors—an optical transcutaneous blood glucose concentration sensor—is yet to be perfected.

Evoked potentials can be readily recorded using external sensors and equipment. For example, the lie detector test—an evoked response by definition—can incorporate an EEG, a facial EMG, voice analysis (based on acoustic sensing), and blood pressure recording (sphygmomanometer). The ECG can be used along with evoking events such as exercise and changes in body posture (sitting, standing, etc.) to assess cardiovascular health of people. Auditory and somatosensory (e.g., pressure or touch) evoked potentials can be tied to the EEG in order to diagnose the behavior in the corresponding temporal and parietal lobes.

All of the medical imaging (i.e., 2D and 3D signal processing) techniques described in the next section belong to the external passive or evoked potential categories, even if the image signal source (X-ray, gamma ray, electrical current, etc.) may have a direct internal effect. However, it is clear from this section that there is a wide array of medical signal processing technologies, making this field suitable to parallel processing. When a particular medical condition needs to be diagnosed, and several signal analysis approaches have been used, it seems clear that parallel analysis approaches—by component, by task, and by meta-algorithmics—can be brought to bear to improve the diagnosis accuracy. By task, the different medical signals can be processed in separate threads. By components, different partitions of the medical signals can be processed in separate threads. By meta-algorithmic, different means of interpreting two of more medical signals can be explored.

3.4.4   Medical Imaging

The distinction between medical signal processing and medical imaging is not strict, nor should it be. Do multiple lead ECGs comprise a set of 1D biosignals or, because they can be combined into a VCG, do they comprise a 2D bioimage? The answer, of course, is that they comprise both. This is an important point. The duality of purpose of multiple biorecordings emphasizes the value of multiple analysis approaches to be performed, in parallel, in order to create a more robust, more accurate diagnosis.

In this section, the medical imaging modalities are categorized based on the type of signal used to create the images. Six primary medical imaging technologies are described: (1) X-rays and the associated computed tomography (CT); (2) magnetic resonance imaging (MRI); (3) single-photon emission computed tomography (SPECT); (4) ultrasound; (5) positron emission tomography (PET); and (6) electrical impedance tomography (EIT). Medical imaging is based on the transmission (X-rays, CT, EIT), reflection (ultrasound), or induced emission (MRI, SPECT, PET) modality.

Table 3.5 lists salient information about the six medical imaging technologies. X-rays—both in 2D and in their 3D form as CT—use electromagnetic waves of high energy as their signal. The waves are penetrative, and only differentially absorbed by either radiopaque biological materials or tissues—such as bone—or by intentionally introduced radiopaque contrast media. The latter, if blood-borne, allows the imaging of the vasculature or specifically targeted tissues. Digital subtraction angiography and temporal subtraction are techniques used to remove background tissue from CT images, increasing the contrast of the final images. This in turn enables better downstream segmentation of diagnostically relevant regions.

MRI is an emissive medical imaging modality. Nuclei release radio frequency (RF) waves during their relaxation from induced nuclear magnetism. Specific nuclei are aligned in the magnetic field based on the Larmor frequency of the nucleus, and the spin-spin relaxation time (to return the nucleus to an unaligned state) is used to determine when the best signal contrast will be obtained. Functional MRI (fMRI) is tuned to specific nuclei (usually hydrogen) with specific chemical bonds. Functional (passive and evoked) measurements include cerebral blood flow, blood volume, blood oxygenation, and various metabolic reactions. Contrast agents such as gadolinium are used to enhance the images of specific brain structures. It is now commonplace to use fMRI to measure evoked responses in the brain in order to identify the brain pathways involved in various thought processes. MRI can also be performed in parallel with other imaging technologies, such as CT or ultrasound, to provide both anatomical and functional/physiological imaging.

SPECT imaging is based on the use of standard radionuclides, and has high resolution since the gamma photons emit directly from the location of the radiopharmaceuticals in the body. Two types of half-lives are involved in SPECT imaging: the half-lives of the radionuclides (the commonly used Tc-99 has a half life of only 6 h) and the half-lives of the radioligands themselves (e.g., radiohalogenated carbohydrates can be used to measure glucose metabolic rates). These latter "half-lives" are tied to the biokinetics of the radioligands; that is, the rate at which they are cleared from the body, body compartment, organ, tissue, and so on. Because radioligand/receptor specificity is tissue dependent, the best contrast may actually be obtained several hours after the introduction of the radioligand, when it will be cleared from all but the target tissue in the body.

Ultrasound is the only reflective medical imaging technology overviewed in Table 3.5, and is based on the partial reflection of acoustic energy at each significant tissue interface; for example, the boundary of the abdominal cavity smooth muscle with the abdominal cavity itself or the boundary of the pericardium with the chest wall. At each interface, a signal is reflected back to the ultrasound imager/recorder that is used to compute the distance from

Table 3.5 Types of medical imaging, the modality of signal measured, a description of the signal measured, and example of evoked potentials measured using these imaging technologies

Medical Imaging Technology	Modality of Signal Measured	Description of the Signal Measured	Important Medical Imaging Considerations
X-ray and CT	Transmission	Electromagnetic waves	Digital subtraction angiography, temporal subtraction, radiopaque contrast media (dyes)
MRI	Emission	Induced nuclear magnetism relaxation-associated radio frequency waves	Spin–spin relaxation time; fMRI for cerebral blood flow, blood volume, blood oxygenation, and metabolism; contrast agents (e.g., Gd)
SPECT	Emission	Gamma photons	Radiopharmaceuticals, biokinetics, ligand/receptor pairing
Ultrasound	Reflection	High-frequency sound	Velocity estimation, compound images
PET	Emission	Simultaneously emitted pair of photons	Positron travel distance before annihilation, biopharmaceutical, tracer molecules (C, N, O, F)
EIT	Transmission	Electric voltage	Inhalation/exhalation, edema

the interface and the recorder. Ultrasound is useful for both anatomical (interface distances) and physiological (e.g., Doppler-based velocity estimation, particular for blood velocity in the heart and major arteries) recordings. Compound images are readily captured in "sweep" mode, such as the familiar prenatal ultrasound images.

PET is another emissive medical imaging technology. Unlike SPECT, however, it is based not on the emission of gamma photons, but instead on the emission of a positron from the nucleus of a commonplace atom such as carbon, nitrogen, oxygen, or fluorine. The positron travels a short distance—which results in some blurring—whereupon it is annihilated when it comes in contact with an electron. Two photons are fired in opposite directions and thus record the location of the annihilation event. Because PET uses biological plentiful atoms, it can be linked to a wide variety of biopharmaceuticals and other tracer molecules. This means it is used primarily for functional imaging.

The last imaging technology overviewed in Table 3.5 is EIT. I was fortunate to work on one of the pioneering systems in EIT, at Rensselaer Polytechnic Institute under the leadership of Drs. Isaacson and Newell (Simske, 1987; Cheney *et al.*, 1990; Cheng *et al.*, 1990). EIT is based on the multiplexed use of an array of surface (external) electrodes to (1) apply an array of currents to the underlying tissue (typically the thorax or abdomen); and (2) read the resulting voltages on the same surface. Some systems use separate sets of electrodes; ours used the same electrodes for both application of current and reading of voltages to improve electrode surface area. The pattern of currents applied to electrodes is optimized for contrast in the conductivity image that is obtained using Ohm's law. Because of the lack of ionizing radiation, this imaging technique can be used for long-term monitoring of physiological conditions. In fact, its current clinically approved and deployed usage is in the monitoring of lung function.

In the medical imaging applications illustrated in later chapters, much of the analysis will be centered on 2D image analysis. In medical imaging, 2D sections through 3D images are called tomographs. Standard image processing techniques used include mean and median filtering, which smooth a noisy image; edge enhancement or sharpening, which is used to highlight structural boundaries in an image; histogram equalization, which can be used to avoid contrast and exposure errors in the image; contrast enhancement, which is used to provide better dark and bright features; image averaging, which is used to remove or at least smooth out noise; image subtraction, which was discussed above for CT imaging; and spectral filtering. Spectral filtering includes lowpass filtering, which is used to remove high-frequency "salt and pepper" noise; highpass filtering, including unsharp masking, which is used to highlight high-frequency features in an image; bandpass filtering, which can eliminate both high- and low-frequency noise simultaneously; bandgap filtering, which is used to remove unwanted signals of a specific frequency or frequency range, such as 50 or 60 Hz noise induced by the electricity used to run equipment; Weiner filtering, which is used to remove a wide variety of noise through comparing an image to a desired "noiseless" image; and hybrid filters that incorporate two or more of the previously described filters.

In short, medical imaging is concerned with both the governing physics—for emission, reflection, and transmission of the image-generating signals—and the mathematics for image-processing-enabled estimation of anatomical and physiological information from the image data. This includes tissue recognition; segmentation of different cellular, anatomical, and so on, structures from the images; quantitative (size, shape, number, distribution, etc.) and qualitative (color, texture, etc.) evaluation of the structures; and diagnosis of specific pathologies from the collective set of image data.

This overview of medical imaging should make it clear that parallel approaches—emissive, transmissive, reflection—of image generation exist, along with parallel types of imaging (passive, evoked). Other opportunities for parallelism include the multiple types of image processing and image restoration that are available. Further parallelism is made possible by the increasing trend for hybridization of two or more medical imaging techniques; for example, CT and MRI to obtain anatomical and functional image information simultaneously. These multiple sources of parallelism—along with its close relationship to image understanding and image segmentation—make medical imaging a suitable domain for this book.

3.4.5 *Natural Language Processing*

NLP is another way of saying "machine understanding of human language." More practically, NLP is concerned with the functional consequence of machine understanding; that is, the ability to perform useful work based on an understanding of the text. Few, if any, domains of machine intelligence research offer such a diverse set of challenges for data extraction, information retrieval, clustering, classification, inference, and tagging as NLP and its closely related, more mathematically driven statistical NLP (SNLP) (Manning and Schütze, 2000).

Not surprisingly, information theory (Shannon, 1948, 1951) and Bayesian mathematics are important for front end data extraction in SNLP. Deviations from expected levels of text entropy are consistent with topics specific to the reduced entropic parts. Entropy can also be computed conditionally for comparing two or more documents or corpora. Related is the concept of mutual information, which indicates how much information one word tells us about another: this value should be highest for synonyms or words with high co-occurrence, high for antonyms, and low for randomly paired words or phrases.

Bayesian mathematics, on the other hand, is concerned with the decisions made during tagging, clustering, and classification. Since Bayesian mathematics is based on conditional probability, they can be used to provide a set of decisions known to minimize the classification error (Duda and Hart, 1973). Conditional probabilities are also the important input data for Markov models and hidden Markov models (HMMs), which are the predominant statistical modeling tool for ASR. They are also widely deployed in OCR and in SNLP. In SNLP, HMMs are used extensively for part-of-speech tagging and sentence parsing. There is still much left to accomplish in this domain, in spite of the development of powerful HMM-based methods such as those of Jelinek and Mercer (1985) and Kupiec (1992). This is understandable based on a consideration of how humans parse speech; that is, estimating the part-of-speech based on what occurs before the word or phrase under consideration, and then correcting it based on context—words, phrases, nuances, and even gestures and other nonspeech cues—after the occurrence. No part-of-speech detector or parsing algorithm has yet incorporated the type of complexity involved in human language understanding. However, parallel approaches—especially meta-algorithmics—may help to bring this complexity to linguistic understanding in the future.

NLP and SNLP output include keyword identification, summarization, part-of-speech tagging, document categorization, and other text analytics. Summarization can be extractive or abstractive: extractive techniques simply replicate the original text that is algorithmically (or meta-algorithmically—see Chapter 10) determined to be the most salient for the summary, and so on (e.g., key clauses, sentences, or paragraphs), while abstraction technique involve paraphrasing sections of the original content, that is, to condense information and provide synopsis and semantic/meaning.

In many ways, abstractive summarization is the key to SNLP, since it can be based on so many different SNLP techniques. For example, extractive summarization sentence weighting approaches reported in the literature are based on the keywords and key phrases extracted; capitalization of text; grammatical case of nouns; word co-occurrences; font formats; sentence position in a paragraph; cue phrases such as "in summary" and "importantly"; correlation of a sentence or phrase with the title, author reported keywords, and so on; sentence length; and sentence centrality or redundancy with other sentences.

Once summarization is obtained, there is an implicit—and usually explicit—ranking of terms, concepts, and loci within the document. This information can be coupled with other analytics—such as curve plotting the rank of terms against their frequencies and the entropy of keywords throughout the text—to categorize the document. Curve plotting the rank against frequency has been the subject of much research, resulting in, for example, Zipf's law and Mandelbrot's formula (Mandelbrot, 1954). Differences in these "power curves" within a document may be indicative of topic, voice, even author change. Entropy, of course, is a measure of the randomness of the "bag of words" for a set of text. Entropy drops when a topic is more focused on a few terms or expressions. Thus, changes in entropy are also potentially indicative of changes in style, topic, or authorship.

NLP clustering techniques benefit from the same approaches used in other areas of machine intelligence (Manning and Schütze, 2000). Expectation-maximization (EM) approaches are widely used, in spite of their sensitivity to initial conditions: inappropriate initial conditions often lead to local maxima. EM algorithms also converge more slowly than other methods, such as gradient methods. Fortunately, however, text categorization approaches are often constraint-rich, meaning good initial conditions and faster-converging methods are often warranted.

In this brief overview of SNLP, one final technique deserves some mention. The vector space model (VSM) is used extensively to compute the similarity of documents. The vector space itself is an N-dimensional space in which the occurrences of each of N terms (e.g., terms in a query) are the values plotted along each axis for each of D documents. The vector $\vec{d}$ is the line from origin to the term set for document d, while the vector $\vec{q}$ is the line from origin to the term set for query q. The dot product of $\vec{d}$ and $\vec{q}$, or $\vec{d} \bullet \vec{q}$, is given by

$$\vec{d} \bullet \vec{q} = \sum_{w=1}^{N} d_w q_w.$$

From this, the cosine between the query and the document is given by

$$\cos(\vec{d}, \vec{q}) = \frac{\vec{d} \bullet \vec{q}}{|\vec{d}||\vec{q}|} = \frac{\sum_{w=1}^{N} d_w q_w}{\sqrt{\sum_{w=1}^{N} d_w^2} \sqrt{\sum_{w=1}^{N} q_w^2}}.$$

Simple to implement, the cosine measure, or *normalized correlation coefficient*, is nevertheless powerful for information retrieval tasks. Sets of documents—and their subsections—can also be compared using the cosine measure, affording additional features for document categorization.

SNLP is highly amenable to parallel processing. Different documents—and different partitions of a document—can be analyzed in separate threads. Within these threads, further parallelism can be implemented; for example, word counts and language identification can be performed independently.

3.4.6 Surveillance

Borrowing from the previous section, it might be said that surveillance is a topic with a large normalized correlation coefficient with both the biometrics and image segmentation domains. While true, this does not limit the utility of exploring surveillance as a topic in its own right. Surveillance is used in this book in its general sense; that is, for monitoring. In all of the examples, the focus will be on image-based monitoring. Surveillance relies on multiple frames; that is, on video. In some ways, this is a 3D imaging approach, but in a different sense than, for example, 3D medical imaging. For the latter, 3D models of the imaged regions are obtained. With surveillance imaging, the individual images are usually 2D. The third dimension for surveillance is time, not the z-axis.

Surveillance consists of image capture, image processing and restoration, object segmentation, object identification, and object tracking. Image capture quality is dependent on the lighting environment, the amount of motion blur, the angle between the camera lens and the subject, and the camera settings. Image processing and restoration is focused on reversing the effects of uneven illumination, underexposure, overexposure, blur, affine distortion, and/or lack of focus. Object segmentation is a task tailor-made for parallelism: See Section 3.4.1 on image segmentation, as the same basic principles apply to much of video object segmentation.

However, the fact that any object under surveillance will appear in multiple—perhaps a very large set of—image frames gives rise to a different set of image processing techniques, collectively called tracking algorithms. Tracking, like segmentation, boasts a number of powerful algorithms, including blob and contour tracking—which compare the locations of connected component centroids and boundaries from image to image, matching the closest location or size of objects that have moved—as well as mean-shift tracking, which can use dynamic programming or other techniques to minimize the overall distance traveled by objects in motion. This minimization provides excellent tracking if the sampling rate is much greater than the time it takes an object to move across the field of view. A fourth type of object tracking is visual feature matching—subsets of which include face recognition and shape recognition—which can be performed on each video frame in parallel, since there may be no dependency of the object segmentation on successive frames.

Since the objects can be individuals or their faces, both face recognition and gait recognition biometric technologies can be part of a surveillance system. Nonbiometric surveillance can provide gesture recognition, including signing, and location identification. The latter benefits from identifying multiple objects and performing downstream image analysis; for example, extracting text from signage that can be used to identify waypoint-narrowing markers such as business names, highway numbers, street names, and so on.

Surveillance can benefit from parallel processing in several ways. Distinct frames in a video segment can be analyzed independently for many tasks, including segmentation and face detection. After segmentation, distinct image analysis tasks—such as facial recognition and object recognition—can also be performed in separate threads.

3.4.7   Optical Character Recognition

OCR comprises a large set of technologies, which, combined, are focused on the digitization, or electronic internalization, of tangible representations of text. This includes printed text associated with documents, magazines, labels, packaging, and so on, as well as various forms of signage. Traditionally, OCR was designed to provide document digitization associated with the use of electronic scanners. Before the advent of multimedia, documents were predominantly text-base, and even the name "OCR" system implies the digitization of text. Documents have evolved to include more graphics such as logos and backgrounds, more images, and more links to additional—usually on-line—content, all of which can be digitally associated with the electronic text. PDF (Adobe's Portable Document Format) is a familiar electronic representation of the OCR engine output, readily allowing for embedded multimedia and link information.

The primary technologies associated with OCR are outlined in Figure 3.5. These technologies, broadly speaking, consist of identification, recognition, and electronic representation. Identification processes are required to frame the downstream recognition tasks. Identification tasks include warp detection, skew detection, page orientation detection, and language detection (Spitz, 1997; Hassan *et al.*, 2011; Lin, Guo, and Chang, 2011), each of which can

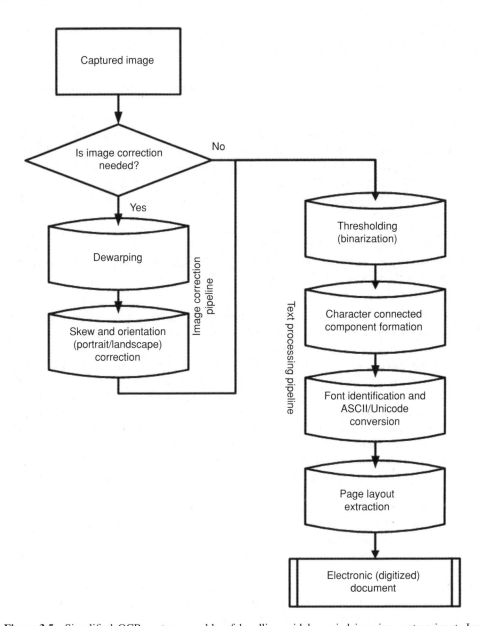

Figure 3.5 Simplified OCR system capable of handling widely varied imaging system input. Image correction is generally not required when documents are scanned using a dedicated scanner or a multi-functional device with a scanning platen. Camera-captured scans will often suffer from warping, especially when low-quality lenses are used. In addition, hand-held cameras will usually have nonzero skew (this can also happen on scanning devices). Documents will often require orientation correction. Once the document is aligned and properly oriented, the text processing pipeline, which converts images into electronic documents for which both the format and content are understood, is performed

be performed before (or after) binarization and before any character recognition has been performed. Warp detection is readily performed, for example, using the following steps:

1. Partition the image into sections, typically 100–400 (e.g., 10–20 rows and columns in the image).
2. For each partition, or subregion, create a histogram of the intensity values. Determine if the histograms are indicative of text or other structured content such as tables, by computing the bimodality of the histogram. Text and tabular regions, which typically have bimodal intensity histograms and relatively short mean run lengths of dark pixels, can be used directly to determine warp. Nontabular regions instead must be converted to their edge maps before determining warp, since the edges of images, logos, and so on, will be used to determine warp.
3. Hough transform methods are used on the intensity values of the text and tabular regions, and on the edges of the nontabular regions, to provide the primary orientation angles of the content. If these are not orthogonal, or if there are not sharp peaks in the angle histograms, then warp is likely present in the image.

If warp is present, dewarping is performed (Ezaki *et al.*, 2005). Next, document skew and orientation are detected and, where appropriate, corrected. Orientation detection accuracy is improved if the language *family* is understood.

The domain-specific work of OCR is then performed. Language identification is used to reduce the relevant character set (in ASCII/Unicode), greatly reducing the errors that occur if the character sets of multiple languages are being simultaneously classified. Font identification can be used to further reduce character errors. Font identification includes the determining the font style (e.g., serif vs. sans-serif) and the font family (e.g., Courier, Times Roman, or Arial). Once this is established, special character (symbol, boldface, italics, underline, superscript, subscript) identification and tagging can be much more accurately assessed. Symbolic fonts are identified and with increasing accuracy distinguished from logos, equations, and other special glyphs.

At this stage, OCR has been successfully performed. However, modern OCR engines have incorporated significant additional analysis technologies over the past decade. For example, OCR engines are increasingly proficient at properly segmenting the text components in mixed-language documents; for example, extracting Spanish terms from predominantly German-language text.

OCR engines also incorporate specialized identification technologies, including table recognition and table field extraction, form recognition and indexing, mathematical equation recognition, logo brand identification, and graphics recognition. OCR engines output this data along with all of the logical zoning information: connected component bounding box and polygon information, column aggregation, and even article extraction (flowed text identification and stitching).

Each of the OCR analyses can be performed with varying degrees of parallelism. For example, upfront the captured images can be processed as separate pages for multi-page documents, or different partitions of a single page image. Once text regions have been identified, they can be morphologically analyzed separately. Tables, document-internal forms, and other specialized document zones can be processed in dedicated parallel pipelines. In addition, document classification and semantic analyses can be performed by multiple algorithms, even multiple OCR engines, in parallel.

3.4.8 Security Analytics

In this text, I will use a broad definition for security analytics, also including privacy and identity analytics. A mistake sometimes made by people concerned with security is to overestimate the value of security components—such as the encryption, digital signing, or certificate generation algorithm—at the expense of the overall system security. Security is only as strong as its weakest link, and so a pragmatic approach to security analytics is to dissect the overall security analytics into relevant subdomains, and use the most relevant methodologies for each of these domains. Technology vulnerabilities are addressed through modeling, simulations, and rugged test scripting. Cryptographic security, for example, can be tested through entropy-based randomness validation and other methods. Usually, tens of megabytes of data must be processed for risk assessment of the technologies.

The risk due to process flaws, however, cannot generally be scripted. Instead, these must be addressed by speculative testing, ethical hacking, and by adapting the litany of known means of breaking processes. A process such as web-based authentication, for example, can be broken by relatively simple means such as a distributed denial of service (DDoS) attack wherein the validity of legitimate mass serialization codes is called into question by the overuse of legitimate and nonlegitimate codes alike. The authentication service must be built in such a way as to recognize, respond, and rollback (the three Rs) such a DDoS. With proper logging, recognition can happen even with a considerable delay. The response is then to check the log looking backward—with 20/20 hindsight here being quite valuable—and then rollback any nonlegitimate authentication attempts, thereby resetting the authentication service to a last known good state and therefore restoring both the utility of, and confidence in, the system.

The risks due to people are often greater, and more readily manifested, than those due to technology or processes. Social engineering, or the extracting of security-compromising information from hapless honest people or from corruptible actors in a system, has long been the preferred method of fraudulent agents. Organizations that underpay or undervalue their security people might find out some of them are willing to earn two paychecks.

It is clear that the best security analytics will not be myopic or highly focused, but will instead consider how multiple vulnerabilities combine to make an environment much weaker than expected from the individual technological, process, or people vulnerabilities. The goal of security analytics is to expect the unexpected. In the extremely complicated, multi-agent systems of today, it is nearly impossible for linear thinking to prevail against the guile, gold, and gutsiness of the modern, corporate fraud. Parallel approaches, operating independently but combining results, offer the best security analytics.

3.5 Summary

In this chapter, a wide range of application spaces, or domains, were overviewed. These primary and secondary domains, while extensive, are shown in this chapter to inter-relate well enough such that lessons learned in one domain—say, medical image processing—will be readily understood to be applicable in a related domain such as biometrics, surveillance, or image segmentation. Since this book aims to introduce and formalize the patterns for meta-algorithms, it must provide some breadth of domain so that the reader is not left unconvinced as to the broad applicability of meta-algorithmics (and parallelism by task or component). I have tried to select domains such that at least one other domain (e.g., security printing for security analytics) is related closely enough for the reader to see a continuum of applicability of the parallel approaches without being unconvinced as to their breadth. The next eight chapters will show whether I have succeeded. I will start with parallelism by task and component, and then spend the bulk of the book on parallelism by meta-algorithmics.

References

Baker, J.M., Deng, L., Glass, J., Khudanpur, S., Lee, C.-H., Morgan, N., and O'Shaughnessy, D. (2009) Research developments and directions in speech recognition and understanding, part 1. *IEEE Signal Processing Magazine*, **26** (3), 75–80.

Baldini, F., Giannetti, A., Mencaglia, A.A., and Trono, C. (2008) Fiber optic sensors for biomedical applications. *Current Analytical Chemistry*, **4**, 378–390.

Bankman, I.N. (chief editor) (2000) *Handbook of Medical Imaging: Processing and Analysis*, Academic Press, San Diego, CA, 901 pp.

Bronzino, J.D. (editor-in-chief) (1995) *The Biomedical Engineering Handbook*, CRC Press, 2862 pp.

Chaudhuri, P., Ghosh, A.K., and Oja, H. (2009) Classification based on hybridization of parametric and nonparametric classifiers. *IEEE Transactions on Pattern Analysis and Machine Intelligence*, **31** (7), 1153–1164.

Cheney, M., Isaacson, D., Newell, J.C., Simske, S., and Goble, J. (1990) NOSER: an algorithm for solving the inverse conductivity problem. *International Journal of Imaging Systems and Technology*, **2**, 66–75.

Cheng, K.-S., Simske, S.J., Isaacson, D., Newell, J.C., and Gisser, D.G. (1990) Errors due to measuring voltage on current-carrying electrodes in electric current computed tomography. *IEEE Transactions on Biomedical Engineering*, **37** (1), 60–65.

Cowie, R., Douglas-Cowie, E., Tsapatsoulis, N., Votsis, G., Kollias, S., Fellenz, W., and Taylor, J.G. (2001) Emotion recognition in human-computer interaction. *IEEE Signal Processing Magazine*, **18** (1), 32–80.

Davis, S. and Mermelstein, P. (1980) Comparison of parametric representations for monosyllabic word recognition in continuously spoken sentences. *IEEE Transactions Acoustics Speech Signal Processing*, **28** (4), 357–366.

Duda, R.O. and Hart, P.E. (1973) *Pattern Classification and Scene Analysis*, John Wiley & Sons, New York, 592 pp.

Enderle, J., Blanchard, S., and Bronzino, J. (eds) (2000) *Introduction to Biomedical Engineering*, Academic Press, San Diego, CA, 1062 pp.

Ezaki, H., Uchida, S., Asano, A., and Sakoe, H. (2005) *Dewarping of Document Image by Global Optimization*. International Conference on Document Analysis and Recognition 2005, Seoul, Korea, pp. 500–506.

Hassan, E., Garg, R., Chaudhury, S., and Gopal, M. (2011) *Script Based Text Identification: a Multi-Level Architecture*. Proceedings of the 2011 Joint Workshop Multilingual OCR Analytics Noisy Unstructured Text Data, article 11, 8 pp.

Jelinek, F. and Mercer, R. (1985) Probability distribution estimation from sparse data. IBM Technical Disclosure Bulletin, Vol. 28, pp. 2591–2594.

Kittler, J. and Illingworth, J. (1986) Minimum error thresholding. *Pattern Recognition*, **19** (1), 41–47.

Krishnamurthy, A. and Childers, D. (1986) Two channel speech analysis. *IEEE Transactions Acoustics Speech Signal Processing*, **34** (4), 730–743.

Kupiec, J. (1992) Robust part-of-speech tagging using a Hidden Markov Model. *Computer Speech and Language*, **6**, 225–242.

Lin, X.-R., Guo, C.-Y., and Chang, F. (2011) *Classifying Textual Components of Bilingual Documents with Decision-Tree Support Vector Machines*. International Conference on Document Analysis and Recognition 2011, pp. 498–502.

Mandelbrot, B. (1954) Structure formelle des texts et communication. *Word*, **10**, 1–27.

Manning, C.D. and Schütze, H. (2000) *Foundations of Statistical Natural Language Processing*, 3rd edn, The MIT Press, Cambridge, MA, 680 pp.

Mohamad, R.A.-H., Likforman-Sulem, L., and Mokbel, C. (2009) Combining slanted-frame classifiers for improved HMM-based Arabic handwriting recognition. *IEEE Transactions on Pattern Analysis and Machine Intelligence*, **31** (7), 1165–1177.

Otsu, N. (1979) A threshold selection method from gray level histograms. *Pattern Recognition*, **9** (1), 62–66.

Rosenberg, A.E., Lee, C.H., and Soong, F.K. (1994) *Cepstral Channel Normalization Techniques for HMM-Based Speaker Verification*. Proceedings of the IEEE ICASSP, pp. 1835–1838.

Shannon, C.E. (1948) A mathematical theory of communication. *Bell System Technical Journal*, **27**, 379–423, 623–656.

Shannon, C.E. (1951) Prediction and entropy of printed English. *Bell System Technical Journal*, **30**, 50–64.

Simske, S. and Adams, G. (2010) *High-Resolution Glyph-Inspection Based Security System*. Proceedings of the IEEE ICASSP, pp. 1794–1797.

Simske, S., Pollard, S., and Adams, G. (2010) *An Imaging System for Simultaneous Inspection, Authentication and Forensics*. IEEE IST 2010, pp. 266–269.

Simske, S.J. (1987) An adaptive current determination and a one-step reconstructive technique for a current tomography system. MS thesis, Rensselaer Polytechnic Institute, 110 pp.

Simske, S.J. (2009) *Dynamic Biometrics: The Case for a Real-Time Solution to the Problem of Access Control, Privacy and Security*. IEEE BIdS 2009, pp. 1–10.

Simske, S.J., Aronoff, J.S., Sturgill, M.M., and Golodetz, G. (2008) Security printing deterrents: a comparison of thermal inkjet, dry electrophotographic and liquid electrophotographic printing. *Journal of Information Science and Technology*, **52** (5), 050201-1–050201-7.

Simske, S.J. and Blakley, D.R. (2012) *Using the Vectorcardiogram to Remove ECG Noise*. IEEE ICIP 2012, pp. 2301–2304.

Simske, S.J., Sturgill, M., and Aronoff, J.S. (2009) *Effect of Copying and Restoration on Color Barcode Payload Density*. Proceedings of the ACM DocEng 2009, pp. 127–130.

Simske, S.J., Sturgill, M., Aronoff, J.S., and Villa, J.C. (2008) *Spectral Pre-compensation and Security Print Deterrent Authentication*. NIP24: 24th International Conference on Digital Printing Technologies and Digital Fabrication 2008, pp. 792–795.

Simske, S.J., Sturgill, M., Everest, P., and Guillory, G. (2009) *A System for Forensic Analysis of Large Image Sets*. WIFS 2009: 1st IEEE International Workshop on Information Forensics and Security, pp. 16–20.

Soria, S., Berneschi, S., Brenci, M., Cosi, F., Conti, G.N., Pelli, S., and Righini, G.C. (2011) Optical microspherical resonators for biomedical sensing. *Sensors*, **11**, 785–805.

Spitz, A.L. (1997) Determination of the script and language content of document images. *IEEE Transactions on Pattern Analysis and Machine Intelligence*, **19** (3), 235–245.

Vans, M., Simske, S., and Aronoff, J.S. (2009) *Barcode Structural Pre-compensation Optimization*. NIP25: 25th International Conference on Digital Printing Technologies and Digital Fabrication 2009, pp. 167–169.

Vaseghi, S., Yan, Q., and Ghorshi, A. (2009) Speech accent profiles: modeling and synthesis. *IEEE Signal Processing Magazine*, **26** (3), 69–74.

Wallace, R., McLaren, M., McCool, C., and Marcel, S. (2012) Cross-pollination of normalization techniques from speaker to face authentication using Gaussian mixture models. *IEEE Transactions on Information Forensics and Security*, **7** (2), 553–562.

Yacoub, S., Simske, S. Lin, X., and Burns, J. (2003) Recognition of emotions in interactive voice response systems. *HPL Technical Report HPL-2003-136*, 5 pp.

4

Applications of Parallelism by Task

I believe that one can indeed work on two or more tasks at once, but in ways yet to be understood.
—Marilyn vos Savant

4.1 Introduction

In this chapter, the first of three major types of parallelism to be addressed in this book—parallelism by task—is considered in depth. Parallelism by task is arguably the most straightforward form of parallelism—a set of N tasks must be completed, and to the extent that they are independent these tasks can be assigned to one of $\{2, \ldots, N\}$ parallel pipelines.

The use of parallelism by task is illustrated with reference to the four primary, or "core," domains of this book: (1) document understanding, (2) image understanding, (3) biometrics, and (4) security printing. This set is chosen because they are a diverse and broad set of domains, and so they exercise many of the possibilities in parallelism by task. My intended approach throughout this book is to favor the pragmatic over the arcane, the general over the specific, and the applicable over the idiosyncratic. This chapter, therefore, is designed to provide a practical guide for implementing parallelism by task. The meaning of "parallelism by task" is, therefore, simple enough: different tasks that would normally be performed in a sequential pipeline are, where possible, performed in parallel.

Document understanding, introduced in Chapter 3, is a complex set of algorithms, knowledge engines, and systems concerned with upgrading the content associated with a document. In the broadest sense, a document is any item in any of a wide variety of media that contains some language content. For example, audio documents require the conversion of speech to text through the process of automatic speech recognition (ASR). Scanned or camera-captured text, such as for mobile sign translation, requires the conversion of image to text using the process of optical character recognition (OCR). In this chapter, OCR-based document understanding systems will be examined in some depth.

For document understanding, many modern engines—such as OCR packages—provide an integrated set of tasks that, internally, rely on a sequential or at most hybrid (parallel-sequential) design. For example, most modern commercial OCR engines provide most or all of the following in one packaged software system: language detection, page orientation

Meta-algorithmics: Patterns for Robust, Low-Cost, High-Quality Systems, First Edition. Steven J. Simske.
© 2013 John Wiley & Sons, Ltd. Published 2013 by John Wiley & Sons, Ltd.

determination, page skew detection, character to ASCII/Unicode conversion, font identification, table recognition, equation recognition, and special character (symbol, boldface, italics, underline, superscript, subscript) identification. This means that the internals of such an engine cannot be "deconstructed" and redesigned for a specific parallelism by task approach. However, within the context of larger, or more complicated, document understanding workflows, the OCR engine can be treated as a single—if complex—task. In this chapter, I consider an integrated document analysis, management, storage, and lifecycle system.

The next topic is image understanding. In this section, the queue-based pattern is applied to emotion detection. The variable-sequence-based pattern is applied to gesture understanding to account for multiple camera angles viewing the same subject. Parallel-separable operations (PSOs) are then considered. The latent introspective PSO (LI-PSO) is discussed in light of quantitative evaluation of images, for example, for object distance and vanishing point determination; the unexploited introspective PSO (UI-PSO) is discussed in light of tracking; and the emergent introspective PSO (EI-PSO) is discussed in light of shape recognition. Finally, the recursively scalable task parallelism approach is applied to image segmentation and restoration pipelines.

The third broad field of interest is biometrics. Simple queue-based patterns are used directly for image-based identification. More complicated queue-based approaches, however, can be used to generate Bayesian inferences. Variable-sequence-based parallelism by task can be used to extend Bayesian-based inference to other means of assigning identity probabilistically. Another probabilistic approach—simultaneous assessment of the likelihood of false positive identification and false negative identification—can be pursued using an LI-PSO in which the top candidates are assessed in parallel. A UI-PSO emerges when different elements of a specific biometric task—for example, gait analysis—are seen to use separable algorithms. Recursively scalable task parallelism is applied to both one-dimensional (1D) (electrocardiogram and speech) and two-dimensional (2D) (iris analysis) biometric signals. The discussion of parallelism by task for biometrics is concluded with the introduction of scaled-correlation and task-correlation approaches.

The chapter concludes with a discussion of parallelism by task approaches for security printing. The queue-based pattern is employed for lot registration and validation. The variable-sequence-based pattern is used to simultaneously authenticate multiple security printing features. Next, the application of an LI-PSO to hybrid security printing features is described. The discussion on security printing is completed with a consideration of probabilistic authentication and the use of multiple variable marks and/or multiple forensic approaches to security printing.

4.2   Primary Domains

In many ways, this chapter presents material that is, conceptually, the most straightforward of any in the book. At risk of celebrating the obvious in this chapter, however, I hope to introduce some important ways of deconstructing even simple tasks that will provide useful insights in later chapters, especially those on meta-algorithmics.

Parallelism by task generally involves the simultaneous parallel usage of multiple instances of the same data set. In this process, a set of sequential tasks that can be performed independently are in fact performed at the same time without respect to the other tasks. A second, distinct, form of parallelism by task is when multiple, dependent sequential tasks are performed in parallel on distinct data sets. This second form is different from parallelism by component (described in the Chapter 5) insofar as the data sets are not componentized (i.e., they are not divided into subcomponents). As described in Chapter 2, this form of parallelism is simply multi-processing where each processor performs the same task.

As a quick recapitulation from Chapter 2, the primary forms of parallelism by task are:

1. queue-based
2. variable-sequence-based
3. LI-PSO
4. UI-PSO
5. EI-PSO
6. recursively scalable task parallelism

The queue-based parallelism by task approach was introduced in Section 2.2.1. This very simple pattern deconstructs a given process into a set of necessarily sequential tasks. For description here, let us suppose there are five tasks $\{A, B, C, D, E\}$, which must be completed in a given order. These tasks, therefore, have built-in dependencies: the input of Task B requires the output of Task A, the input of Task E requires the output of Task D, and so on. This constitutes a pipeline, shorthanded as A I B I C I D I E. In a queue, any other pipeline, for example, D I C I A I B I E, will nonpredictably provide different results. Pipeline architects are skilled in determining the correct order for the individual tasks (algorithms, transformations, etc.), and this knowledge should not be abrogated by the parallel processing architect. In some cases, however, the pipeline can be broken into subpipelines that can be performed in variable order. Domain expertise—often deep domain knowledge—is generally needed to know how to break up queues into parallel queues. Fortunately, however, for most nontrivial systems, this form of parallelism is in general a waste of effort.

The effort of breaking queues, or pipelines, into parallel subpipelines is often unnecessary because, for most high-throughput systems, the increased efficiency of such parallelism by task is more than made up for by the increased efficiency of assigning (complete) pipelines to parallel paths. That is, assigning $\{A_1$ I B_1 I C_1 I D_1 I $E_1\}$, $\{A_2$ I B_2 I C_2 I D_2 I $E_2\}, \ldots ,$ $\{A_N$ I B_N I C_N I D_N I $E_N\}$ to parallel paths provides nearly all of the increased throughput that would be possible in assigning, for example, $\{A_1$ I B_1 I $C_1\}$, $\{D_1$ I $E_1\}$, $\{A_2$ I B_2 I $C_2\}$, $\{D_2$ I $E_2\}, \ldots , \{A_N$ I B_N I $C_N\}$, $\{D_N$ I $E_N\}$ to parallel paths. This is because each step, or transform, in a high-throughput pipeline is completed quickly.

These latter sets—$\{A_1$ I B_1 I $C_1\}$, $\{D_1$ I $E_1\}$, $\{A_2$ I B_2 I $C_2\}$, $\{D_2$ I $E_2\}, \ldots , \{A_N$ I B_N I $C_N\}$, $\{D_N$ I $E_N\}$—are examples of the variable-sequence-based approach introduced in Section 2.2.1. That is, the larger pipeline $\{A_1$ I B_1 I C_1 I D_1 I $E_1\}$ is split into two smaller pipelines $\{A_1$ I B_1 I $C_1\}$ and $\{D_1$ I $E_1\}$, and so on for each larger pipeline $\{A$ I B I C I D I $E\}$. While scheduling of these individual pipelines may result in improved throughput, the overhead of splitting and reassembling the pipelines will generally only be warranted when the number of parallel processors is relatively sparse.

The PSOs include the LI-PSO, which arises when an algorithm or process is deconstructed and readied for parallelism. As an example, image convolution operations, typically used for sharpening and other forms of image filtering, are readily amenable to parallel processing operations. Images can be tessellated into subimages with minimal pixel redundancy that is proportional to the ratio of the width of the convolution kernel to the width of the subimage. After tessellation, each subimage can be sharpened in parallel. Additionally, each sharpening operation itself can be performed using parallel processing approaches. This is an UI-PSO, which is born of deployment.

A more complex pattern for parallelism by task is the *recursively scalable task parallelism* pattern, as shown in Section 2.2.2 (Figure 2.3). This "Parallelism by Task" pattern is somewhat blurred with parallelism by component—the main topic of Chapter 5—inasmuch as a task is reduced to subtasks. It is appropriate to include it here, however, to illustrate how to optimize a parallel processing approach. The aforementioned example—tessellating an image to run parallel operations such as convolution or other filtering on subimages—will be explored further here as the system design for performing this filtering.

First, a simple application using a filtering kernel is introduced. We wish to lowpass filter an image, which effectively blurs it. A simple operation for blurring an image is to replace each pixel with a weighted average of the pixel and its surrounding pixels. If the pixel of interest is labeled $p(x,y)$, then a lowpass filter is used to create an altered, new pixel, $p_{new}(x,y)$. If we choose to weight the new pixel 2/3 based on its original value and 1/3 based on its 8-nearest neighbor pixels, then we obtain the following formula: $p_{new}(x,y) = 2/3p(x,y) + 1/24p(x - 1,y - 1) + 1/24p(x - 1,y) + 1/24p(x - 1,y + 1) + 1/24p(x,y - 1) + 1/24p(x,y + 1) + 1/24p(x + 1,y - 1) + 1/24p(x + 1,y) + 1/24p(x + 1,y + 1)$. We can imply this transformation explicitly with the notation

$$I_{new} = I_{old} \times \mathbf{C}.$$

Meaning that I_{new} is I_{old} convolved with the filtering matrix $\mathbf{C}$. In this case, $\mathbf{C}$ is given by

$$\mathbf{C} = \begin{bmatrix} 1/24 & 1/24 & 1/24 \\ 1/24 & 2/3 & 1/24 \\ 1/24 & 1/24 & 1/24 \end{bmatrix}.$$

The effect of this transformation on a mixed (text, photo) image is shown in Figures 4.1 and 4.2.

In order to perform convolution on a subset of the image I, which is delimited by a rectangular bounding box $\{xmin, xmax, ymin, ymax\}$, we need a subimage, S, that is sized $(xmax - xmin + 2) \times (ymax - ymin + 2)$ with the extra two pixels in the x- and y-directions enabling the convolution to extend to the pixels immediately surrounding the pixels on the edge of the subimage. The convolution kernel, C, used in the example is a 3×3 kernel. Usually, kernels are symmetric $N \times N$ matrices where N, usually called the kernel size, is an odd integer. This means that the border around the subimage S will need to be B pixels, where $B = (N - 1)/2$. As B increases, the relative size of the subimage together with its necessary boundary pixels increases in comparison to the subimage, expressed by the equation

$$\text{subimage convolution overhead ratio} = \frac{(S + 2B)^2}{S^2}.$$

On the other hand, as S, the original subimage size, increases, the relative impact of a fixed value for B decreases. These trade-offs are illustrated in Table 4.1.

Figure 4.1 Original image with the filter "Lowpass Filter" selected. The image as shown has not been filtered yet. The lowpass filter is described in the text. Reproduced by permission of Cheyenne Mountain Zoo

The value $(S + 2B)^2/S^2$ in Table 4.1 represents the overhead of preparing the convolution operation for parallelism. For imaging applications, since the convolution operation is generally a function of real resolution, B tends to be small when resolution is low, and large when resolution is high. Thus, larger images, which would tend to have large values of S, are also the ones with large values of B—as a result the *subimage convolution overhead ratio* is typical less than 1.1.

A second type of overhead is associated with the recursively scalable approach to image filtering—the overhead required to tessellate the original image and then restitch it after performing the parallel operations. This overhead grows as the number of subimages created increases; that is, as the number of subimages grows. Table 4.2 provides the relative overhead due to subimage formation and reversal. As is true for all timings reported in this book, processing times were obtained using built-in timing capabilities (e.g., using the QueryPerformanceCounter() Win32 API function).

These two types of overhead penalize the creation of subimages. However, they can be offset by the increased throughput provided by performing filtering on the smaller images.

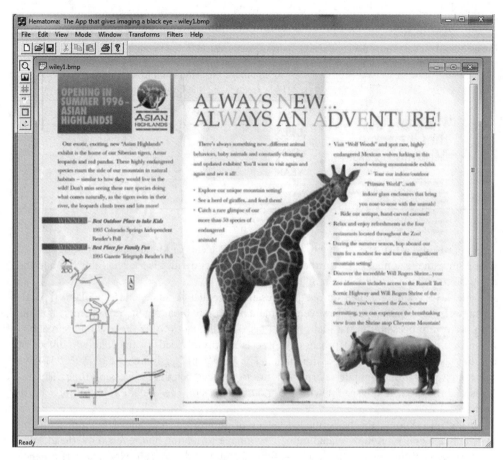

Figure 4.2 The image of Figure 4.1 after the lowpass filtering has been performed as described in the text. Reproduced by permission of Cheyenne Mountain Zoo

Table 4.1 Subimage convolution overhead as a function of subimage size (S), which is the same in the x- and y-directions, and as function of the convolution

		Subimage Convolution Overhead Ratio $= (S + 2B)^2/S^2$			
N	$B = (N - 1)/2$	$S = 200$	$S = 100$	$S = 50$	$S = 25$
3	1	1.02	1.04	1.08	1.17
5	2	1.04	1.08	1.17	1.35
7	3	1.06	1.12	1.25	1.54
9	4	1.08	1.17	1.35	1.74
11	5	1.10	1.21	1.44	1.96

Table 4.2 Subimage formation and reversal (splitting and rejoining) impact on throughput for image filtering applications. Overhead is relative to the timing required for loading and unloading the original image. Based on the timing data generated, most of the overhead is due to repetitive access to memory

N	$B = (N-1)/2$	Subimage Formation and Reversal Overhead (Relative to Using the Original Image in its Entirety)			
		$S = 200$	$S = 100$	$S = 50$	$S = 25$
3	1	1.01	1.02	1.06	1.15
5	2	1.01	1.02	1.06	1.15
7	3	1.01	1.02	1.06	1.15
9	4	1.01	1.02	1.06	1.15
11	5	1.01	1.02	1.06	1.15

This improved performance is due to the efficiency, in terms of higher relative throughput, of processing smaller images (for further illustration of this trend, please see Table 5.6 and the discussion around it), and is given in Table 4.3.

Table 4.3 accounts for the improved processing time that occurs even if the planned parallelism were not to take place. However, the main purpose in breaking up the task into subtasks is to perform processing in parallel on each task. The impact of this processing approach is given in Table 4.4. Here, the processing time is simply divided by the relative number of parallel tasks that are performed simultaneously. Thus, smaller images (e.g., $S = 25$) have substantial gain in throughput—corresponding to reduced relative processing time. This increased throughput is consistent with the modified version of Amdahl's law introduced in Section 2.3.2.

Taking into account the performance values of Tables 4.1, 4.2, and 4.4—corresponding to "subimage convolution overhead ratio," "subimage formation and reversal overhead," and "relative processing time," respectively—Table 4.5 shows the overall expected performance for the combinations of (N,S) investigated. The sum of (subimage convolution overhead ratio -1.0) + (subimage formation and reversal overhead) + (relative processing time) is used to

Table 4.3 Relative processing time for each image. Times are computed by summing up the processing time for each subimage multiplied by the number of subimages. Further performance improvement, of course, is obtained using parallel processing (see Table 4.4) in addition to smaller subimages. This table illustrates the contribution of structural reframing on the improved throughput

N	$B = (N-1)/2$	Relative Processing Time for the Entire Image (Subimage Processing Time) × (Number of Subimages)			
		$S = 200$	$S = 100$	$S = 50$	$S = 25$
3	1	0.94	0.87	0.83	0.81
5	2	0.94	0.87	0.83	0.82
7	3	0.94	0.88	0.84	0.82
9	4	0.94	0.88	0.84	0.83
11	5	0.95	0.89	0.85	0.83

Table 4.4 Relative processing time for each image, accounting for the parallel processing improvement. Here, the number of parallel processors is relatively limitless, meaning throughput directly scales from Table 4.3 based on the relative area of the subimages

N	$B = (N-1)/2$	Relative Processing Time for the Entire Image (Subimage Processing Time) × (Number of Subimages)			
		$S = 200$	$S = 100$	$S = 50$	$S = 25$
3	1	0.94	0.22	0.05	0.02
5	2	0.94	0.22	0.05	0.02
7	3	0.94	0.22	0.05	0.02
9	4	0.94	0.22	0.05	0.02
11	5	0.95	0.22	0.05	0.02

mark performance. The value of subimage convolution overhead ratio has 1.0 subtracted from it so as not to count the processing of the subimage twice.

Table 4.5 does not account for relative differences in processing time for these three metrics. However, in performing these experiments on a set of 50 digital camera-captured (mean size 1.6 MB) 3-channel color images of natural scenes, it was found that the relative processing time of each of these three factors is given by the following:

1. Subimage convolution overhead ratio $= 1.01 = k_1$
2. Subimage formation and reversal overhead $= 0.07 = k_2$
3. Relative processing time $= 1.00 = k_3$.

Using these three values $\{k_1, k_2, k_3\}$, the overall performance timing (OPT) can be exactly calculated from OPT $= k_1 \times$ (subimage convolution overhead ratio $- 1.0$) $+ k_2 \times$ (subimage formation and reversal overhead) $+ k_3 \times$ (relative processing time). These values are shown in Table 4.6.

The relative results for Table 4.6 are very similar to those of Table 4.5. This is an important point to note, as it is an example of a result that will be observed in many of the (especially

Table 4.5 Overall relative performance for each convolution kernel size (N) and subimage size (S) combination, where the performance is simply the sum of Tables 4.1 4.2, and 4.4, or (subimage convolution overhead ratio $- 1.0$) $+$ (subimage formation and reversal overhead) $+$ (relative processing time). The optimal configuration for each value of B is boldfaced in the data cells of the table. For $B = 3$, the scores are identical for $S = 100$ and $S = 50$

N	$B = (N-1)/2$	Total Performance = Sum of Values in Tables 4.1, 4.2, and 4.4			
		$S = 200$	$S = 100$	$S = 50$	$S = 25$
3	1	1.97	1.28	**1.19**	1.34
5	2	1.99	1.32	**1.28**	1.52
7	3	2.01	**1.36**	**1.36**	1.71
9	4	2.03	**1.41**	1.46	1.91
11	5	2.06	**1.45**	1.55	2.13

Table 4.6 Overall relative performance for each convolution kernel size (N) and subimage size (S) combination, where the performance is $k_1 \times$ (subimage convolution overhead ratio $- 1.0) + k_2 \times$ (subimage formation and reversal overhead) $+ k_3 \times$ (relative processing time), where k_1, k_2, and k_3 are the timing weights for the values in Tables 4.1, 4.2, and 4.4, respectively. The optimal configuration for each value of B is boldfaced in the data cells of the table

N	$B = (N-1)/2$	Total Performance = Weighted Sum of Values in Tables 4.1, 4.2, and 4.4 (multiplied by k_1, k_2, and k_3, respectively)			
		$S = 200$	$S = 100$	$S = 50$	$S = 25$
3	1	1.03	0.33	**0.21**	0.27
5	2	1.05	0.37	**0.30**	0.45
7	3	1.07	0.41	**0.38**	0.65
9	4	1.09	**0.46**	0.48	0.85
11	5	1.12	**0.50**	0.58	1.07

imaging-related) parallel processing systems in this book. The theme is that the *preparation for parallel processing* often results in significant improvement in processing efficiency, even before the parallel processing is performed. The range of subimage sizes is clearly sufficient to find an optimal strategy: for the most common convolution kernel sizes, $S = 50$ provides optimal throughput. For larger kernel sizes, $S = 100$ provides the optimal throughput.

With this simple introductory example, we now turn our attention to the four primary domains (document understanding, image understanding, biometrics, and security printing) considered in this book for further illustration of the features and deployment recommendations for parallelism by task.

4.2.1 Document Understanding

As noted in the introduction to this chapter, modern document understanding systems are quite complex, comprising multiple intelligent systems such as (1) OCR, (2) document workflow management, (3) distributed document storage, and (4) document lifecycle management, as overviewed in Table 4.7. Different document understanding tasks require different sets of these engines, and the distinctions between them are not always clearly defined; for example, many modern enterprise resource management (ERM) systems comprise two or more of these four systems. In general, however, the integrated ERM systems allow the deployment settings to be customized at the level of these four large systems (or even more discretely).

Table 4.7 identifies the primary purposes of each of these four systems. The OCR and document recognition system is primarily concerned with digitization of text. However, graphics (logos, business graphics, etc.) and image-based information (photos, drawings, etc.) are also converted to digital form by OCR systems. PDF is a familiar output format for OCR engines, and its format allows for embedded images, graphics, and text.

Document workflow management systems provide all read, write, and editing access to all participants in the document workflow, including reminders or automation for distribution of the document after each access to the document. For secure documents, the access rights may include connection to an authentication service.

Table 4.7 Four large document understanding systems and their primary purposes

Document Understanding System	Primary System Purpose
Optical character recognition and document recognition	Convert image-based information into electronic information, in particular the digitization of text
Document workflow management	Provide/control access to documents, including read/write/edit access, ensure proper document distribution among participants
Distributed document storage	Determine location and settings (including compression/encryption, redundancy, and fast look-up) of all document storage
Document lifecycle	Perform important workflows checkpoints, including authorization, authentication, notification, and validation

In addition to workflow management, document content—data, meta-data, usage, and all logging information—must be stored. Distributed document storage systems are used to determine where and how to store the documents. This includes specification of storage location and settings: whether and how to composite the data, compress the information, and how and where to use encryption.

The fourth document understanding system defined in Table 4.7 is a document lifecycle system. A document lifecycle is the set of operations governing all stages/checkpoints in the document's pathway from creation to completion. A document lifecycle systems is used to perform important workflows checkpoints, including authorization, authentication, notification, and validation.

Given the definition of these document understanding systems, I now consider each in more depth to prepare the discussion of them in context of parallelism by task.

The first of these systems, OCR, has incorporated many other document understanding technologies over the years. Modern OCR systems, such as those produced by numerous commercial OCR vendors such as Nuance, Abbyy, and Iris, provide a wide array of features focused on the digitization of paper-based (document image) information. In the past, OCR engines were not particularly adept at automatically determining the language of the scanned document: users were usually prompted to select the language manually. Now, language identification is a prominent feature upfront during the OCR process. Increasingly, OCR engines are also adept at mixed-language documents; for example, extracting French phrases from English-language documents.

OCR engines have also incorporated many of the technologies formerly associated with stand-alone page restoration and analysis products. These technologies include page orientation (landscape vs. portrait, left vs. right detection), page skew detection and correction, contrast enhancement, color detection, and page size determination. Additional capabilities of OCR and other digitization technologies are listed in Table 4.8. From a text perspective, the most important function of an OCR engine is the conversion of raster images of text characters to ASCII/Unicode. To aid in downstream text interpretation, though, OCR engines additionally perform font identification, including the tagging of special text such as superscript, subscript, underlining, boldface, and italics. Symbolic fonts are identified and with increasing accuracy distinguished from logos, equations, and other special glyphs.

Table 4.8 Four large document understanding systems and sample tasks performed by each system

Document Understanding System	Example System Tasks
Optical character recognition and document recognition	Language detection, page orientation determination, page skew detection, portrait/landscape detection, character to ASCII/Unicode conversion, font identification, table recognition, equation recognition, special character (symbol, boldface, italics, underline, superscript, subscript) identification, page layout extraction, semantic extraction, text tagging, logical region segmentation, and location extraction.
Document workflow management	Document version control, document access control, definition of logical participants at each step of the workflow, definition of the workflow steps, allowance of different document operations at each stage of the workflow, including save, send, print, and archive.
Distributed document storage	Map document sensitivity to storage regimen. Perform document composition and decomposition as necessary to store the document and/or the document elements with the proper access rights at each stage of the workflows. Determine where storage should occur (local file structure, local shared drive, private cloud, public cloud, secure repository, etc.). Perform appropriate document compression and/or encryption, including to the different logical parts, as deemed appropriate by the document workflow management system.
Document lifecycle	Perform all security- and privacy-related tasks. Validate receipt of the documents at each stage of the workflow by each party in the workflow. Authorize every participant as they validate, read, or edit the document. Authenticate the changes using a public key infrastructure (PKI) or other security infrastructure. Provide logging, notification, and auditing.

OCR engines also perform advanced region segmentation and identification, including table recognition, form recognition, mathematical equation recognition, logo identification, and graphics recognition. OCR engines output this data along with all of the logical segmentation information: region bounding box and polygon information, page layout description, and text tagging for downstream text mining and semantic extraction. OCR engines have been sold for more than two decades, and rely on a substantial amount of legacy code. Therefore, all of the OCR engines for which I have had the opportunity to view the source code rely on a sequential or at most hybrid (parallel-sequential) design. Unless one has access to the source code of the OCR engine (possible for open source OCR packages such as the HP Labs-spawned, Google-supported Tesseract OCR engine), one cannot "deconstruct" and thereafter redesigned an OCR engine for a specific parallelism by task approach. Based on the thousandfold improvement in processing speed, memory access, and storage capacity since the early days of OCR, however, this "limitation" is of little consequence. Most modern parallel systems—from multi-core to multi-processor to fully virtualized architectures—provide enough resources in each core, processor, or virtual machine (VM) to perform OCR on a sizeable document or document portion. This means that the three orders of magnitude relative diminution in the scale of the OCR task over the years has allowed the relative "perspective" of parallelism to move from, say, the word, sentence, or paragraph level to the level of an image, a page, or a document. From a document engineering perspective, this difference is simply one of scale. From a parallelism by task standpoint, however, this difference is highly significant, since the elements in the parallelism are now more conveniently exactly the familiar storage primitives such as documents or images, or logical subelements within these primitives, such as pages or single frames.

Salient document understanding tasks can make use of the primary set of the Parallelism by Task patterns introduced in Chapter 2. For example, in order to make use of the *queue-based* approach originally shown in Figure 2.1, we need to identify sequential document understanding tasks that must take place in a specific order. One such sequential system (or set of tasks) using each of the four major systems of Tables 4.7 and 4.8 is the following pipeline:

A. Determine the document identity from a scanned or electronic version of the document. This uses OCR, if necessary, followed by document recognition technologies such as described in earlier chapters. This corresponds to the "optical character recognition and document recognition" block in Tables 4.7 and 4.8.
B. Determine the security level required for the document based on its document type and the associated policies. This is performed by the "document workflow management" block in Tables 4.7 and 4.8.
C. Based on the size of the document—including its original content and the additional content created in Step A—and the storage policies, determine the storage settings for the document and locate the document accordingly. This is handled by the "distributed document storage" system.
D. Finally, all relevant security and privacy algorithms are employed for each of the appropriate authentication, authorization, validation, logging, and/or notification tasks. These tasks are performed according to the appropriate policies, which are also associated with the document workflow management system, and together comprise the "document lifecycle" system tasks for the document. For example, logging of all tasks performed in Steps A, B, and C above may be required, and completion of certain tasks triggers notification of salient participants; for example, the document owner/creator/administrator. Based on policy, authentication—for example, in the form of a digital signature—may be required for one or more participants/steps in the document lifecycle. Finally, the appropriate tasks are performed in anticipation of document storage—for example, compression and backup.

Steps A, B, C, and D, in combination, comprise a classic queue-based parallelism by task approach, where a given (sequential) pipeline can be assigned to each of the documents processed in a parallel path for each document. The pipeline is simply A | B | C | D, where the output of A is required as input to B, the output of B is required input for C, and the output of C is required input for D. There are some shared policies—for example, for the security and storage protocols—but these policies do not allow task order variance in the order of the steps.

However, there are document understanding tasks for which the steps can be varied in order or composition. An example of a *variable-sequence-based* document understanding process is a document workflow that undergoes a split and merge stage. For example, suppose a particular document of value is sent to three different participants for their approval. Also suppose that

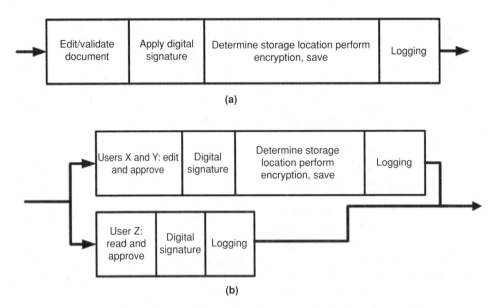

Figure 4.3 Variable-sequence-based document understanding example. Here, the Task A is different for the two types of users, and as such indicated by A and A′, where A results in a changed (edited) document, but A′ only requires User Z to correctly digitally sign the document. Since User Z does not change the document, the storage/saving Step C does not occur in User Z's pipeline. Figure (a) shows the general pipeline A | B | C | D, which is performed for Users X and Y in the upper path for (b). The lower path in (b) consists of the pipeline A′ | B | D

two of the participants (X and Y) are allowed to edit the document and the third participant (Z) is only allowed to view (read) the document. Then the pipeline for X and Y may require a different set of access right security algorithms, encryption algorithms, notifications, and logging requirements than is the case for participant Z. These separate types of participants, therefore, can be assigned to different variable-sequence subpipelines, as illustrated in Figure 4.3 in which the pipelines are A | B | C| D and A′ | B | D.

Next, the PSOs are considered in the context of document understanding systems. These operations cannot be defined by a single pattern per se. Instead, they depend on domain knowledge to prepare for parallelism. A LI-PSO is parallelism by task wherein the original operation is deconstructed for the purposes of preparing it for subsequent parallel processing. Each of the four large systems described in Table 4.2 are likely highly amenable to LI-PSO if, in fact, they are not already explicitly designed internally for parallelism by task. In the case of OCR and Document Recognition, for example, the skew detection task is advantageously performed before any of the other tasks. Next, portrait/landscape detection is advantageously performed, meaning a sequential pipeline suffices at the front end of this system. However, once skew correction and orientation have been performed, the following operations may be performed in parallel:

1. Table recognition and page layout extraction
2. Character segmentation and language detection

After these parallel task pipelines are completed, further pipelines can be performed in parallel:

1. Semantic extraction
2. Special character and font identification

Each of these two parallel sets of pipelines emerges from a reconsideration of the larger system. There are several benefits of restructuring the system in accordance with this parallelism by task. One benefit is that, in many cases, some of the analyses to be performed by the OCR engine are not required: language detection, for example, if it has already been determined for the document from which a page is taken. Another benefit is that such flexibility in the OCR and document recognition tasks allows the system designer to leave otherwise large document elements—such as an entire clip of video or a many-page document—unpartitioned, and in so reducing the organizational-, logging-, and partitioning-associated overhead.

As the complexity of system architecture increases in future chapters of this book, feedback patterns—in which the results of a later step in a sequential or sequential-parallel hybrid architecture are used to affect the earlier steps in associated or even future jobs performed—will become very familiar. It is worth pointing out, then, that a PSO becomes far more adaptive when we do allow parallel branches to operate with schedulers ensuring that certain parallel processes are completed before others for a given task. In the system considered in this section, for example, language detection can be used to improve the accuracy of the orientation detection. This is because there are different language-specific rules that can be used to improve upside-down versus right-side up classification of text. The character sets for the five prevalent Western European "EFIGS" (English, French, Italian, German, Spanish) languages, for example, have a much higher occurrence frequency of ascenders (elements of characters that extend above the top of the small letter "x"), such as occur in all capital letters and the common letters "l," "t," "h," and "d," than the occurrence frequency of descenders such as occur with the highest probability in the less common letters "p" and "y." This type of font-shape difference does not occur in many other languages; for example, Japanese, Mandarin or Cantonese Chinese, or Korean. For this reason, we may wish to have language detection and its associated tasks scheduled for completion before orientation detection is scheduled. From an architectural standpoint, this means that parallel pipelines for processing—one associated with at least language detection and the other associated with at least orientation detection are put in the design: the scheduler then ensures that the former task occurs before the latter for each document.

The second type of PSO, the UI-PSO, is parallelism by task that is born of deployment. An example of a UI-PSO is taken from performing OCR on multiple pages or sections of a larger document in parallel. There are a number of valid reasons to not analyze sections of a document in parallel, two of them being (a) a large enough number of documents are analyzed by the overall system that even a multi-page document is a reasonably sized atomic element for the scheduling of the parallel processing; and (b) the OCR engine can "learn" important settings about later pages or sections, such as language, orientation, fonts used, and so on—from the first or first few pages. The UI-PSO pattern may be delayed, therefore, when (a) the document size is large relative to the number of documents analyzed, such that a more refined atomicity is desired; and/or (b) the document is associated with a known document type—for example, it is built from a template, identified as a specific document in a workflow, and so on—such that most of the "global" document settings such as font families used, boilerplate text, and so on, can be readily deployed to partitions of the document rather than the entire document itself without loss of OCR accuracy.

Another example of deploying a UI-PSO for document understanding is when security algorithms are deployed in parallel to provide different access rights to different parts of the document. In this case, the otherwise monolithic security policy (applied to all portions of the document) is replaced with an adaptive, variable security policy appropriate to each portion of the now-partitioned document.

The third type of PSO is the EI-PSO. As mentioned previously, the EI-PSO, unlike a UI-PSO, is not obvious in the original system; and unlike an LI-PSO, does not become obvious in the deconstruction of the original system. It therefore takes even more domain expertise to uncover an EI-PSO than a UI-PSO. An EI-PSO that may be useful for document understanding is one in which different features for identifying the font typefaces being used in a document are broken up into parallel-friendly tasks such as convolution. Then, multiple glyph recognition can be performed in parallel and the highest match selected as the likely font for each character.

The last parallelism by task approach, overviewed in depth in Chapter 2, is quite robust: *recursively scalable task parallelism*. This parallelism is based on the ability to create and optimize a cost function to choose between processing a given Task A and a subset of tasks $\{A_1, A_2, \ldots, A_N\}$ that together comprise A. This cost function is of course dependent on the overhead involved in the unwrapping of Task A into subtasks $\{A_1, A_2, \ldots, A_N\}$ and then rewrapping $\{A_1, A_2, \ldots, A_N\}$ into A, and recursively into smaller subtasks, as possible/appropriate (see Figure 2.3). In document understanding, there is usually a very small amount of recursion possible for the individual classification and identification tasks. For example, character recognition usually begins at the connected component level, which results in primary classification tasks focused on distinguishing two glyphs—for example, two characters such as "l" and "1" or a character such as "m" from two consecutive characters such as "r" and "n." Usually, document classification builds upward—into words, phrases, and sentences. However, in the case of specific comparisons such as "m" versus "r" and "n," some recursive scalability to focus on the image more closely at the end of the first "hump" in the "m" proves valuable.

4.2.2 Image Understanding

The next domain of interest to us for parallelism by task is image understanding. Many, if not most, image understanding tasks require sequential operations in a specific (fixed) order. This type of processing, termed a pipeline, implements *queue-based* parallelism by task. Queue-based imaging tasks require a set of image operations to be performed in a specific order. For example, suppose that we wish to down-sample an image, then sharpen it for better image quality of the lower-resolution image. To down-sample the image, we wish to perform an image convolution operation. For image convolution, the convolution kernel is centered on each pixel in turn, and the value of the center pixel is updated to be the sum of the kernel multiplied by the image values. Down-sampling kernels, on the other hand, commonly have even dimensions. A down-sampling kernel that performs well for $2\times$ down-sampling (e.g., reducing a 600×600 pixels/in.—or ppi—image to 300×300 ppi), is given here:

$$\mathbf{D} = \begin{bmatrix} 1 & 3 & 3 & 1 \\ 3 & 9 & 9 & 3 \\ 3 & 9 & 9 & 3 \\ 1 & 3 & 3 & 1 \end{bmatrix}.$$

To ensure that uniform areas do not have a change in image intensity or saturation, the $\mathbf{D}$ kernel is normalized so that the coefficients sum to 1.0, as shown here by $\mathbf{D}'$:

$$\mathbf{D}' = \begin{bmatrix} 1/64 & 3/64 & 3/64 & 1/64 \\ 3/64 & 9/64 & 9/64 & 3/64 \\ 3/64 & 9/64 & 9/64 & 3/64 \\ 1/64 & 3/64 & 3/64 & 1/64 \end{bmatrix}.$$

The down-sampling kernel is centered over pixel $P(x,y)$ and then multiplied by each of its neighbors to obtain a new pixel, $P'(x,y)$.

In anticipation of sharpening, however, we also wish to find a down-sampling kernel that is 3×3 instead of 4×4. One such kernel is

$$\mathbf{D} = \begin{bmatrix} 1 & 2 & 1 \\ 2 & 4 & 2 \\ 1 & 2 & 1 \end{bmatrix}.$$

Normalizing, this down-sampling kernel becomes $\mathbf{D}'$ as shown:

$$\mathbf{D}' = \begin{bmatrix} 1/16 & 1/8 & 1/16 \\ 1/8 & 1/4 & 1/8 \\ 1/16 & 1/8 & 1/16 \end{bmatrix}.$$

Note that this $\mathbf{D}'$ kernel provides an equal representation of every pixel in the original image P in the down-sampled image P', since the four corners of the kernel have coefficients of (1/16) and are used in four down-sampled pixels each, the four side coefficients of the kernel have coefficients of (1/8) and are used in two down-sampled pixels each, and the center coefficient is (1/4) and is used in only one down-sampled pixel. Thus, exactly one-fourth of the value of every pixel in P is used in P', which has exactly one-fourth the pixels of P.

After down-sampling, we wish to sharpen the down-sampled pixels. One 3×3 sharpening kernel is the Laplacian kernel as described in Gonzalez and Woods (2008), defined as

$$\mathbf{L} = \begin{bmatrix} -1 & -1 & -1 \\ -1 & 8 & -1 \\ -1 & -1 & -1 \end{bmatrix}.$$

Figure 4.4 (a) The original image (2048 × 1536 pixels). (b) The (1024 × 768 pixels) image after performing convolution using the Laplacian kernel **L** as described in the text. (c) The (1024 × 768 pixels) image after down-sampling using the modified down-sampling kernel, **D′**, as described in the text, followed by convolution using the Laplacian kernel. Note the relatively sharper detail in the edges caused by reducing the resolution using **D′**. (d) The sharpened (2048 × 1536 pixels) image obtained by convolving the original image with the original sharpening kernel, **S**, as described in the text. Much of the blur in the original image has been removed

The Laplacian kernel as shown will convert an image into an edge image. If the intensities are randomly distributed in an image, then half of the pixels in a Laplacian image should be black. The impact of the Laplacian on the original image (Figure 4.4a) is shown in Figure 4.4b. The Laplacian creates a black image wherever the centered pixel has a mean intensity below that of the surrounding pixels, which is important in order to keep the edges from expanding (only the "bright" side of the edge shows up in the Laplacian-transformed image). The edges are therefore much more pronounced in Figure 4.4c, which is the Laplacian of the image after down-sampling (as described below). The reason the down-sampled image has brighter edges in Figure 4.4c in comparison to Figure 4.4b is that the down-sampling process generally reduces edges to half their former thickness, meaning the differences between neighboring pixels along edges wider than one pixel in the original image will be sharper in the Laplacian of the down-sampled image.

The next process considered is sharpening. Sharpening is intended to overcome blurring, and so it is very closely related to edge detection, with the difference being that the expected (mean intensity) value of the sharpened image is the same as the original image. The simplest sharpening approach, perhaps, is directly based on the Laplacian kernel. The center coefficient is simply incremented to obtain the sharpening kernel, $\mathbf{S}$:

$$\mathbf{S} = \begin{bmatrix} -1 & -1 & -1 \\ -1 & 9 & -1 \\ -1 & -1 & -1 \end{bmatrix}.$$

The sharpened version of the original image is shown in Figure 4.4d. The image is certainly sharp, but careful evaluation reveals some high-frequency noise in the image. This is expected, since the sharpening kernel itself has high-magnitude, high-frequency content (transitioning from nine times a pixel to the negative of all surrounding pixels).

Next, the effects of down-sampling are shown. The first, Figure 4.5a, uses the 3×3 $\mathbf{D}'$ kernel described above. A "sharper" down-sampled image is obtained in Figure 4.5b through a much simpler down-sampling method. Figure 4.5b is created by simply keeping only those pixels for which the x- and y-locations are both even—for example, $P(0,0)$, $P(0,2)$, ..., $P(2,0)$, $P(4,0)$, ..., $P(2n,2n)$. This down-sampling operation creates a sharper image than does the $\mathbf{D}'$ kernel for the simple reason that no pixel averaging is performed. A second reason is that the original image is actually oversampled (i.e., at 2048×1536 pixels), and so dropping three-fourth of the pixels does not significantly drop image information.

The next step is to combine the operations of sharpening and down-sampling. The first method, a true queue-based approach, follows up convolution with $\mathbf{S}$ with convolution with the 3×3 $\mathbf{D}'$ (down-sampling). The resulting image is shown in Figure 4.5c, and is characterized by high-frequency noise that is not reduced by the down-sampling operation (in fact, the image noise is much more noticeable than for Figure 4.4d). Since this sequential pipeline did not provide excellent results, another approach, in which down-sampling and sharpening are combined into a single convolution kernel, is defined. This "pipeline" kernel $\mathbf{P}$ is given by

$$\mathbf{P} = \begin{bmatrix} -1 & -2 & -1 \\ -2 & 13 & -2 \\ -1 & -2 & -1 \end{bmatrix}.$$

$\mathbf{P}$ is then normalized to have the same off-center coefficients as the $\mathbf{D}'$ matrix, forcing the center coefficient to become 1.75:

$$\mathbf{P}' = \begin{bmatrix} -1/16 & -1/8 & -1/16 \\ -1/8 & 7/4 & -1/8 \\ -1/16 & -1/8 & -1/16 \end{bmatrix}.$$

The effect of convolving the original image with $\mathbf{P}'$ is shown in Figure 4.5d. This image is both sharp and ostensibly noise-free—and certainly better than the approach used to create Figure 4.5c.

Figure 4.5   The two types of down-sampling considered in this example are shown in (a) and (b). In (a), the $\mathbf{D}'$ kernel is used and so there is some pixel averaging and a slightly greater blur in the down-sampled image than in (b), which simply drops three-fourths of the pixels from the original image. In (c), the sharpened image of Figure 4.4d is down-sampled using the $\mathbf{D}'$ kernel. Although convolution with $\mathbf{D}'$ should introduce some blur to the image, the effect of the dramatic sharpening is only exacerbated. The relatively poor results indicate that sharpening should not occur before down-sampling for the kernels defined and the particular image. In (d), the $\mathbf{P}'$ kernel, which combines the down-sampling and sharpening into a single operation, is shown. This high-quality image was created in half the time as (c), with better results

It should be noted that the results here are not expected to be universal. As noted, the image tested was likely oversampled, and so down-sampling by dropping pixels (Figure 4.5b) and combining sharpening with down-sampling using a 3×3 kernel both provided excellent results. The key point to this explanation is one that will be noted throughout this book: the very act of preparing an imaging task for potential parallel processing leads to a better system design, irrespective of whether parallel processing is actually performed. This *structural* aspect of architecting parallel processing systems will be even more important for parallelism by component in Chapter 5. However, it is clear that considering a plurality of architectures results in better final system design. This observation, too, will have tremendous implications for the later meta-algorithmics-focused chapters.

Another example of applying queue-based parallel processing by task to image understanding is that of extracting facial information from an image. The set of steps involved are readily amenable to a pipeline, or sequence:

1. Identify candidate skin color patches.
2. Detect possible faces.
3. Perform face recognition.
4. Perform emotion detection.

In this system, parallelism by task can occur at a number of possible points. The entire set of {1, 2, 3, 4} can be run in parallel, with the task being "emotion detection in an image" where each parallel pipeline processes one image at a time. The parallel branches could also be assigned only Steps {2, 3, 4} if Step 1 is fast and a good "triage" step for scheduling. For example, if there is a strong correlation between the relative amount of candidate skin pixels and the processing time required for Steps {2, 3, 4}, then performing Step 1 upstream is a good architectural decision.

Potential skin pixels are usually identified based on the hue of the pixel. Hue is used since human skin hue is consistent across human subgroups, even though intensity can be quite different. The pixel at (x,y) location (i,j) is therefore written as $P(\Theta_H, i, j)$, where Θ_H is the hue angle of the pixel, and $0 \leq \Theta_H \leq 360$. If the target skin hue is Θ_S, and the allowable error to either side of Θ_S is Δ_S, then the pixel P is assigned to the candidate skin pixel set, S, under these conditions:

$$\{(\theta_S - \Delta_S) \leq \theta_H \leq (\theta_S + \Delta_S)\} \Rightarrow \{(P(\theta_H, i, j) \in S)\}.$$

Once all of the candidate skin hue pixels are identified, connected components are formed using operations familiar to image segmentation proficients: erosion operation(s) first to get rid of noise; dilation operation next to return legitimate patches to their original sizes; run-length smearing (Wahl, Wong, and Casey, 1982) and/or convex hull operations to eliminate overly complex connected component perimeters; and connected component formation consisting of the following:

1. Bounding box information (xmin, xmax, ymin, ymax). This is predominantly used for sizing—are the regions the right size for face identification?
2. Polygon vertices, xvertices[] and yvertices[]. These are used to render the image.
3. Scan line segment representation, which includes the set of all pixels in the image by row in the image. These allow overlapping connected components.

Next, the now-formed connected components are segmented and separately evaluated using all the relevant downstream image understanding processes: facial feature identification, face matching, emotion detection, and if sufficient image frames are available, lip reading for the video sequence.

These two examples—intelligent image down-sampling and face understanding—illustrate the application of queue-based parallel processing by task to image understanding. We next consider the other types of parallelism by task for intelligent image analysis tasks.

Next, *variable-sequence-based* parallel processing by task is considered. From the above discussion about face recognition and facial expression interpretation, it is clear that the queue-based pipeline can be made more complicated. That is, it can elaborate more parallelism by architecture, not just by running the entire pipeline in multiple parallel processors. For example, as noted above, the initial image segmentation—extracting the potential facial image areas—can be performed upfront, with the scheduling of this task for different images maintained independently of a separately scheduled set of parallel processing for the specific face-understanding tasks. This architecture, a variable-sequence-based approach, is illustrated in Figure 4.6.

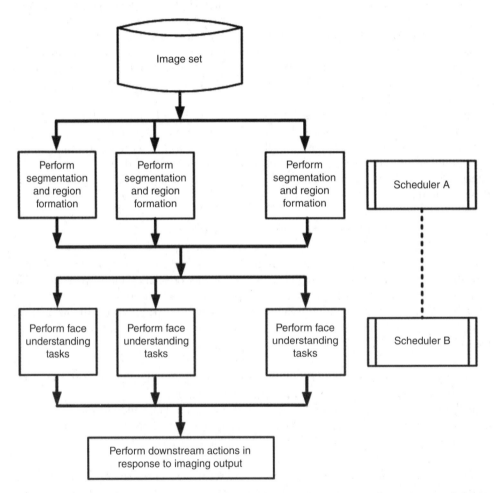

Figure 4.6 Example of variable-sequence-based image understanding architecture, here used to perform a face recognition-related task. Scheduler A assigns different images to one of multiple parallel processing pipelines used to extract the possible face regions. The relative area of the image face regions is used to predict the time for secondary processing (as input to Scheduler B). Scheduler B then assigns the images to the face recognition and information extraction processing tasks in a separate set of parallel processing pipelines

An additional example of a variable-sequence-based architecture could be used to interpret gesture and associate it with a specific individual in a scene. In this image understanding system, the architecture could be based on that shown in Figure 4.6, with a third parallel processing step (managed by a Scheduler C) used to interpret gestures. Gesture interpretation would require the output from several images, making the scheduling task of Scheduler C more complicated than that of either Scheduler A or B. In this example, the parallel processing by task is strictly enforced by assigning each of the three main task pipelines to a separate scheduler.

We now move to applying an LI-PSO to image understanding. Here, we determine the vanishing point of an image. Without parallel processing, a simple set of steps for this operation is as follows:

1. Create edge image from the original image using an edge detector such as Canny (1986) or Haralick (1984), followed by a thresholding algorithm (binarizing algorithm) such as Kittler and Illingworth (1986) or Otsu (1979).
2. Perform a Hough transform on the edge image to obtain the primary angles of line segments in the edge image (Duda and Hart, 1972).
3. Perform projection profiles along the primary angle in the Hough transform histogram to obtain the major line segments in the image.
4. Create line segments and compute the major intersection. One approach to this is outlined in Gallagher (2002). This is the vanishing point.
5. Determine the distance and/or size of other objects based on relative location with respect to the vanishing point. This requires one or more calibrating (known-dimension) items, or fiducials.

We now consider an example of UI-PSO used on an image understanding task: identity tracking. Traditional tracking is based on mapping key features of images sequentially, but in tying tracking to a specific model of the item tracked, it can be opened up to massive parallelism by task: each image can be independently analyzed and the tracking put together through a best fit to the set of images in the video frames. This approach is also robust to errors in individual frames.

We next turn to the EI-PSO. Moving an image understanding task such as shape identification from a model-based approach amenable to CPU processing to a pattern-recognition-based approach amenable to GPU processing is an example of an image understanding EI-PSO.

Finally, we consider *recursively scalable task parallelism*, which is readily applied to a wide variety of image understanding tasks. Compression approaches, including JPEG compression, are defined by the size of the blocks an image is split into. For JPEG compression, images are broken into 8×8 pixel blocks, or subimages, suitable for the computation of the discrete cosine transform coefficients that allow subsequent image compression. In a more general image compression, the block size may be allowed to be variable such that the following occur: (a) tessellation of the image, or breakdown of the original image into a set of squares with no overlap between squares and 100% coverage of the original image by the squares; and (b) the image can be restored with a specified quality, Q, after restoration from the subimages. This variable-block size approach requires some overhead (as is typical of all recursively scalable task parallelism) to tessellate and recompose the image, but also provides huge advantages inasmuch as large areas of spot colors are very efficiently compressed.

This section showed how each of the task parallelisms can be applied to image understanding tasks. It is clear that queue-based approaches are the ones most readily applied to imaging, since they are pipelines.

4.2.3   Biometrics

Biometrics, at first blush, is an application domain in which parallelism by task would seem to have little benefit, at least in comparison to the benefits of parallelism through scheduling. Most individual biometrics have been created and remain based on a pipeline, or sequential set of tasks. Thus, scheduling for large biometric systems is relatively uncomplicated—biometric pipelines as integral processes (i.e., untessellated) are assigned to the available parallel processors by a simple scheduler.

However, as with other processes described in this chapter, a reconsideration of biometric analysis in light of the increasing availability of parallel processing resources, is certainly worth pursuing. The *queue-based* pattern for parallel processing by task is directly applicable to face recognition. Parallelism occurs at the individual face level, while the recognition itself is performed using a sequential set of steps such as the following, modified from Nanavati, Thieme, and Nanavati (2002):

1. *Image acquisition*: Most cameras suffice for the task—the keys to high-accuracy face recognition are usually the lighting and having the face of the person to be identified properly facing the camera.
2. *Image processing*: As described in the previous section, the primary purpose of the image processing is to identify possible face regions, and prepare them for later recognition. This includes cropping to the outline of the face, conversion of the image to grayscale, rotation of the face to be perfectly upright, and scaling the image for downstream processing.
3. *Locating distinctive characteristics on the face*: For face recognition, the key distinctive features are, for example, cheekbone outlines, sides of the mouth, nose outline/shape defining points, upper ridges of the eye socket, distance between midpoints of the eyes, and other relative locations and distances between features on the face. This comprises a *generic set* of features to evaluate, and is useful for distinguishing an individual versus the population of all other individuals.
4. *Template creation*: Templates are created from as many (decent-quality) images as can be captured and assigned to the same individual. The templates vary depending on the face recognition system, but the attributes of the templates are typically the distinctive characteristic features described in Step 3. Templates are typically no less than 100 bytes and no more than 3 kB in length.
5. *Matching*: The final stage of the face identification process involves determining the probability of matching the individual. Since this can be performed on a multiplicity of images, including, for example, video streams, a statistical model rather than a binary yes/no model for identification is required.

Thus, in any real-world case, biometrics requires addressing multiple sources of identification at a time. Even simple surveillance requires the determination of both identification (which person is that) and credibility (could that person have actually been at each of these points in the timeframe allotted). Hybridization (combining two or more decision streams) is therefore implicit in any biometric task, meaning biometrics is innately preadapted to parallelism by task. Extending this further, though, we need to consider how multiple biometric assays, computed in parallel, can be combined to provide improved biometrics.

Generally, biometrics in combination provide higher-accuracy identification, and that is the focus of this section. However, hybrid biometrics can also be used to provide improved location and/or interaction history for an individual. One example is where surveillance information is hybridized with voice information—the former coming from a system of video cameras and the latter from the mobile phone company to which an individual subscribes.

When two or more biometric data streams are combined, there are at least three different means in which they may be combined, or three different types of hybridization we need consider:

1. *Latency, or relative timing of the events used for identification*: Latency hybridization is concerned with the relative timing of events; for example, the response time of an individual to a challenge. Example challenges include lie detector test questions followed by salient physiological measurements. Latency hybrid temporal parallelism, wherein the identified events (e.g., sudden change in blood pressure) are tagged temporally, describes the process of evaluation following such a "challenge-based" biometric. Since multiple sensors may be used—for example, microphone and sphygmomanometer—this approach is also termed a *multi-sensor* approach. In terms of parallelism, the parallel channels may be multiplexed into the same signal recorder, so that this approach can also be termed MUX (or "multiplexed") hybridized parallelization. As with any multiplexed sampling, these multiple sensor events can be windowed, which allows these signals to be processed with a single processor. Parallelism by task can be then deployed at the individual (person) level, with each user assigned to an individual parallel path.
2. *Absolute identification*: Here, two or more events are either directly correlated (both occur) or the identification fails. This "absolute hybrid biometric identification" in general requires two or more signals to be monitored in parallel (not multiplexed). This parallel simultaneous approach is thus readily amenable to parallelism by task, where each task is one of the channels being analyzed (audio, video, etc.). The decision on identity is thus binary, and dependent on the output of the multiple parallel channels: either the two or more events correlate and are both discovered or they are not/do not. This in general requires overlap in time, and so is not generally multiplexed.
3. *Predictor/predicted event*: This is the term for paired events, one of which happens beforehand or afterward in comparison to a primary (or priming) event. In this type of hybrid biometric parallelism, one event is the trigger for the other, rather than the two events simply being correlated. The predictor event can also be termed the "eliciting" event. The predicted event can also be termed the "elicited" event. One example is when a particular word is flashed on a screen, followed by the user having to perform a gesture in response. The elicited event, the gesture, either proceeds as needed for identity, or it does not. Among the factors used in positively identifying the individual are the latency between the two events and the similarity between the expected and measured elicited events.

These three classes of hybrid biometrics approaches are outlined in Table 4.9. It is clear from these examples that parallelism applied to many tasks simultaneously does not benefit

Table 4.9   Three hybrid biometric approaches and the types of parallelism by task involved

Hybrid Biometric	Types of Parallelism Salient
Latency	Hybrid temporal, multi-sensor, multiplexed (MUX)
Absolute	Parallel simultaneous, nonmultiplexed (non-MUX)
Predictor/predicted	Elicitor/elicited, triggered event

from a "one-size-fits-all" solution. Indeed, when we look beyond the hybridization to the implementation, we see that we can perform parallelism by task by performing all of the tasks associated with an individual as one parallel pipeline; alternatively, the individual tasks comprising the hybrid biometric procedure can all be performed in parallel.

Scheduling of these tasks or sets of tasks can get particularly complicated when we do not know beforehand the number of channels to which we can assign the tasks. One situation that must be avoided is when the overhead required to keep track of the scheduling and the multiple channels associated with a hybrid biometric exceeds the overhead of just looking for an equally relevant set of events in a single channel. If identity is established with equal or superior statistical significance simply by looking for multiple voice results, for example, in comparison to searching for specific voice, video, and fingerprint events, then the additional simplicity engendered by only investigating a single (audio) channel should be considered.

Queue-based parallel processing by task is applied directly to the simple single-channel, multi-event task described above. The queue-based approach is, additionally, consistent with the latency and the predictor/predicted approaches outlined in Table 4.9. Simple queue-based patterns are also used directly for image-based identification.

More complicated queue-based approaches are supported mathematically by the use of Bayesian inference models. As described in Chapter 1, the general form of Bayes' theorem is given by

$$p(A|B) = \frac{p(B|A) \times p(A)}{p(B)},$$

where $p(A|B)$ is the conditional probability of event A given that the event B has already occurred. In many cases, we wish $p(A)$ and $p(B)$ to be very low (rarity of identifying event/events) but $p(A|B)$ to be very high (specificity of event/events). Rearranging Bayes' theorem to show the ratio of $p(A)$ and $p(B)$, we can see that if the probability of events A and B are both low, $p(A|B)$ and $p(B|A)$ are similar but not necessarily high ratios:

$$\frac{p(A)}{p(B)} = \frac{p(A|B)}{p(B|A)}.$$

In general hybrid biometrics, however, we are concerned with events that occur with some probability when one or more of multiple other events occur. In fact, we may wish to find the probability of event A when considering all of a set of mutually exclusive events, denoted as B here. We thus rearrange to solve for $p(A)$:

$$p(A) = \frac{p(A|B) \times p(B)}{p(B|A)}.$$

Let us now consider the case where B can occur as one of N outcomes. If the sum of all $p(B_i)$ = 1.0, which is true by definition for mutually exclusive events, then we obtain the following for $p(A)$:

$$p(A) = \frac{\sum\limits_{i=1,\ldots,N} p(A|B_i) \times p(B_i)}{\sum\limits_{i=1,\ldots,N} p(B_i|A)}.$$

This latter equation also governs the probability of event A for any subset of the outcomes of B; for example, under constrained situations in which one or more of the B_i is not allowed to, or cannot, occur. This is important in queue-based approaches in which each of the various B_i is tested for in a sequence. Additionally, if there are other event sets, say C and D, which are partially or completely independent from event set B, then we can test for which set results in a disproportionately high value for $p(A)$. In this way, Bayesian approaches can be used to identify useful event sets for biometric identification.

Parallelism by task for biometrics using the *variable-sequence-based* pattern (originally shown in Figure 2.2) includes cases in which increasingly refined assessment is performed until identity is established with a certain statistical probability (p-value). This type of parallelism by task can be used to extend Bayesian-based inference to other means of assigning identity probabilistically, including but not limited to voting and weighted voting. A series of tests, for example, event sets B, C, and D as described above, can be performed in sequence until a given confidence in the outcome is achieved.

Parallelism by task for biometric tasks can use a LI-PSO approach. Consider a retinal scan in which honing in on specific regions of the retina is valuable for more confidently identifying the person. Suppose the initial retinal scan, which can use a traditional image analysis pipeline (a set of sequential operations such as binarization, segmentation, and feature identification), provides a statistical confidence of 0.92. Now suppose that in subdividing the retinal scan into three regions of interest (e.g., the optic disk, the macula lutea, and the remainder of the retina), the three regions, respectively, provide an identification statistical confidence of 0.94, 0.88, and 0.85. From this, it is clear that the retinal scan can and should be made a parallel operation, wherein previously combined regions are analyzed in parallel.

Another LI-PSO emerges from a consideration of performing two multiple (in this case complementary) identity approaches in parallel. The first approach proceeds as outlined above, where evidence (and statistical confidence) of positive identification is accumulated in one parallel processing pipeline; in contrast, evidence (and perhaps also statistical confidence) of negative identification is accumulated in the second parallel processing pipeline. As an example of the latter, suppose an individual is known to be 6 ft in height, and image analysis performed in the second parallel processing pipeline establishes with sufficient confidence that the individual is 5 ft 8 in. in height (and not 6 ft tall!), then the negative identification can override (and terminate) both itself and the positive identification pipeline. This LI-PSO thus constitutes a dual probabilistic approach in which simultaneous assessment of the likelihood of false positive identification and false negative identification can be pursued.

The UI-PSO, or PSO that is born of deployment, can also be applied to biometric identification. A UI-PSO emerges when different elements of a specific biometric task—for example, gait analysis—are seen to use separable algorithms. There are at least three types of parallelism by task that can be deployed for biometric gait analysis:

1. Multiple video streams can be analyzed in parallel. This allows different views of each limb and joint, helping ensure at least one good perspective for every salient joint.
2. The combination of limb/body velocity and joint dynamics can be assessed using computational techniques known as inverse dynamics. The solution of Newton–Euler equations of motion are used to combine the analysis of the net forces and the net moments of force about each salient joint at every stage of the gait cycle. This set of computations can be performed in parallel for each video stream, as well as for segments of each video stream.
3. Individual metrics of importance to gait analysis—cadence, dynamic base, foot angle, progression line, speed, step-length, and stride—can be computed for each of the parallel video streams and for each of the parallel segments extracted from each stream. Task-specific image analysis techniques can be used for the calculation of each of these metrics.

Recursively scalable task parallelism is applied to both 1D (electrocardiogram and speech) and 2D (iris analysis) biometric signals. For the 1D signals, relatively long arrays of signal, or "streams," can be quickly analyzed for the presence of specific spectral content—or, more saliently to long-term monitoring, for the presence of specific differential spectral content. This is especially salient to voice identification. If there is a given level of confidence that specific spectral content is in the longer stream, it can be subdivided (e.g., split into N equal length substreams) and then each substream analyzed for this spectral content. In this way, the longer stream can be quickly analyzed, the task parallelized, and the best substream for biometric voice characterization identified.

Two final types of approaches to parallelism by task are relevant to biometric parallelism. The *scaled-correlation* pattern is similar in nature to the recursively scalable pattern, except that it generally involves a different "effective sampling" rate. With scaled-correlation, we are concerned with signal analysis at different scales—for example, sampling frequencies for 1D signals or image resolution for 2D signals—and how well the results can be made to correlate. For example, consider the voice identification task described above. Suppose the longer stream comprises 100 s of data with a sampling frequency of 8 kHz. This requires the analysis of 800 kSB of data, where SB is the bytes in each sample of the signal. If, however, the effective sampling rate is made 2 kHz using lowpass filtering, compressive sampling or windowed sampling, then the 800 kSB is scaled to 200 kSB prior to analysis. Then, the original stream is split into two substreams, for which the effective sampling rate can now be doubled, for example, to 4 kHz. Each of these substreams therefore uses 50 s of 4 kHz sampling resulting in the same 200 kSB of data to be analyzed by the two (now parallel) pipelines. The substream(s) for which the spectral characteristics of interest correlate well with those of either the original stream or the desired characterizing spectral components can then be further scaled—for example, to 25 s of 8 kHz sampling, for analysis of the full spectrum.

In contrast to the scaled-correlation approach, the *task-correlation* approach is focused on substituting one task (usually one requiring more resources) with another task (requiring less resources, often substantially less). As a consequence, task-correlation is often used during system training and optimization for deployment. Scaled algorithms as just described are one means of identifying substitute tasks. However, there are others. For example, an algorithm or engine can also be *scaled back*. This means that certain functionality can be omitted (turned off, not paid for, etc.) because it will not deleteriously affect the outcome of the information generated. An example in biometrics may be in the substitution of a full fingerprint detection algorithm with a faster, simpler algorithm that performs a 2D Fourier transform (2DFT) of the image and uses the spectra of the 2DFT to identify the person. For a large group of people, the 2DFT will be less specific than a full-fledged fingerprinting biometric; however, for a smaller set of people, it may provide equal or nearly equal accuracy. If the task provides high enough correlation with the full-fledged system—that is, if there is task-correlation—then it can be substituted within the context of the application it is being used for.

Task-correlation approaches are related to the meta-algorithmic pattern of Constrained Substitution (see Section 6.2.2). In that pattern, one algorithm, engine, system, or other means of knowledge generation is substituted for another. The substitute is constrained by a minimum accuracy and/or robustness metric, and is approved as a suitable substitute if it both achieves the minimum constraint and (typically) reduces the cost of the system.

4.2.4 Security Printing

Many of the Parallelism by Task patterns can be deployed for security printing problems. The *queue-based* pattern, for example, is employed when a set of variably printed items are simultaneously registered and validated. For example, suppose a unique barcode is printed on each of a large set of labels. Then, the inspection process (successful reading of the barcode after printing and validating that it should be in the database) is a queue-based approach.

In addition, the *variable-sequence-based* pattern can be used to simultaneously authenticate multiple security printing features. Suppose that, in addition to the barcode, a variable sequence of text (e.g., microtext), serialized alphanumerics, graphical alphanumerics, and so on, are printed, each with a unique representation. The inspection of these two or more printed sets can be performed in parallel. An LI-PSO pattern can be used to separate multiple layers of information into a series of operations, each run in parallel. Security printing offers a parallelism by task possibility not at all obvious at first sight, based on a scalable data representation possible with three-dimensional (3D) (color 2D) barcode security printing features. The so-called staggered 3D barcode approach relies on scaled-correlation parallelism, in which image analysis at different scales—for example, sampling frequencies or image resolution—are used to provide identification for a wide variety of imaging devices.

The basic 3D barcode consists of a 2D barcode array—like the QR, Aztec, or Data Matrix barcode—which uses multiple colors. The standard configuration uses the six colors red (R), green (G), blue (B), cyan (C), yellow (Y), and magenta (M), as exemplified in Table 4.10. Since $\log_2(6) = 2.585$, the third dimension adds 1.585 bits per module by the inclusion of

Table 4.10 Simple 4×4 module 3D (or 2D + color) barcode representation, with the sequence in English reading order being "GMCGRCBYYBMGRRGB"

G	M	C	G
R	C	B	Y
Y	B	M	G
R	R	G	B

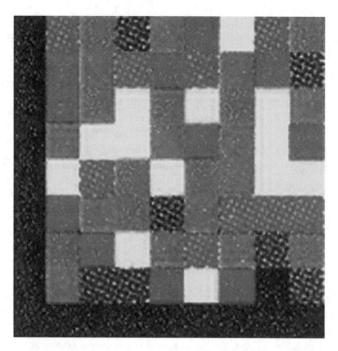

Figure 4.7   Sample 3D (color 2D) barcode that uses a calibrating (nonpayload) indicia approach borrowed from the 2D Data Matrix barcode. All of the perimeter tiles are calibrating. Along the left and bottom sides (with one additional tile near the lower right), the black tiles are used to set the black point and determine orientation and skew. Along the top and right, the six colors used in the payload section—R, G, B, C, M, and Y—are repeated in reading order. The payload consists of 48 color tiles, which in reading order are MCRMRCMMGYGYGYCYYCMCYYRGYGCYMCGBCRRGMYCRCBCBBMYM

variable colors. Table 4.10 illustrates a simple 4×4 module color tile that holds $16 \times 2.585 = 41.36$ bits of information.

The content in a color barcode is therefore dependent on the number of payload tiles. These are tiles (or modules) that contain data, as opposed to those tiles that are used for calibrating the overall barcode feature. In Data Matrix barcodes, the perimeter tiles on top, bottom, left, and right are used to convey calibration information—calibrating black and white point, orientation, skew, and nonaffine warping. Borrowing from this approach, color tiles use the perimeter for all of these tasks and, additionally, color calibration. An example is shown in Figure 4.7.

In Figure 4.7, the left and bottom sides are black and can be used to calibrate for black point, for tile thickness, for orientation (they identify the left and bottom sides), and for skew: the pixels along their boundaries can be processed with linear regression and the slope of the linear regression line of best fit is the skew angle. Along the top and right sides are tiles whose colors repeat the sequence of the allowable payload tile colors—in this case R, G, B, C, M, and Y. Collectively, these calibrating or "nonpayload" indicia comprise 33 tiles (40.7% of the 81 tiles in the complete barcode). The remaining 48 color tiles, if read in reading order, comprise the sequence, when shorthanded by the color symbol, of   MCRMRCMMGYGYGYCYYCMCYYRGYGCYMCGBCRRGMYCRCBCBBMYM. Since there are six different color possibilities for each of the payload tiles, there are $6^{48} = 2.245 \times 10^{37}$ possibilities. This represents just over 124 bits of data.

Unlike black and white tiles, color tiles can have intermediate values. They, therefore, can be used in different aggregations (single tiles, 2×2 pairings of tiles, 3×3 pairings of tiles, etc.) to provide data that is readable by a wide array of imaging devices. High-end devices such as scanners and line cameras will accurately read each individual (small) module, or tile, in the barcode, while less expensive cameras, such as cameras in mobile phones, will only be able to accurately read clusters of the tiles (e.g., 2×2 module "aggregations") at a time. We address this by using a novel type of error-correcting code (ECC), the chroma-enhancing tile (CET), which reduces the payload density by 25% for a 2×2 pairing, but allows 2×2 pairings to reliably map to the same set of colors—usually {RGBCMY}—as the original (single) modules.

The payload approach described above is based on two assumptions: (1) all individual tiles can be read, and (2) the colors are consistent across the deterrent (see below). The former is aided by using predetermined tile dimensions, while the latter is aided by uniform lighting and relatively compact deterrent size (or high-quality capture, such as with a scanner or vision system). However, these assumptions often fail in the mobile world.

For the purposes of describing the use of parallelism by task, we need the following definitions:

1. A *tile* is a uniformly colored glyph, nominally a square, from which the overall deterrent is constructed.
2. A *cell* is the largest set of tiles that can be individually authenticated by any reading device. In the simple example of Table 4.11, a 2×2 set of tiles is this cell.
3. A *deterrent*, or mark, is the complete set of cells, combined to form the color tile security feature. For purposes of illustration, we define the deterrent to be an $N \times N$ array of cells. For further illustration, we make N an integral multiple of 2, so that the deterrent can be entirely tiled by 2×2 and 1×1 sized clusters of tiles as in Table 4.11.
4. A *cluster* is any $P \times P$ set of tiles from the size of an individual tile (1×1 cluster) to the size of a cell (e.g., 2×2 cluster). Power-of-two cluster sets like the ones illustrated here will line up with the cells such that no clusters overlap more than one cell.

In terms of size, tile $\leq$ cluster $\leq$ cell $\leq$ deterrent. In Table 4.11, the R, G, B, C, M, and Y labeled tiles are the payload indicia (PI). These PI tiles, when deployed, would contain one of the allowed sets of colors that convey information, for example, {RGBCMY}. The $X =$ marked tiles in Table 4.11 represent what we term the CETs. CETs are used to guide the hue at

Table 4.11   Representation of 16 color tiles organized as four 2×2 color tile sets

R	G	B	C
M	X	Y	X
R	C	XX	XX
G	X	XX	XX

which the successively larger clusters will be authenticated. The 2×2 tile cluster in the upper left, for example, consists of three PI tiles and one CET. The three PI tiles are red, green, and magenta (R, G, M). In terms of the red, green, and blue channels $\{r,g,b\}$, $R = \{255,0,0\}$, $G = \{0,255,0\}$, and $M = \{255,0,255\}$, so the sum is $\{510,255,255\}$. To enhance the chroma of the 2×2 cluster, therefore, we set the CET to $R = \{255,0,0\}$ and so the 2×2 cluster comes to $\{765,255,255\}$, which is overwhelmingly red. The same approach is used for the larger 4×4 and 8×8 clusters (the 2×2 and 4×4, respectively, CETs, are also a single color).

In general, when the final deterrent is an $N \times N$ deterrent, and $N = 2M$ for some integer M, then the following are true:

1. The final number of independent tiles when the deterrent has been specified at every power of 2 from 0 to M is

$$(3/4)^M * 2^{2M}.$$

2. All remaining CETs are the final authority for the cells they monitor. Thus, remaining X and XX CETs in Table 4.11 enhance the chroma for their respective 2×2 and 4×4 cells, irrespective of the presence of the larger CETs added to the deterrent. For the (B, C, Y) tiles, which sum to $\{255, 510, 510\}$, the associated CET (marked "X") should be C; for the (R, C, G) tiles, which sum to $\{255, 510, 255\}$, the associated CET (marked "X") should be G. Finally, the CET tiles marked "XX" are determined by the sum of the other 12 tiles, which is $\{1265, 1785, 1265\}$, and so the "XX" tiles are G, bringing the overall 4×4 tile set to $\{1265,2805,1265\}$, or a not fully saturated G.

Parallelism by task enters into the evaluation of these 1×1, 2×2, and 4×4 cells when image analysis is performed simultaneously for each cell size. If the 1×1 cells are correctly interpreted, they should read "RGBCMRYCRCGGGGGG" in reading order (left to right in each row, and each row in order top to bottom). If the 2×2 cells are correctly read, they should read "RCGG" in reading order. Finally, the 4×4 cell should read "G." After parallel processing, each of these sequences can be compared for logical consistency. Any failures in consistency remove the smaller size (e.g., 2×2 compared to 4×4) from consideration for full authentication. In this way, the maximum data density that can be reliably read is automatically determined.

Having described the parallelism by task associated with the staggered color tile representation, we conclude the section on security printing with a consideration of one of the simplest forms of parallelism. From a parallelism-by-task standpoint, the greatest advantage provided to security printing is the ability to use multiple (hybridized) security printing marks (known as deterrents) for probabilistic authentication. Suppose, for example, we are using the following set of security printing marks:

1. A 2D (e.g., QR or Data Matrix) barcode
2. An alphanumeric string of microtext, for example, "L42GJK8D"
3. A graphical alphanumeric, for example, guilloche marks or mixed-color/mixed-symbol mark, for example, θ4βTfc□ds⊥

In general, the amount of information we wish to uniquely print—for example, in a mass serialized set—is dependent on how many unique variable-data items we are going to print throughout the lifetime of the set of marks. We must exceed the number of bits needed to represent every individual item in a related set by the number of bits needed to achieve the desired statistical improbability of correctly guessing a legitimate data representation.

Let us define B_L as the lifetime number of bits required to represent all of the printed items associated with a plurality of security printing marks, or "deterrents." There are two factors involved. The first is the bits required to reach security level SL, which we designate B_{SL}. Next, there is the number of bits required to reach statistical confidence level SCL, designated B_{SCL}. From these definitions, it is clear that

$$B_L = B_{SL} \times B_{SCL}$$

Suppose, for example, that we wish to provide security for 100 million (10^8) printed items, with less than 1 in 10^{18} probability of guessing a correct identifier. So, $B_{SL} = \text{ceiling}(\log_2 10^{18})$ = ceiling(59.795) = 60; that is, 60 bits are required to provide the required security level for an individual item. Since there are 10^8 printed items, not 1, $B_{SCL} = \text{ceiling}(\log_2 10^8)$ = ceiling(26.575) = 27. Thus, another 27 bits are required to provide sufficient additional variability for all of the items associated with the related set of products. Thus, if we can embed 87 bits into our one or more security deterrents, we can provide sufficient security for this large set of printed items. In the above set of deterrents, suppose we have the following limitations:

1. The 2D barcode can hold between 16 and 64 bits of data in the space allocated.
2. The alphanumeric microtext can hold 5 bits/character and up to 8 characters.
3. The graphical alphanumeric uses 8 colors, 256 different characters, and can be up to 10 characters long. Thus, each character holds 11 bits, since $11 = \log_2(8 \times 256)$.

From this, we have a maximum of 64 bits from (1), 40 bits from (2), and 110 bits for (3). We can see that graphical alphanumerics are especially dense means of embedding data. The three deterrents can hold up to 214 bits, more than twice the needed bits. A good strategy for this set of security printed marks, then, may be to encode portions of the payload data (87 bits) in each of the three deterrents, and relate them to one another in a fashion proprietary to the specific product. This allows the overall product security printing approach to be changed without changing the set of deterrents used. For example, for product A, we might choose to place 12 bits in the (6×6 module payload section) 2D barcode with 200% ECC; 20 bits with 50% ECC in the microtext; and 55 bits with 50% ECC in the graphical alphanumeric. For product B, however, we may choose to place 32 bits within the (8×8 module payload section) 2D barcode with 100% ECC; 0 bits in the microtext (using them instead as a decoy for would-be counterfeiters); and 55 bits with 50% ECC in the graphical alphanumeric.

4.3 Summary

In this chapter, we have explored some of the ways in which parallel processing can be used to break up tasks into sets of tasks. The chapter emphasizes the use of parallel processing for serialized tasks (e.g., queue-based parallelism and pipelines) and for further parallel-ready tasks (recursively scalable task, scaled-correlation parallelism, and task-correlation parallelism). It is clear that parallelism can be applied to virtually any task. Where the task itself is able to be restructured into parallel operations, these subtasks can be scheduled in parallel. Where the task cannot be restructured for parallelism, the task itself is the primitive for parallel scheduling. Regardless, it is clear that if there are enough tasks in a population, assigning tasks to parallel processing pipelines is an efficient means of improving throughput. When a small set of tasks comprise the population, the restructuring and scheduling overhead may obviate the advantages of parallelism.

References

Canny, J. (1986) A computational approach to edge detection. *IEEE Transactions on Pattern Analysis and Machine Intelligence*, **8**, 679–714.

Duda, R.O. and Hart, P.E. (1972) Use of the Hough transformation to detect lines and curves in pictures. *Communications of the ACM*, **15**, 11–15.

Gallagher, A.C. (2002) A ground truth based vanishing point detection algorithm. *Pattern Recognition*, **35**, 1527–1543.

Gonzalez, R.C. and Woods, R.E. (2008) *Digital Image Processing*, 3rd edn, Prentice Hall, 954 pp.

Haralick, R. (1984) Digital step edges from zero crossing of second directional derivatives. *IEEE Transactions on Pattern Analysis and Machine Intelligence*, **6** (1), 58–68.

Kittler, J. and Illingworth, J. (1986) Minimum error thresholding. *Pattern Recognition*, **19** (1), 41–47.

Nanavati, S., Thieme, M., and Nanavati, R. (2002) *Biometrics: Identity Verification in a Networked World*, John Wiley & Sons, Inc., New York, 300 pp.

Otsu, N. (1979) A threshold selection method from gray level histograms. *Pattern Recognition*, **9** (1), 62–66.

Wahl, F.M., Wong, K.Y., and Casey, R.G. (1982) Block segmentation and text extraction in mixed/image documents. *Computer Vision Graphics and Image Processing*, **2**, 375–390.

5

Application of Parallelism by Component

The least flexible component of any system is the user.

—Lowell Jay Arthur

5.1 Introduction

In this chapter, specific applications of parallelism by component are considered in depth. While the set of "core" domains is in some ways idiosyncratic, they are also broadly different and so in composite provide an acceptably sweeping perspective of the value of parallelism by component. This chapter is meant to provide a pragmatic guide to implementing parallelism by component. The four core domains considered are the same as the previous chapter: (1) document understanding, (2) image understanding, (3) biometrics, and (4) security printing.

The meaning of "parallelism by component" is, of course, related to the use of different parts of a particular item, or data set, in parallel. However, as this type of parallelism is applied to document understanding, I show that much—in the particular example the vast majority, in fact—of the improved throughput coincident with converting the process to parallelism by component is due to the *structural reframing* of the process, and not the parallelism per se. This emphasizes the importance of domain expertise in each parallelization process.

The next topic is image understanding. As with document processing, image understanding benefits from efficient preprocessing. This includes the structural reframing benefits observed in the document understanding parallelism, but extends to a potentially circuitous consideration: knowing what we are looking for in an image helps us look for it more efficiently in at least three ways I will discuss: (a) *model down-sampling*, in which an image and the means to understand specific features within the image are simultaneously and, from the standpoint of recognition accuracy, uniformly down-sampled; (b) *componentization through decomposition*, in which, for example, different relative or absolute image planes are separated and separately processed; and of course (c) *structural reframing*, the act of intelligently dividing a large, unwieldy, and potentially inefficiently processed component into a plurality of readily processed subcomponents.

Meta-algorithmics: Patterns for Robust, Low-Cost, High-Quality Systems, First Edition. Steven J. Simske.
© 2013 John Wiley & Sons, Ltd. Published 2013 by John Wiley & Sons, Ltd.

Next, biometric applications are considered. As opposed to the hybrid biometrics described in Chapter 4, biometric features that are parallelizable by component focus on elements and subelements of the same biometric measurement. *Componentization through decomposition* is an important means of addressing biometrics-related machine intelligence. In this section, I also address two other means of parallelization by components for biometrics: (1) *temporal parallelism*, with obvious extensions to video analysis, surveillance, and motion tracking; and (2) *overlapped parallelism*, with obvious extensions forward to meta-algorithmics. Finally, the "parallelism by component" flavor of the scaled-correlation parallelism first described in Chapter 4 will be applied to the field of biometrics.

The fourth "core domain" to which parallelism by component is applied is the broad, interdisciplinary science of security printing. Because security printing itself draws on the expertise of multiple fields, this topic provides an excellent one to show how an expert in parallelism by component can utilize her entire bag of tools to solve multiple, related problems of the highest possible value—serialization, inspection, authentication, forensics, and access rights validation. In addition to the six patterns of parallelism by component described for one or more of the other three core areas, security printing introduces two new patterns for parallelism that naturally arise as a consequence of the modern printing technologies: (a) variable data printing (VDP) and (b) print on demand (POD). VDP naturally leads to *variable element parallelism*. POD, on the other hand, leads to *search parallelism*, which is simple conceptually but assuring globally unique identifiers requires *prelocking of content*. This means that the globally unique identifiers are adding to a pool of noncolliding numbers before they are allocated to an item with which they will thence be associated. The interweaving of these eight patterns of parallelism will be shown to benefit the robustness, accuracy, and efficiency of security printing ecosystem architecture.

5.2 Primary Domains

The four primary domains are (1) document understanding, (2) image understanding, (3) biometrics, and (4) security printing. As we progress from one domain to the next, we build up a repertoire of parallel approaches that comprise a toolset for secondary domains covered elsewhere in the book.

5.2.1 Document Understanding

In Section 2.3.2, I overviewed an important factor in deciding on how to best perform image understanding (e.g., image segmentation, object identification, object extraction, and other image processing) tasks. When an image is broken up into multiple subimages, there may be improvements in throughput *even without performing parallelism by component*. That is, smaller images are often processed more efficiently relative to the size of the image than larger images, even when using only a single processing resource.

This image-splitting approach is more likely to be advantageous the larger the original image is. As an example, I consider document image segmentation. In Figure 5.1, the original scanned document image is shown. This document (Cheyenne Mountain, 1996) is a mixed-content document, comprised text, line art, and photos. A quick binarization, or thresholding, operation, such as those described in Kittler and Illingworth (1986) or Otsu (1979), and shown

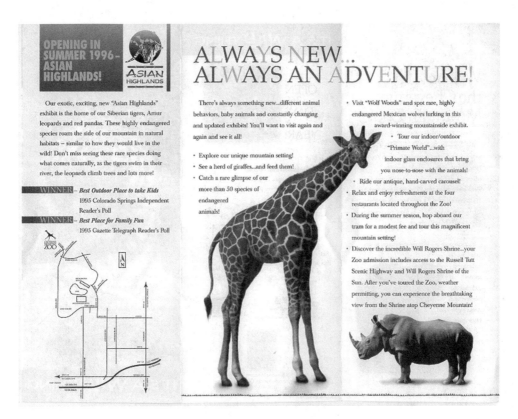

Figure 5.1   The original image (the "small image"), an advertisement for Cheyenne Mountain (1996). The image is 24 bits/pixel (8 bits/channel, RGB (red, green, blue) channels) with dimensions 825 × 638 pixels. Reproduced by permission of Cheyenne Mountain Zoo

earlier in Figure 3.3, is used to separate foreground (document content) from background. Projection profiles (Zramdini and Ingold, 1993) in the horizontal and vertical directions are used to find optimal cuts in the document image—minima in the projection profile generally correspond to the best locations for dissecting images to create subimages. One such dissection into two (unequally sized) partitions is shown in Figure 5.2.

In order to explore the effectiveness of the parallelization by component approach to document image understanding, I start with the simple document image shown in Figure 5.1. This image was originally scanned at 75 × 75 pixels/in. (ppi), and the overall image size is thus 825 × 638 pixels. This is an RGB 3-channel, 8 bits/channel image, which when loaded into memory is just over 1.5 MB of image data. Several operations were then performed on the image. The first (designated "thresholding") is the binarization process, examples of which are shown in Figure 3.3. Thresholding is itself a broad field of science—both global (one single threshold for the entire image) and local (different threshold for different sections of the image) thresholds are valuable depending on the nature of the content. For example, in Figure 5.1, the box in the upper left containing the text "Opening in Summer 1996 – Asian Highlands!" requires a darker threshold than the overall image threshold in order to extract the

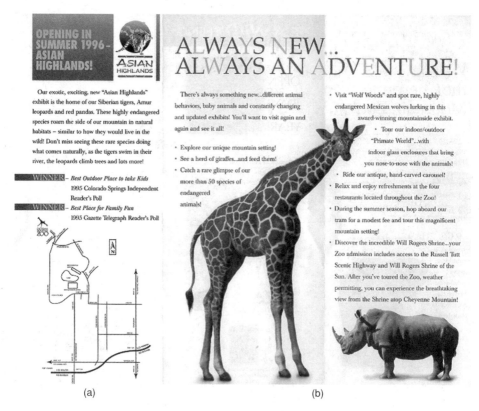

(a)   (b)

Figure 5.2   The original image of Figure 5.1 after quick presegmentation into two parts. Reproduced by permission of Cheyenne Mountain Zoo

text. A lighter, whole-image threshold results in the text and the spot color background both being binarized to black.

The whole-image thresholding is based on a modification of Kittler and Illingworth (1986), designated ModKitt, which uses the image luminosity histogram (as shown in Figure 3.4) and performs a dual-Gaussian fitting. After generating the histogram for the image, the ModKitt algorithm assumes a single large peak in the image luminosity histogram corresponding to the background of the image (typically white) and a less cohesive set of luminosities, distinct from the background peak and generally separated by a trough in the histogram.

After thresholding, the document skew is determined. The subthreshold black pixels form a binary threshold image. We then find solid areas of nonuniform hue or intensity, if they are present in the document. These represent photos. We first remove any black shadows along the edge of the image (often concomitant with scanning or camera-introduced vignetting), then adaptively determine the proper amount of run-length smearing (introduced in Wahl, Wong, and Casey, 1982) to generate a sufficient number of regions to estimate skew correctly, but not so many that we sacrifice performance. This estimate is based on the number of black runs rather than using full region formation, and thus is independent of later document image segmentation. After run-length smearing the threshold image, we form regions from the black pixels on the image. Solid (nontext) regions are, for example, greater than 1.0 in^2 in area and contain 70% or more black pixels, among other possible experimentally determined factors. In general, solid regions have areas with long black runs while "nonsolid" (usually text) regions do not have a high percentage of long black pixel runs. These solid regions are then scrubbed from the threshold image, but their boundary information is stored for secondary skew analysis.

Skew analysis is performed, not surprisingly, using a parallel-architecture approach. Two sets of points are collected and analyzed for their primary set of skew angles. The first set of points is the set of centroids of the small connected components, typically individual characters like "c" or "s" or else parts of characters for ones such as "i" or ":". The angles between these centroids are accumulated in a *centroid–centroid angle histogram*, for example, a 901-element histogram with bin values of $\{0.0°, 0.2°, 0.4°, \ldots, 180.0°\}$. The histogram is then smoothed with a moving average filter, and peaks are found. The highest peaks—typically 3–5 of them—are kept as candidate skew angles. The second, parallel, set of points are the boundaries (eroded to a single pixel edge thickness) of any of the "solid" regions uncovered during the quick segmentation stage described in the previous paragraph. Since the number of edge pixels is relatively small, the angles between them can be readily computed and used to create another histogram, the so-called *image edge histogram*. After that, the procedure is the same as for the centroid-centroid angle histogram.

After these two sets of skew candidates are compiled, they are weighted based on integrating under the peak curve in the histogram. Typically, each set's weight is normalized to sum to 1.0. The peak sets are then compared. If there are matching skew candidates from the centroid-centroid and image edge histogram sets, their weights are combined. Regardless, the two sets are now merged, and if there is no clear individual candidate that stands out as superior to the rest, then the set of highest weighted candidates are evaluated further. If there are substantial text areas, then projection profiles—described above—are computed parallel and perpendicular to the skew angle for each candidate. Text regions will have significantly different projection profile entropy in comparing the parallel and perpendicular profiles when the parallel profiles are best aligned with the text. Thus, when sufficient text is present, the actual skew angle is therefore set to the skew candidate with the maximum entropy difference in comparing the parallel and perpendicular projection profile histograms.

When there is insufficient text, the boundary pixels themselves are used to compare the skew candidates. Projection profiles are computed, again parallel to and perpendicular to the angle of skew. The correct skew will have the minimum overall (sum of parallel and perpendicular) projection profile entropy, since when the correct skew angle is found, the projection profiles will feature the sharpest peaks (and thus the lowest entropy).

The third document understanding process is segmentation and classification. Segmentation and classification begin with the thresholding and skew detection processes described above. After the skew angle is determined, the image is deskewed using the shear-based rotation approached shown in Chapter 3. Much of the segmentation is performed as part of skew detection. The remaining segmentation tasks are associated with compound region segmentation; for example, extracting text over images, identifying the text regions in line art, drawing, and table regions, and extracting specific shapes from images. The individual connected components, or document regions, are then labeled as text, line art, image/photo, table, business graphic, and so on. In the final step, clustering of like (e.g., lines of text) or associated (e.g., graph labels with the graph) regions to form the relevant composite regions is performed.

Table 5.1 Processing time for the thresholding (Kittler and Illingworth, 1986), skew detection and correction (Simske and Baggs, 2004), and segmentation and classification (Simske and Arnabat, 2003) operations on a large and small image (the large image was not available during the time I owned the OLD computer). The small image is 825×638 pixels; the large image is 2816×2112 pixels, or 11.3 times the size of the small image

	OLD Computer, Small Image (ms)	NEW Computer, Small Image (ms)	NEW Computer, Large Image (ms)
Thresholding	55	31	156
Skew detect	815	406	1 420
Segmentation and classification	1 022	421	39 249
Total document image processing	1 892	858	40 825

The total document image processing time is the sum of the thresholding, skew detect, and segmentation and classification processing times. Processing times were measured using two different laptop computers: the first (herein designated "OLD computer") a vintage Pentium P-III, 1.13 GHz, 256 MB RAM, Windows 2000 system; the second (herein designated "NEW computer") an Intel i7 Q740, 1.73 GHz, 8.00 GB RAM, Windows 7 Enterprise system. All of the image processing software was written in C++. The software was executed without threading on a single processor. Table 5.1 gives the processing times for each of the tasks for the image shown in Figure 5.1 (*small image*). The NEW computer was also used to process a separate *large image* (More in-depth consideration of this larger image is given later in Figure 5.5).

The results of Table 5.1 confirm that the NEW computer processes the image faster than the OLD computer. The timing ratio (858 ms/1892 ms = 0.453) is less than the inverse ratio of clock speed (1.13 GHz/1.73 GHz = 0.653), indicating a concomitant improved architecture in the NEW computer. The nonlinear relationship between image size and total document image processing time is shown in comparing the large and small images. While the large image is only 11.3 times the size of the small image, the total document image processing time (lowermost row) for the large image is 47.6 times the processing time for the small image. The increase in processing time is not uniform for the image processing tasks. For (1) thresholding, (2) skew detection, and (3) segmentation and classification, the increase in processing time is (1) 5.03, (2) 3.50, and (3) 93.23 times for the large image compared to the small image. Since the latter value is roughly 20 times the first two values, it is clear that the imaging task has a larger effect on processing time than the image size per se. Thresholding and skew detection processing improve or are not as adversely affected with increasing image size since only one pass through the image is required, after which all operations are in common. Segmentation and classification, however, require multiple passes through the image, and so processing time is exacerbated by increasing image size.

After splitting the image of Figure 5.1 into two unequally sized subimages (Figures 5.2a and b), the document image processing tasks were performed on both sections (Table 5.2). The left image was half the size of the right image, and thresholding (19 ms compared to 38 ms) and skew detection (278 ms compared to 559 ms) processing times were also half as much for the left image compared to the right image. Segmentation and classification (289 ms compared to 647 ms), however, more than doubled (to 2.24 times as much) when the image size doubled. These data support the general principle that splitting simple (e.g., single

Table 5.2 Processing time for the thresholding (Kittler and Illingworth, 1986), skew detection and correction (Simske and Baggs, 2004), and segmentation and classification (Simske and Arnabat, 2003) operations on the bisected small image. The processing time of the two images was 1830 ms, slightly less than the 1892 ms for the entire (unbisected) image

	OLD Computer, Left Image (ms)	OLD Computer, Right Image (ms)
Thresholding	19	38
Skew detect	278	559
Segmentation and classification	289	647
Total document image processing	586	1244

pass) imaging tasks into subimage tasks does not improve performance. Thresholding time actually slightly increased from 55 ms for the whole image to 57 ms for the two subimages. Skew detection time also increased slightly, from 815 to 837 ms, when the image was split into the two subimages. However, the processing time for the more complicated, multi-pass image processing task—segmentation and classification—was noticeably reduced (from 1022 to 936 ms) when the image was split into two subimages. This highlights a second general principle: when a complicated (e.g., multi-pass) imaging task is split into subimage tasks, there is improved overall performance.

Aside from the image processing times provided in Tables 5.1 and 5.2, there is additional information of interest we can glean from the document image processing of the images comprising Figures 5.1 and 5.2. The results of the document processing are shown for the intact image in Figure 5.3, and for the two subimages in Figure 5.4. These images show the output of the document image analysis using indicative region boundary polygons: dashed outlines indicate the boundary polygons of text regions, dotted outlines indicate the boundary polygons of photo regions, dash–dot lines indicate the boundary polygons of color line art regions, and solid lines indicate the boundary polygons of binary line art regions (and tables). Interestingly, the results obtained for the left subimage in Figure 5.4 differ from the results obtained for the left third of the intact image in Figure 5.3. In Figure 5.3, the black and white map in the lower left of the image is classified as a black and white line art region. In Figure 5.4 (left), this same region, even though segmented nearly identically (the boundary polygons differ only slightly, predominantly in the lower left portion of the region), is classified as color line art. In addition, the two solid text boxes labeled "Winner" in the left center of the images are classified differently in the two figures. Finally, there are differences in segmentation and classification of the text regions in the upper right of the left image of Figure 5.4. These differences illustrate a finding of perhaps even more importance to our purposes than the differences in performance—the fact that performing image processing tasks on subimages can produce different results than performing image processing tasks on the intact images. This means that the same image can be processed by the same algorithm or intelligent system and yield a plurality of results. The reason for this is that the thresholds and other variables used in the image processing are based on different data sets—even though the image processing is deterministic it will form different segmented regions on cropped versions of a larger image. In effect, this subimage dependency is a form of *meta-algorithmic parallel processing*, which is the focus of the next four chapters. It is also a form of sensitivity analysis—if image

Figure 5.3 The original image of Figure 5.1, with the classified scanned regions indicated. Lightest outlines indicate text regions, second lightest outlines indicate photo regions, darkest lines indicate color line art regions, and second darkest indicate binary line art regions. Reproduced by permission of Cheyenne Mountain Zoo

processing output differs for a subimage in comparison to the original image, then it is likely that the algorithm or intelligent system has less than full confidence in the output—and perhaps multiple algorithms should be used. We leave this topic until the next chapter to consider the right side image of Figure 5.4.

The results obtained for the right subimage in Figure 5.4 are identical to those obtained for the right two-thirds of the intact image in Figure 5.3. This is likely due to the fact that the images are more alike in size than the left subimage and the original image. Regardless, because the thresholds were the same for both of these images, the results—skew angle detected and segmentation and classification—are also the same. These identical results allow for a more direct assessment of the effect of image size on the image processing times—if size has no impact, the ratios should all be an exactly 2/3, or 0.667. Indeed, for thresholding, the ratio (comparing the last data column in Table 5.2 to the first data column in Table 5.1) is 0.691. For skew detection, the ratio is 0.686. These are both approximately 0.667, and thus scale more

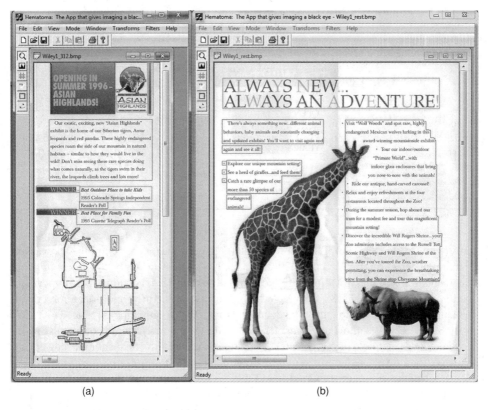

(a)   (b)

Figure 5.4 The original image of Figure 5.1, after quick presegmentation into two parts, with the classified scanned regions indicated. Lightest outlines indicate text regions, second lightest outlines indicate photo regions, darkest lines indicate color line art regions, and second darkest lines indicate binary line art regions. Reproduced by permission of Cheyenne Mountain Zoo

or less directly with size. For segmentation and classification, however, the ratio is 0.633, less than 2/3. This ratio supports the interpretation that when a complicated imaging task is split into subimaging tasks, there is improved overall performance.

Next, document image processing of bisections of the small image and, separately, the large image was performed using the NEW computer (Table 5.3). Document image processing time was not improved for the small image when the two parts were processed. Total thresholding time did not change (31 ms), but both skew detection (452 ms compared to 406 ms for an intact small image) and segmentation and classification (499 ms compared to 421 ms for an intact small image) actually *increased* when the small image was broken into two parts. These results indicate that the smaller image was already small enough that multi-pass processes like region segmentation and classification did not challenge the resources of this particular processing device. In general, the results obtained will be device dependent.

The results for document image processing of the large image (Table 5.3), however, strongly justify the bisecting of this image prior to performing document image processing. While the combined processing time for the thresholding of the left and right subimages (halves) increases

Table 5.3 Processing time for the thresholding (Kittler and Illingworth, 1986), skew detection and correction (Simske and Baggs, 2004), and segmentation and classification (Simske and Arnabat, 2003) operations on the bisected large and small images

	NEW Computer, Left Small Image (ms)	NEW Computer, Right Small Image (ms)	NEW Computer, Left Large Image (ms)	NEW Computer, Right Large Image (ms)
Thresholding	15	16	93	94
Skew detect	156	296	702	577
Segmentation and classification	141	358	11 202	14 602
Total document image processing	312	670	11 997	15 273

slightly (from 156 to 187 ms), this is more than made up for by the reduced combined processing time for skew detection (reduced to 1.279 s from 1.420 s). More strikingly, the processing time for segmentation and classification is reduced by more than 13 s, equating to a proportional reduction of one-third (to 25.80 s from 39.25 s). Because of the disproportionate processing time required for segmentation, the total document image processing time is reduced to 27.27 s from 40.83 s (33.2%), without any consideration of the (additional) benefits of processing the two subimages in parallel—there, the processing time is reduced to 15.27 s for the slower right side image, for a 62.6% reduction in overall processing time.

Next, each of the subimages reported on in Table 5.4 were segmented into two parts, resulting in four equally sized (1408 × 1056 pixel) subimages of the original 2816 × 2112 pixel image. The results for those four subimages vary greatly since, for example, the upper right quadrant and lower right quadrant of the original image vary significantly in image information (e.g., in terms of image entropy), the upper right being mainly sky and the lower right being information-rich cliffs and foliage. Here, the combined processing time for the thresholding of the four subimages (quadrants) increased more noticeably (from 156 to 235 ms). The combined processing time for skew detection also increased significantly from the intact original image (to 2.450 s from 1.420 s). However, the processing time for segmentation and classification

Table 5.4 Processing time for the thresholding (Kittler and Illingworth, 1986), skew detection and correction (Simske and Baggs, 2004), and segmentation and classification (Simske and Arnabat, 2003) operations (and the sum of the three) for the image of Figure 5.5 when divided into four equal-sized (1408 × 1056 pixel) subimages

	Upper Left (ms)	Upper Right (ms)	Lower Left (ms)	Lower Right (ms)
Thresholding	93	47	63	32
Skew detect	359	265	858	968
Segmentation and classification	1 888	1 123	4 664	9 624
Total document image processing	2 340	1 435	5 585	10 624

Figure 5.5 The "large image." The image is 24 bits/pixel (8 bits/channel, RGB channels) with dimensions 2816 × 2112 pixels. This is the image later broken into two subimages (1408 × 2112 pixels each); four subimages (1408 × 1056 pixels each); and eight subimages (704 × 1056 pixels each)

is reduced by more than half, and by nearly 22 s (to 17.30 s from 39.25 s). Again, because of the disproportionate processing time improvement in segmentation and classification, the total document image processing time is reduced to 19.98 s from 40.83 s (51.1%). If the four subimages are processed in parallel, the processing time is reduced to 10.62 s for the slowest quadrant (lower right) subimage, for a 74.0% reduction in overall processing time.

In this and the following related examples, I have ignored the negligible overhead time for splitting the image into subimages. In these examples, I have in each case locked the complete image into memory and the pixels are then streamed directly from the locked memory to the image processing algorithms. Processing time was computed in-code with 1 ms sampling resolution, and no difference in access time was measured. This is because there was no overhead for splitting, just the dimensions (xmin, xmax, ymin, ymax) of the memory access changed for subimages compared to the main image. This access time is anyway included in the processing time of the thresholding.

As a final illustration in this example, I next bisected each quadrant of the original image into two equally sized 704 × 1056 pixel subimages. These (now eight) subimages then underwent document image processing as before, with the results given in Table 5.5. Not unexpectedly

Table 5.5 Processing time for the thresholding (Kittler and Illingworth, 1986), skew detection and correction (Simske and Baggs, 2004), and segmentation and classification (Simske and Arnabat, 2003) operations (and the sum of the three) for the image of Figure 5.5 when divided into eight equal-sized (704 × 1056 pixel) subimages

	Upper Left		Upper Right		Lower Left		Lower Right	
	Left (ms)	Right (ms)	Left (ms)	Right (ms)	Left (ms)	Right (ms)	Left (ms)	Right (ms)
Thresholding	31	31	31	31	47	31	47	47
Skew detect	250	109	172	94	640	609	780	359
Segmentation and classification	640	390	686	421	1482	2652	4087	2292
Total document image processing	921	530	889	546	2169	3292	4914	2698

given the trends reported in the earlier tables, in Table 5.5 the summed processing time for thresholding rises to 296 ms; the summed processing time for skew detection rises to 3.01 s; and the summed processing time for segmentation and classification drops even further to just 12.65 s. The total document image processing time decreases from 40.83 to 15.96 s (60.9%), and if all eight subimages could be processed in parallel, the slowest of the eight subimages will complete in 4.91 s, an 88.0% reduction in processing time completion.

The results for the large image (Tables 5.1, 5.3, 5.4, and 5.5) are summarized in Table 5.6 and plotted in Figure 5.6. The total document image processing time closely follows a quadratic curve that asymptotes to 12.9 s as the number of subimages into which the original image is dissected increases to a very large number. This is a simple, but nevertheless important, finding. For the large image shown, these data indicate that the improvement in throughput, where throughput is inversely proportional to the processing time, is largely due to the act of *converting* the process to parallelism by component and not due to the downstream parallelism

Table 5.6 Processing time for the thresholding (Kittler and Illingworth, 1986), skew detection and correction (Simske and Baggs, 2004), and segmentation and classification (Simske and Arnabat, 2003) operations (and the sum of the three) for the image of Figure 5.5 on the original image (Table 5.1), the image bisected into two subimages (Table 5.3), the image broken into quadrants (Table 5.4), and the image broken into eighths (Table 5.5)

	NEW Computer One Image (ms)	NEW Computer Two Subimages (ms)	NEW Computer Four Subimages (ms)	NEW Computer Eight Subimages (ms)
Thresholding	156	187	235	296
Skew detect	1 420	1 279	2 450	3 113
Segmentation and classification	39 249	25 804	17 299	12 650
Total document image processing	40 825	27 269	19 984	15 959

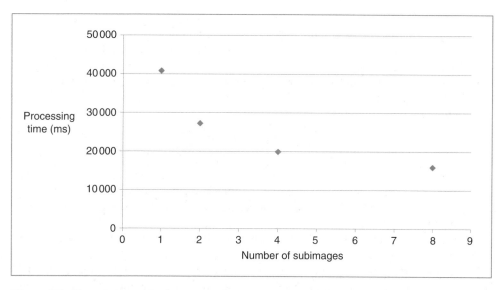

Figure 5.6 Plot of the processing time (in ms) versus the number of subimages for the image of Figure 5.5. Based on the quadratic best fit of the curve, processing time levels off at 12 900 ms as the number of subimages reaches a very large number

itself. That is, the *structural reframing* of the process results in a processing time reduction of more than 60% before any actual parallel process can occur. The advantages of parallel processing for the large image example thus cannot be more than 40% of the processing time.

The *relative* additional impact of actually processing the subimages in parallel, of course, can be as high as $100\% \times$ (number of processors $-$ 1)/(number of processors); that is, 50%, 75%, and 87.5%, respectively for two, four, and eight subimages. The variability in processing time for each of the subimages, however, reduces the actual impact from this ideal value. The throughput for this simple example is delimited by the maximum subimage processing time. As shown in Table 5.7, it is this maximum processing time that sets the "effective parallelism," or EP, of the parallel processing of the subimage set:

$$EP = n_{\text{subimages}} \left(\frac{t_{\text{proc}}(\text{mean})}{t_{\text{proc}}(\text{max})} \right),$$

Table 5.7 Relationship between the square root of the number of subcomponents (here subimages), or Q, and processing time proportionality value, or P, for the data in Tables 5.3, 5.4, and 5.5

	Mean Processing Time (s)	Maximum Processing Time (s)	Effective Parallelism (EP)	Percent Throughput Improvement $(\%TI) = [(EP - 1)/EP] \times 100\%$ (%)
Two subimages	13.64	15.27	1.786	44.0 (max 50)
Four subimages	5.00	10.62	1.881	46.8 (max 75)
Eight subimages	2.00	4.91	3.248	69.2 (max 87.5)

where $n_{\text{subimages}}$ is the number of subimages, $t_{\text{proc}}(\text{mean})$ is the mean processing time, and $t_{\text{proc}}(\text{max})$ is the maximum processing time.

Table 5.7 shows that this ratio of mean/maximum processing time reduces EP from 2, 4, and 8 to 1.786, 1.881, and 3.241, respectively. This means the percent throughput improvement, designated %TI, is significantly less than the predicted value for each set of subimages. The value %TI is defined as

$$\%\text{TI} = \left(\frac{\text{EP} - 1}{\text{EP}} \right) \times 100\%.$$

The actual values for %TI are 44.0%, 46.8%, and 69.2%, respectively, for two, four, and eight subimages. These values are only 10.2% higher, on the mean, than the *structural-reframing*-related throughput improvement percentages of 33.2%, 51.1%, and 60.9%, respectively, for two, four, and eight subimages (Tables 5.3, 5.4, and 5.5). Thus, in this example, structural reframing and actual parallel processing itself have similar relative improvements on throughput.

I have gone into great detail in this example for another reason. In illustrating the different relative impact of structural reframing on the processing time for different document image processing tasks—for example, the generally negative effect of subimage processing on thresholding time versus the highly positive impact of subimage processing on segmentation and classification time—I wish to emphasize the importance of domain expertise in the parallelization process. The need for domain expertise extends to the decision of how many subimages to form in order to not affect the accuracy of the important document image processing output. Applying domain knowledge and the data obtained to the current example, the following overall design recommendations are made:

1. Thresholding and skew detection will be performed on the entire image before any subimage formation. When a global threshold is employed, it makes sense to use all of the image data for the threshold determination, since by the *central limit theorem* a better estimate of the means of each of the peaks will be obtained. The same argument holds for skew determination, and is appropriate since we assume the individual document has only one skew angle. In addition, performing skew detection on the entire image actually improves throughput over the four and eight subimage cases.
2. Before segmentation and classification, subdivide the image into eight subimages and then perform the associated document image processing. We do not further subdivide (e.g., into 16 subimages) as the data in Table 5.3 indicate that this will not further improve throughput. Moreover, at some point, the subimages will be small enough that segmentation and classification will become less accurate.

It should be noted that a global threshold will not always be the right approach. In cases of nonuniform illumination or image background, a local threshold may be employed. It should also be noted that the optimum subdivision is in general a function of the computing device's capabilities and the image size. For images of approximately the same size processed on the same device, the optimal subdivision is consistent.

Given these recommendations, the architecture for the document image processing task is given in Figure 5.7. Using this architecture, all document image processing tasks are

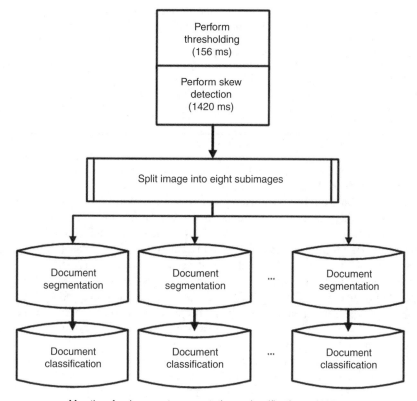

Max time for document segmentation + classification = 4087 ms

Figure 5.7 Document image processing architecture chosen. Thresholding and skew detection are performed on the whole image, which is then subdivided into eight subimages, for each of which segmentation and classification are performed in parallel. The document image processing task is complete in 5663 ms (compared to 40 825 ms when there is no subimage formation or parallel processing)

complete in 5663 ms, an 86.1% reduction in processing time and a 7.2× improvement in throughput.

Structural reframing shows that subcomponent processing can take far less processing time than for a larger component. We can formalize the impact of structural reframing by considering a document image processing task in which M passes through the image are required. The processing time is thus proportional to $M(HW)^P$, where H is the height of the image, W is the width of the image, and P, a coefficient to be determined that describes the relationship, is often $\gg 1.0$. Breaking the image into Q^2 subcomponents (i.e., $Q \times Q$ subimages), the processing time is then proportional to $MQ^2(HW/Q^2)^P$, or $M(HW)^P/Q^{2P-2}$. The ratio of processing time is therefore proportional to $1/Q^{2P-2}$. If $Q = 4$ and $P = 2$, then the proportion is 1/16, a huge impact on processing. For the "large image" example illustrated in this section, the values for Q and P are given in Table 5.8. The data are consistent with the chart shown in Figure 5.6; that is, as the number of subimages is increased, the relative value of the subdividing the images further decreases. This means that P is negatively correlated with Q.

Table 5.8 Relationship between the square root of the number of subcomponents (here subimages), or Q, and processing time proportionality value, or P, for the data in Tables 5.3, 5.4, 5.5, and 5.6

	Image Size (MB)	Q	P
Original image	17.02	1.0	N/A
Two subimages	8.51	1.414	1.582
Four subimages	4.25	2.0	1.514
Eight subimages	2.13	2.828	1.451

There are a number of assumptions made in this section, mainly for the purpose of providing a fully transparent worked example. The first is that the data obtained for the large image are representative of what would be obtained for a large number of images that require document image processing. That is, the architecture finally deployed in Figure 5.7 is suitable for a large set of data. Fortunately, that is the case, as I have tested such a system on literally hundreds of thousands of document page images. The second assumption used in this section is perhaps more important to note. I have assumed that once the subimages are assigned to parallel pipelines, the throughput is limited by the maximum processing time required for any of the subimages. In practice, many images will be parallel processed at a time and so the impact of EP and %TI will generally be minimal—the more important predictor of system throughput will be the mean processing time. As such, the relative impact of parallel processing will be larger than shown in this example.

Given these caveats, then, this example shows that much of the improved throughput coincident with converting a document image processing serial process to parallelism by component is due to the *structural reframing* of the process, and not just the parallelism by itself. The importance of domain expertise for deciding on the final parallel system architecture (as in Figure 5.7) is also made apparent.

5.2.2 Image Understanding

5.2.2.1 Introduction

The second "core" application area to be considered in this chapter is image understanding. As described in Chapter 3, there is a plethora of important image understanding tasks, ranging from the mundane (image differencing to open automatic entrance doors) to the germane (face recognition, surveillance) to the arcane (medical image analysis, generalized scene recognition, augmented reality, etc.). In this section, rather than focus on the specific image understanding algorithms, I will focus on how parallelism by component can be deployed to simplify the downstream analysis.

In the example above (Section 5.2.1, document understanding), I showed that image segmentation plays a central role in extracting useful content from the document image. In the same manner, parallel processing by component for image understanding benefits from efficient preprocessing. However, unlike documents, which often use a Manhattan layout (meaning that successive horizontal and vertical "cuts" can break the document into its logical parts, or "regions"), general images have irregular (nonlinear) boundaries between objects of interest. As a consequence, the relatively simple, global thresholds such as those described in Kittler and Illingworth (1986) and Otsu (1979) will not be as useful. Instead, image segmentation is often performed using one or more of the following to define segment boundaries:

1. Texture
2. Chroma (hue)
3. Edges
4. Background (areas of low variance)

In the case of texture, significant changes in texture often accompany the transition from an object in the foreground to an object in the background, due to differences in focus. Chroma changes significantly when we move from one natural object to the next—for example, a blue sky to a tan beach, although it is relatively insensitive to sharpness of focus. This allows us to "regroup" three-dimensional (3D) objects that have been too aggressively segmented using one of the other methods. Edges are well understood. When we move from one object to another, edges define the boundaries of both objects. Thus, edges are shared. This has importance when considering the direction of the edge. Real edges—meaning edges that are not false positives—should match or blend the edge periphery statistics of two objects they border. Finally, background-based boundary determination is quite simple, since the background is defined based on binarization of the image. Wherever the image is "above threshold" we define it as background. If there are multiple peaks in the histogram above the threshold, we are likely to have multiple background colors and so this segmentation approach may be less applicable to that image.

Image understanding benefits from structural reframing along the lines used for document understanding parallelism by component. Once an image has been segmented, each of the segments, or regions, are now treated as a separate image. Polygonal boundaries, or even line segment boundaries, as described in Chapter 3, are used to access only the pixels associated with each segment. The region bounding box (x-minimum, x-maximum, y-minimum, and y-maximum) can be used to efficiently lock only this part of the large image in memory. If the image must be saved as a new (rectangular) image, then the pixels outside of the polygon or line segment, but inside the bounding box, definition of the region, must be "zero padded" with the appropriate background value. If the background is white, and the image is a 3-channel RGB image, then these extra-polygonal pixels will be set to (255,255,255). These "subimages" are then directly analogous to the subimages explored in Section 5.2.1, and the image understanding processing can be performed directly on each subimage independently of the other subimages.

At the highest level, there are two sources of variance in the image: image creation variance and image capture variance. If an image is man-made, there are differences in how the image is printed, or displayed on a screen, since different hardcopy (e.g., printers) and softcopy (e.g., displays) devices have (sometimes widely) different resolution, color, and permanence characteristics. As printers and displays have become ubiquitous, their manufacturers have in many cases moved to simplified engineering in order to retain some net profit in a commoditized market. Electronic displays, moreover, have to compete with ambient light since they project rather than reflect light. Nevertheless, it is still safe to say that the image capture process is more challenging. The quality of camera and scanner technologies is more widely variable, and in addition there is extensive variation in illumination and scene framing. These latter variations cannot be controlled by the manufacturer. What this means to the image understanding engineer is that she must—for practical, "real-world" applications—build some amount of insensitivity to such wide variation in captured image quality.

One manner in which to use parallelism by component to build in this robustness, of course, is the structural reframing provided by intelligently presegmenting the image. Presegmentation can be performed at low (often very low) resolution, meaning it can be used in common across all imaging devices. For scanned documents, I have historically used fixed or variable resolution input in the range of 30–75 ppi from the preview scan to perform document image analysis. Carrying this forward, a low-resolution "preview" camera image, such as the LCD image viewed by the camera user, is almost always sufficient for this segmentation, allowing subsets of the overall image to be analyzed more quickly as described in Section 5.2.1.

Structural reframing provides the intelligent division of a large, unwieldy, and potentially inefficiently processed component into a plurality of readily processed subcomponents. The benefits of this structural reframing, however, are not a given for every type of image understanding process, nor is it generally recursive. In addition, it can do nothing about the image understanding algorithms themselves. In order to aid the parallelism by component approach, we turn our attention to two approaches that provide real-time predictive insight into how well an image-understanding task is—or is capable of—proceeding. These approaches can, in fact, be used to provide feedback to the algorithm designer.

5.2.2.2 Model Down-Sampling

The first of these is *model down-sampling*, in which an image and the means to understand specific features within the image may be simultaneously and, from the standpoint of recognition accuracy, uniformly down-sampled. This means, at the highest level, that the model for analyzing an image is largely independent of the image characteristics—that is, the model is based on characteristics that can be computed for all input images. For example, suppose images to be analyzed will have compression sizes from 10% to 100% of the original image size, include grayscale and color images, and comprise resolutions ranging from 72 to 1200 ppi. Ideally, then, the model for these images will be based on an image that is (a) compressed to 10% of its original size, (b) grayscale (color information discarded), and (c) 72 ppi resolution. This is because every image used in the set will meet these minimum requirements. However, this "down-sampled" model only makes sense if the accuracy of the overall imaging task using the highest quality, non-down-sampled, images is maintained despite the down-sampling.

The complement of model down-sampling is *model up-sampling*, which scales every image to the greatest size, number of channels, and resolution. Model up-sampling, however, has several disadvantages. Artifacts are introduced by scaling up lossy-compressed images. The mean size of the images increases, which, as shown in Section 5.2.1, leads to a nonlinear increase in processing time, which works against the intended processing advantages of parallelism by component.

In order to test the effect of model down-sampling and model up-sampling on image understanding tasks, a set of 100 images belonging to one of four classes {nature, city, faces, documents: $n = 25$ each} was analyzed separately using each of the three approaches to image modeling (unchanged, model down-sampling, model up-sampling) followed by image classification accuracy (Simske, Li, and Aronoff, 2005) determination. A separate set of 100 images (also 25 images in each of the four classes) was used for training. Here, accuracy is defined as the percent of images assigned to the correct one of the four classes. Processing

Table 5.9 Task accuracy with different approaches to image modeling. Unprocessed images were a mean of 260 ppi, and a median of 250 ppi. Down-sampled images were all 150 ppi, and up-sampled images were all 400 ppi

	Unprocessed Images	Model Down-Sampling	Model Up-Sampling
Image classification	0.71	0.74	0.69
Barcode reading	0.88	0.88	0.88

time was recorded for each image classification. Model down-sampling was simple image down-sampling, while model up-sampling was simple image up-sampling. Both sampling approaches used cubic interpolation. Of the 25 images in each set, 5 each were captured at 150, 200, 250, 300, and 400 ppi. Down-sampled images were all resampled to 150 ppi; up-sampled images were all resampled to 400 ppi. Unprocessed images were a mean 260 ppi (median 250 ppi).

A second simple experiment was performed to illustrate the effects of model down-sampling on functional image processing tasks. In this case, image readability was assessed by looking at barcode reading success. In this second, separate experiment, a set of 100 Aztec barcode images was created, printed, and then read with an InData Systems LDS-4600 barcode reader. Only 88% of the barcodes were readable after image noise was added (coffee spills, abrasion, folding, etc.), for which the reading times (to the nearest millisecond) were recorded. Readability of the barcodes was not affected by model down-sampling or up-sampling.

The accuracies of the image classification and barcode reading results are shown in Table 5.9. The same 88 barcodes were readable for each approach, yielding an accuracy of 88% for all three image sets (unprocessed, down-sampled, up-sampled). Image classification was improved by model down-sampling (from 71% to 74% accuracy), likely because down-sampling applied to the training set resulted in less variable or less noisy training data. Model up-sampling decreased the accuracy slightly (to 69%), likely a consequence of image scaling adding noise to the training data.

Processing time, of course, was more noticeably affected by the imaging operations. The processing times required for image classification and, separately, barcode reading, for each of the three image sets are shown in Table 5.10. For this simple example, both accuracy and processing are optimized when the training and later classification use model down-sampling as opposed to unprocessed (mixed-resolution) or model up-sampled images.

The findings of Table 5.10 are not surprising. The mean image size of the model down-sampled images is smaller than the mean unprocessed image size, and far smaller than the mean

Table 5.10 Mean processing time with different approaches to image modeling. Both image classification and barcode reading tasks required the minimum processing time using model down-sampling. The processing time shown includes the time to down-sample or up-sample the image, if needed

	Unprocessed Images (ms)	Model Down-Sampling (ms)	Model Up-Sampling (ms)
Image classification	1734	567	2941
Barcode reading	226	194	346

up-sampled image size. The processing time, as shown in the previous section, is proportional to the mean image size. However, the accuracy results in Figure 5.9 may seem counterintuitive. Why do smaller images lead to higher classification accuracy? Perhaps it is due to the down-sampling normalizing—or smoothing out—some of the noise that is present at the higher initial resolution. Perhaps it is due to the significant content of the image being reliably represented at lower resolution (i.e., the significant content has an aliasing frequency far below the higher-resolution sampling frequency). Nevertheless, the results shown in Table 5.9 have their precedence. In previous work, in fact Simske, Sturgill, and Aronoff, (2010), down-sampling by a factor of as much as 3600 yielded improved classification results for a binary classification problem.

In this previous work, we obtained packages for a set of authentic and a set of counterfeit HP inkjet cartridges. The counterfeit cartridges were shipped in counterfeit packaging, from which the images were taken. Four different types of images I will discuss in this section were each scanned at 600×600 dots/inch (dpi) horizontal $\times$ vertical resolution using a desktop scanner (HP Scanjet 8200). The image types include a set of two barcodes (hereafter "Barcode"), a blue spot color region (hereafter "Blue"), a set of five branding images separated by white space (hereafter "Images") and a single large image (hereafter "Meadow"). The image areas were approximately 2.7, 2.8, 4.2, and 10.1 in^2, respectively. These images were originally scanned at high resolution (600 ppi) and then down-sampled to as little as 10 ppi (continuous tone ppi).

Ten image processing measurements, comprising the feature set, were computed for each of these images. All image sets were 50% from scans of authentic, and 50% scans of counterfeit packaging. Simple binary classification was performed. The individual features were weighted inversely proportional to their error rates and then combined to create a single binary classifier (Simske, Sturgill, and Aronoff, 2010).

Table 5.11 summarizes some of the results presented in Simske, Sturgill, and Aronoff (2010), highlighting the different relative effect of image down-sampling on the classification

Table 5.11 Original and model down-sampled (here as "pure" image down-sampling) images and the accuracy of classification. Original classification accuracy (top data row) is in italics. Any down-sampled image sets with higher classification accuracy than the original images are shown in boldface in the other rows

Down-Sampling Factor	"Barcode"	Blue Spot Color, or "Blue"	Composite Image (Set of Five Images), or "Images"	Single Meadow Image, or "Meadow"
Original (600 ppi)	*0.896*	*0.708*	*0.816*	*0.743*
4 (2 × 2)	0.832	**0.773**	**0.834**	**0.801**
9 (3 × 3)	0.828	0.674	**0.819**	**0.880**
16 (4 × 4)	0.774	0.675	**0.954**	**0.961**
36 (6 × 6)	0.739	0.682	**0.954**	**0.893**
64 (8 × 8)	0.734	0.669	**0.944**	**0.836**
100 (10 × 10)	0.722	0.636	**0.917**	**0.882**
144 (12 × 12)	0.710	0.648	**0.945**	**0.880**
225 (15 × 15)	0.732	0.642	**0.960**	**0.865**
400 (20 × 20)	0.731	0.668	**0.896**	**0.834**
900 (30 × 30)	0.740	0.662	**0.866**	**0.768**
3600 (60 × 60)	N/A	N/A	0.718	**0.806**

results for four different image types. The first, labeled "Barcode," is in fact a pair of black and white, one-dimensional (1D) barcodes segmented as a single image. The barcodes were classified with 89.6% accuracy at their original scanned resolution (600 ppi), and any level of down-sampling sharply reduced the classification accuracy. For a blue spot color region, the so-called Blue image set, down-sampling by a factor of two in both the x- and y-directions improved the classification accuracy from 70.8% at 600 ppi to 77.3% (at, effectively, 300 ppi). However, any greater down-sampling resulted in decreased accuracy in comparison to the original images. Thus, image down-sampling was not in general a good method for model down-sampling for either of the "Barcode" or the "Blue" image sets.

A completely different set of results was obtained for the continuous tone, or *contone*, images used in the study, labeled "Images" and "Meadow." These images were down-sampled by as much as 3600: for every down-sampling ratio up to 900, improved classification accuracy was observed for both images. Indeed, for the "Meadow" image, the classification accuracy was 80.6% (above the 74.3% on the original image) even when both the x- and y-directions were down-sampled by a factor of 60–10 ppi. Certainly, some of this is due to the elimination of noise—at least for the lower down-sampling ratios—but more likely the continued improved classification accuracy is due to a fundamental difference in the image sets that favors lower frequency coefficients of the image spectrum.

Regardless, these results add yet another item to the list of reasons for considering parallel processing. Because the parallel processing of images often does not benefit from the parallel analysis of a spatial subset of the image—for example, in determining the image skew—the system algorithm designer may wish to consider *resolution down-sampling* (which is a very straightforward example of model down-sampling, as well as perhaps the simplest exemplar of compressive sampling) by resolution when preparing an image processing task for parallel processing. Thus, it is possible that preparing an image for parallel processing will not only improve throughput—as shown for structural reframing in Section 5.2.1—but also improve the accuracy of the image understanding task, as shown in Table 5.11. A savvy algorithm architect, therefore, will be interested in exploring both of these forms of preparation for parallel processing when defining the overall imaging approach.

The results of Table 5.11 illustrate that, for contone images captured with a fixed imaging device (e.g., a scanner, not a camera), scanning at a resolution higher than 150 ppi is likely unnecessary for many image understanding tasks. At 4×4 down-sampling (resulting in the same resolution and similar image entropy as scanning at 150 ppi), we see that the mean accuracy of the two contone images increases from 78.0% to 95.8% (error is reduced by more than fivefold), while the system additionally benefits from a 16-fold decrease in image size. Based on the structural reframing results provided in Section 5.2.1, we anticipated a more than 16-fold improvement in throughput. This means that if overall system performance is considered a function of the product of the inverse of the error rate multiplied by the inverse of the image processing time, then we expect a more than 80-fold improvement in overall system performance for images like these.

This simple form of model down-sampling is part of parallelism by component since each image component can be treated with model down-sampling prior to applying parallel processing. The example shown is extremely simple, but more complex approaches—in which the images and the models for analysis are not identically down-sampled—can be readily derived from it. Suppose, for example, that we consider an alternative to performing a single down-sampling approach for the entire image. Among the alternatives are any of, or any combination

Figure 5.8 Sample of the authentic "girl with ball" image (left) and counterfeit "girl with ball" image (right). Note that the counterfeit image, in addition to being lower quality, also has a different halftone implementation

of, the following: (1) differential down-sampling of image regions based on their frequency, entropy, and so on, measurements; (2) palettization of different image regions based on their histograms, chroma variance, entropy, and so on; and (3) filtering (e.g., sharpening, blurring, unsharp masking, etc.) of different image regions based on their regional characteristics.

It is clear that such approaches naturally lead to meta-algorithmic patterns such as Predictive Selection (see Section 6.2.4), but I will not follow these possibilities here. Instead, I will focus on an example of how the components of the model itself vary as the image size is varied. For this new experiment, several hundred images were scanned from counterfeit packages that were seized during a raid on a counterfeit printing/manufacturing facility. A sample original image (scanned at 600 ppi) is shown on the left of Figure 5.8. A sample counterfeit image (also scanned at 600 ppi) is shown on the right of Figure 5.8. To the human eye, there are noticeable differences in the image quality. To the trained eye, the completely different halftoning approach employed by the counterfeiters is striking. To an image understanding expert, then, it is hardly surprising that a wide array of quantifiable image features will be effective at disambiguating between authentic images and counterfeit images taken from legitimate and nonlegitimate packaging.

To illustrate this, 10 features were computed for the images from both authentic and counterfeit packaging when scanned at 600 ppi ("A"), 300 ppi ("B"), and 150 ppi ("C"). The 10 features are relatively straightforward features, and are described in brief here:

1. *Entropy* is standard image entropy:

$$-\sum_{x=1}^{N} p_x \log_2(p_x),$$

where p_x is the percent of image intensities in bin "x" of the histogram, and the histogram comprises N elements. In this case, the images were captured in 8-bit *RGB*, so $N = 256$. Intensity I is defined as

$$I = (\|R\| + \|G\| + \|B\|)/3.$$

2. *Mean intensity, μ_I,* is defined by

$$\mu_I = \left[\sum_{j=1}^{H} \sum_{i=1}^{W} \frac{(\|R_{i,j}\| + \|G_{i,j}\| + \|B_{i,j}\|)}{3} \right] \bigg/ (WH),$$

where W is the width of the image in pixels and H is the height of the image in pixels, $\|R_{i,j}\|$ is the magnitude of the red channel for pixel $P(i,j)$, $\|G_{i,j}\|$ is the magnitude of the green channel for pixel $P(i,j)$, and $\|B_{i,j}\|$ is the magnitude of the blue channel for pixel $P(i,j)$.

3. *Image % edge* is determined by computing the edges in an image using an edge detector (e.g., Laplacian, Sobel, Roberts, etc., method) followed by an edge thresholding operation (e.g., Otsu (1979) performed on the edge histogram). The percent of the edge histogram that is above the threshold is considered the "image % edge."

4. *Mean edge magnitude* (MEM) is the mean value of the edges left over after thresholding to compute the image % edge in the previous step. If the threshold T is defined so that $1 \leq T \leq$ ME, with ME being the maximum edge value, then the MEM is defined as

$$\text{MEM} = \frac{\sum_{i=T}^{\text{ME}} i \times E(i)}{\sum_{i=T}^{\text{ME}} E(i)},$$

where $E(i)$ is the number of elements in the edge histogram for edge value i.

5. *Pixel variance* is typically performed in the intensity domain, and is simply a measure of neighborhood variability in the image.

6. *Mean region size, intensity-based segmentation* is the mean size of the connected components formed after thresholding in the image intensity plane.

7. *Region size variance, intensity-based segmentation* is the variance in the size of the connected components formed after thresholding in the image intensity plane.

8. *Mean image saturation* is mean value of all the pixels for saturation. There are several definitions for saturation, but all are based on how far from white any of the image channels—usually red, green, and blue, or *RGB*—stray. Saturation S is defined as

$$S = (255 - \min(R, G, B))/(R + G + B),$$

where $S = 255$, if $R = G = B = 0$. As a consequence, mean saturation μ_S is defined as

$$\mu_S = \left[\sum_{j=1}^{H} \sum_{i=1}^{W} \frac{(255 - \min(\|R_{i,j}\| + \|G_{i,j}\| + \|B_{i,j}\|))}{(\|R_{i,j}\| + \|G_{i,j}\| + \|B_{i,j}\|)} \right] \bigg/ (WH).$$

9. *Mean region size, saturation-based segmentation* is the mean size of the connected components formed after thresholding in the image saturation plane.

10. *Region size variance, saturation-based segmentation* is the variance in the size of the connected components formed after thresholding in the image saturation plane.

Table 5.12   Image features and the accuracy of the classification provided by each for the original 600 ppi scanned images (A), the images when scanned at 300 ppi (B), and the images when scanned at 150 ppi (C)

Image Metric	A (600 ppi)	B (300 ppi)	C (150 ppi)
1. Entropy	0.998	0.936	0.708
2. Mean intensity	0.993	0.930	0.823
3. Image % edges	0.977	1.000	0.927
4. Mean edge magnitude	0.789	0.988	0.999
5. Pixel variance	0.761	0.992	0.694
6. Mean region size, intensity-based segmentation	1.000	0.941	0.877
7. Region size variance, intensity-based segmentation	0.999	1.000	0.790
8. Mean image saturation	0.999	1.000	1.000
9. Mean region size, saturation-based segmentation	1.000	1.000	0.831
10. Region size variance, saturation-based segmentation	1.000	1.000	0.879

Taken together, these 10 metrics comprise a rather eclectic combination of image features, which are in general useful for comparing printing and/or scanning differences among image sets. Entropy, pixel variance, mean intensity, and intensity-based region metrics, for example, are sensitive to changes in image luminosity, both globally and locally. Image % edge and MEM are highly sensitive to image sharpness and contrast. Mean image saturation and saturation-based region metrics are sensitive to changes in image contrast and color balance. Table 5.12 provides the classification accuracy using the Simske, Li, and Aronoff (2005) classifier. The features providing the highest accuracy vary with the scanning resolution. For set A, the highest accuracy features are $\{1, 6, 7, 8, 9, 10\}$; for set B, the highest accuracy features are $\{3, 7, 8, 9, 10\}$; and for set C, the highest accuracy features are $\{4, 8\}$. Only one feature, mean image saturation, provides the highest accuracy for each resolution investigated. On the other hand, entropy is far less useful for image sets B and C than for set A; image % edge is most useful for set B; and MEM is most useful for set C. This means that the set of features optimal for classification varies with image size.

Thus, the experiment shows a second, more complex form of model down-sampling. In this form, the set of features is changed, and thus the model for classification changes with change in the image size. In effect, the model's components are varied with resolution. Note that the example shown here involves selecting an optimal set of global image features to provide image classification—by analogy, regional image features, such as different approaches to segmentation, could also be optimized. The application to parallel processing by componentization architectures such as shown in Figure 5.7 is strong: model down-sampling can potentially be used to ensure that structural reframing does not result in decreased accuracy. The data in this section indicates that this approach will work for some image types—for example, contone images in this example—but not for others—for example, barcodes and spot colors in this example.

5.2.2.3   Componentization Through Decomposition

We have seen that reducing the size of an image greatly improves image processing throughput even before parallel processing is deployed, generally possible through structural reframing. We also saw that model down-sampling provides the means to maintain image understanding accuracy, even when significant image compression or down-sampling is employed. In this section, we investigate another important approach for streamlining image understanding tasks for later parallel processing. This approach, termed *componentization through decomposition*, employs the use of separating an image into different relative or absolute image planes for separate, usually parallel processing.

There are a number of means of performing componentization through decomposition. An obvious one is JPEG compression, which decomposes image blocks into their discrete cosine transform (DCT) coefficients. Certain types of image analysis (e.g., shape matching and signal processing) can be performed directly on the DCT coefficients. Other examples include decomposing the image into distinct channels of information, including, for example, the following:

1. Red, green, and blue (RGB) channels
2. Cyan, magenta, and yellow (CMY) channels
3. Hue, saturation, and lightness (HSL) channels
4. Luma, blue-difference chroma, and red-difference chroma (YCbCr) channels

After decomposing an image into its channels, each channel can be processed separately (i.e., in parallel) and generally much more rapidly, since each is only one-third the size of the original image. One such image decomposition is shown in Figure 5.9: 5.9a is the original image, 5.9b is the hue image, 5.9c is the saturation image, and 5.9d is the intensity, or luminance, image.

Image decomposition instantly prepares a single image—even an image already prepared for parallelization by component—for improved parallel processing by trifurcating an image, with an expected disproportionate (i.e., greater than a factor of three) improvement in image processing throughput. Such an approach can also be used to improve overall image understanding accuracy. For example, consider separation of foreground from background. The three maps—for example, hue, saturation, and luminance—can each be binarized (thresholded) upfront to define the background and foreground parts of an image. Only where all three maps are above their appropriate threshold values do we define the overall image background. This prevents the determination of false positives for the background.

5.2.2.4 Image Understanding Recapitulation

Image understanding is a very broad field, and this section has only scratched the surface on a limited set of image analysis approaches—primarily image segmentation. I have omitted entirely such complex image understanding technologies as face detection and recognition, object tracking, scene recognition, and medical imaging. Nevertheless, focusing on simple image segmentation allowed a more limpid illustration of how to employ *model down-sampling* and *componentization through decomposition*, the two new parallelism-by-component-enabling approaches introduced in this section.

The discussion on model down-sampling (Section 5.2.2.2) was initially focused on image down-sampling, since it is clear that reducing the size of an image has disproportionate improvement on processing throughput. The results of Table 5.11 show, perhaps surprisingly, that this simplest of model down-sampling approaches can result in increased accuracy even as it aims for improved throughput. As a second form of model down-sampling, the data in

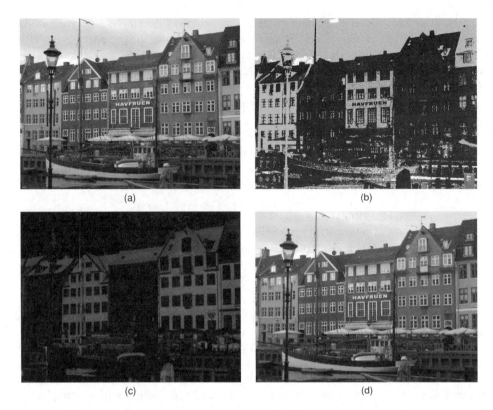

Figure 5.9 Sample image (a) and its hue (b), saturation (c), and luminance (d) channel. The image (a) is as large as (b), (c), and (d) combined, since it comprises three channels

Table 5.12 illustrate that certain image features can be used for high accuracy classification across a wide range of image resolutions. Specifically, the mean image saturation feature affords 99.9% or higher classification accuracy for image resolutions across the range of 150–600 ppi—it is "robust" to change in image resolution. Thus, mean image saturation should be part of the classification "model" irrespective of image resolution.

The example for componentization through decomposition involves the decomposition of an image into its hue, saturation, and intensity channels, each one-third the size in memory of the original image. These channels-as-images afford different segmentation of the image—by color and brightness, for example—than simple binarized image segmentation (which corresponds to the intensity image) as described earlier. Effectively, this process of preparing an image for parallelism by component also enhances the segmentation: two new images are available to provide salient region extraction.

5.2.3 Biometrics

In Chapter 4, which was focused on parallelism by task, hybrid biometrics were discussed. Hybrid biometrics rely on multiple input streams to validate identity. In this section, component-based biometrics are introduced. As opposed to hybrid biometrics, component-based biometric features focus on elements and subelements of the same biometric measurement.

Componentization through decomposition, introduced in Section 5.2.2 for image understanding approaches, will be shown to be an important means of addressing biometrics-related machine intelligence. In this section, I also introduce two other means of parallelization by components for biometrics: (1) *temporal parallelism*, with obvious extensions to video analysis, surveillance, and motion tracking; and (2) *overlapped parallelism*, with obvious extensions forward to meta-algorithmics—the focus of the next four chapters and this book in general. Finally, scaled-correlation parallelism—first described in Chapter 4 in relationship to parallelism by task—will be reconsidered for parallelism by component and applied to the field of biometrics.

5.2.3.1 Componentization Through Decomposition and Biometrics

Image-based biometric analyses—such as face detection and recognition, fingerprint recognition, iris recognition, and gait analysis—are often performed on a greatly simplified version of the original image. Fingerprint recognition, for example, is usually performed on a binary (thresholded) image. Typically the ridges (high points) of the fingerprints are binary black, and the valleys between the ridges are binary white. As described in Nanavati, Thieme, and Nanavati (2002), fingerprint image processing usually (for 80% of the fingerprint reading systems) consists of erosion (thinning) of the ridges to a single pixel in width, followed by the identification of the finger-scan minutiae such as crossovers, cores, bifurcations, ridge endings, islands, deltas, and pores. Minutiae matching—which can be assessed using a rules-based, or expert, system—is a data-parsimonious approach, although it is sensitive to wear and tear of the fingertips. A second method, pattern matching, generally requires more data for analysis and is also more sensitive to finger placement during reading (Nanavati, Thieme, and Nanavati, 2002).

Fingerprint analysis, therefore, seems readily suited to benefit from the application of componentization through decomposition. The original image can be sent to two parallel pipelines as shown in Figure 5.10. The left pipeline is used for minutiae matching. The image is thresholded and the set of minutiae that match for the individual are collected. A convex hull is formed around these minutiae. This convex hull is shared with the right parallel pipeline to be described shortly. After the set of matching minutiae are determined, the confidence value (probability of a match based on the minutiae) is reported as p_M. The right parallel pipeline uses a grayscale (nonbinarized) image, which must be prepared for analysis in a different fashion. For example, the contrast should be made uniform across the image and any damaged areas within the convex hull should be excluded. The remaining pattern within the convex hull can then be matched to the person's stored fingerprint pattern using correlation. The matching probability corresponding to the correlation value is then reported as p_C. The pair of probabilities (p_M, p_C) is then used to determine an overall matching probability, p_{C+M}.

Generally, the overall probability of matching, p_{C+M}, will be higher than either individual probability when there is a true match, and lower when either of the two individual metrics fails or has relatively low accuracy. Table 5.13 shows one such result set, obtained using my 10 fingerprints. In Table 5.13, the values of p_{C+M} are reported for 49 combinations of (p_C,p_M),

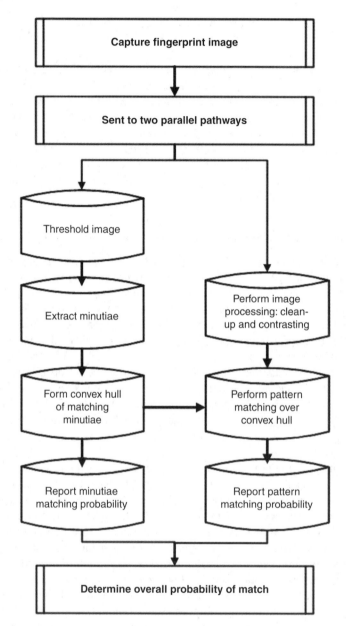

Figure 5.10 Componentization through decomposition design for fingerprint analysis. The fingerprint image is binarized (thresholded) and then analyzed using minutiae matching. The convex hull of the matching minutiae can be used to crop the pattern for matching with the original image. Minutiae matching and pattern matching pipelines can be assigned to separate parallel hardware optimized for these two different processes

Table 5.13 Probability of a fingerprint biometric match when only minutiae (p_M), only pattern matching correlation (p_C), or the combination (p_{C+M}, as shown in the design of Figure 5.10) are considered. Note the italicized, boldface values in the lower right of the table, wherein the parallel combination of minutiae and correlation provide higher accuracy than either of the two metrics by themselves

p_{C+M}		p_M						
p_C		0.50	0.80	0.90	0.95	0.975	0.99	0.999
	0.50	0.50	0.75	0.82	0.85	0.87	0.89	0.90
	0.80	0.72	0.79	0.85	0.90	0.95	0.97	0.99
	0.90	0.78	0.86	0.89	0.94	*0.98*	*0.992*	*0.997*
	0.95	0.83	0.92	0.95	*0.97*	*0.99*	*0.994*	*0.998*
	0.975	0.86	0.96	*0.98*	*0.99*	*0.995*	*0.997*	*0.9996*
	0.99	0.88	0.98	*0.994*	*0.9992*	*0.996*	*0.998*	*0.9998*
	0.999	0.91	0.99	*0.9992*	*0.9994*	*0.9997*	*0.9998*	*0.9999*

where p_C and p_M vary from 0.50 to 0.999. The combinations of (p_C,p_M) for which $p_{C+M} > \max(p_C,p_M)$ are shown in italicized boldface. In this example, $p_{C+M} > \max(p_C,p_M)$ as both p_C and p_M approach 1.0. In fact, the results from Table 5.13 indicate the following approach to fingerprint classification, with the assumption being that the two parallel pipelines in Figure 5.10 each report their confidence values (p_M and p_C, on the left and right, respectively):

1. When $p_C + p_M \geq 1.875$, choose the classification as provided by the combination of p_C and p_M; that is, of p_{C+M}.
2. Otherwise, choose the classification as provided by the minutiae if $p_M \geq p_C$.
3. Otherwise, choose the classification as provided by the pattern matching correlation.

Other means of employing componentization through decomposition for biometrics include separating signals into distinct frequency bands—for example, for speech or ECG recognition. These can be nonoverlapping partitions of the spectrum, or different transformations—for example, linear, logarithmic, log-linear, filtered—to create domain-specific spectrums, or "cepstrums." For example, if voice recognition is the biometric in use, then transformations of the normal mel-frequency cepstral coefficients—namely, cepstral mean subtraction—can be used to provide speaker identification (Rosenberg, Lee, and Soong, 1994), while correlation of the untransformed audio spectrum can be used for speaker verification.

5.2.3.2 Temporal Parallelism and Biometrics

Biometric analysis comprises both static and dynamic biometric measurements, as described in Chapter 3. A simple means of parallelism by component, therefore, is to assign a static biometric assay to one component of the parallelism, and a dynamic biometric assay to another component. At first blanch, this would seem to be an unbalanced parallel processing design, since the static biometric usually requires a single image or signal for analysis, and the dynamic biometric requires a sequence of images and/or video for analysis. However, the parallelism is often readily balanced since many of the dynamic biometrics can be computed using significantly smaller images.

Consider, for example, a static biometric that requires an image W pixels wide and H pixels high. Fingerprint recognition, for example, may require an image size of 355×390 pixels for high accuracy (corresponding to 500 dpi). To track the same finger for gesture recognition may require video analysis of a much smaller image, say 80×100 pixels. This is 17.3 times smaller. If the video frame rate is 24 frames/s, and the biometric gesture is complete within 1 s, then the amount of processing time required for the static and dynamic biometrics are roughly equivalent. It is feasible to subsample the video stream in time to very closely match the throughput of these two parallel pipelines.

Temporal parallelism is especially useful when a large signal stream—for examples, an audio stream or a large video stream from touch screen capture of fingerwriting—is available and can be streamed to multiple pipelines simultaneously. Face detection, for example, can be used to determine the minimum cost tracking of a person, while the same stream is used in a parallel path for scene recognition, or slightly delayed for facial emotion recognition.

Other forms of temporal parallelism are when different aspects of motion are analyzed at the same time in two or more parallel paths. In the simple surveillance example, one path can be used for facial detection and recognition, and the other path for walking speed, gait analysis, and other kinetics calculations.

5.2.3.3 Overlapped Parallelism and Biometrics

Overlapped parallelism is the spatial analog to temporal parallelism. Overlapped parallelism prevents the loss of context through the use of sharp boundaries, such as those between arbitrary tessellations of images such as Figure 5.5. The benefits of overlapped parallelism are obvious: larger images are tessellated into much smaller images, which benefit from the disproportionate improvement in throughput as summarized in Table 5.6. However, the improved throughput may come at a price—reduced accuracy of segmentation due to the artificial image boundaries introduced by tessellation cutting through objects that need to be recognized.

The trade-off between throughput and accuracy is illustrated by the example of Tables 5.14 and 5.15. In Table 5.14, a set of surveillance images are analyzed (face detection) and the overall accuracy is defined as (IP – FP – FN)/TP, where IP is the identified positives, FP is the false positives, FN is the false negatives, and TP is the true positives. The images are analyzed using 10 different cases. For Case 1, the images are analyzed without subsegmentation. For Cases 2–4, the images are subsegmented into 4, 9, or 16 equally sized subimages, respectively, with no overlap between the subimages. For Cases 5–7, the images are subsegmented into 4, 9, or 16 equally sized subimages with 20% neighboring overlap. Subimages against the edge of the image must overlap 20% to one side; subimages not against the edge of the image overlap 10% in each direction. Finally, Cases 8–10 are subsegmented into 4, 9, or 16 equally sized subimages with 50% neighboring overlap. For each case, the percentage of the image spanned by the subimage (%Span) and the percent overlapped by the subimages (indicative of the amount of redundancy of search caused by subimage overlap, designated %Overlap) are computed. Overall accuracy (A) is defined as described above:

$$A = \left(\frac{\text{IP} - \text{FP} - \text{FN}}{\text{TP}} \right),$$

Table 5.14 Effect of overlapped parallelism on a biometric surveillance process. The images are analyzed without subsegmentation (Case 1); subsegmented into 4, 9, or 16 equally sized subimages (Cases 2–4) without overlap; subsegmented into 4, 9, or 16 equally sized subimages with 20% neighboring overlap (Cases 5–7); or subsegmented into 4, 9, or 16 equally sized subimages with 50% neighboring overlap (Cases 8–10). The percent image span and area overlap are given by "%Span" and "%Overlap" columns, respectively. The throughput improvement for each subsegment, overall accuracy (A) and overall throughput improvement (OTI) are given in the remaining columns. See text for details

Case	Subsegments	%Span	%Overlap	Throughput Improvement by Subsegment	A	OTI
1	1	100.0	100	1.0×	0.931	1.000
2	4	50.0	100	7.2×	0.857	1.800
3	9	33.3	100	27.4×	0.829	3.044
4	16	25.0	100	54.6×	0.798	3.413
5	4	60.0	144	4.2×	0.907	1.050
6	9	40.0	144	15.6×	0.889	1.733
7	16	30.0	144	35.4×	0.877	2.213
8	4	75.0	225	2.1×	0.926	0.525
9	9	50.0	225	7.2×	0.914	0.800
10	16	37.5	225	19.6×	0.903	1.225

Table 5.15 Optimizing the parallelism by component approach of Table 5.14 through the use of three different figures of merit (FoMs). The first, $A \times$ OTI, is concerned with the product of accuracy and throughput. The second, $(A/\text{OTI}) + E$, is an estimate of the time to successfully analyze all of the faces in the image. The third, OTI $\times$ ($E/Emin$), penalizes the system in direct proportion to throughput and to relative error rate. See text for details

Case	A (Overall Accuracy)	E (Error Rate)	OTI (Overall Throughput Improvement)	FoM, $A \times$ OTI	FoM, $(A/\text{OTI}) + E$	FoM, OTI $\times$ ($E/Emin$)
1	0.931	0.069	1.000	0.931	1.000	1.000
2	0.857	0.143	1.800	*1.543*	*0.619*	0.869
3	0.829	0.171	3.044	*2.524*	*0.443*	*1.228*
4	0.798	0.202	3.413	**2.724**	**0.436**	*1.166*
5	0.907	0.093	1.050	*0.952*	*0.957*	0.779
6	0.889	0.111	1.733	*1.541*	*0.624*	*1.077*
7	0.877	0.123	2.213	*1.941*	*0.519*	**1.241**
8	0.926	0.074	0.525	0.486	1.838	0.490
9	0.914	0.086	0.800	0.728	1.229	0.642
10	0.903	0.097	1.225	*1.106*	*0.834*	0.871

and the overall throughput improvement (OTI) is defined as

$$OTI = TI(SS)/N_{SS},$$

where TI(SS) is the throughput improvement by subsegment and N_{SS} is the number of sub-segments.

As expected, there is a disproportionate improvement in throughput with decreasing subimage size (Table 5.14, fifth column from the left). For example, when the image is divided into four equally sized subimages, each subimage is processed 7.2 times as quickly. Dividing into 9 and 16 equally sized subimages results in 27.4 and 54.6 times the throughput. Accuracy, as expected, drops in going from 1–4 to 9–16 subimages (93.1% to 85.7% to 82.9% to 79.8%), largely due to the splitting up of faces by the tessellation-imposed subimage boundaries. Accuracy is improved with increasing subimage overlap: from 85.7% to 90.7% to 92.6% for the four subimage cases; from 82.9% to 88.9% to 91.4% for the nine subimage cases; and from 79.8% to 87.7% to 90.3% for the 16 subimage cases.

The OTI improves the most when there is no overlap. For the 9 and 16 subimage cases without overlap, OTI is greater than 3.0. This is an important input to Table 5.15, which provides analysis of the accuracy (and its complementary error) and the OTI in three important figures of merit (FoMs) (effectively, these are cost functions).

Table 5.15 provides the error rate, E, which is defined as $1.0 - A$. The minimum error rate, designated *Emin*, is the error rate on the original (unsegmented) images, or Case 1. *Emin* is therefore 0.069. Three FoMs are then used to compare the nine parallel approaches to Case 1. The First FoM, the simplest, is the product of accuracy and OTI (a higher FoM is better here). The FoM makes no assumptions about the system's ability to recognize errors when they occur, and thus assumes the output of the system will be used "as is." For this FoM, most of the parallelism by component approaches outperform the nonparallel Case 1. The best FoM is obtained for Case 4, followed closely by Cases 3 and 7. In general, the improved throughput of ever-smaller subimage tessellations outweighs the loss of accuracy for this FoM.

The second FoM, defined by $(A/OTI) + E$, is an estimate of how much time will be required to obtain the highest accuracy output (a lower FoM result is better here). It is assumed that if the correct faces are not found, the same subimages would be analyzed as complete images (Case 1). The results closely mirror those of the first FoM, with Cases 4, 3, and 7 providing the best results.

The third FoM assumes that the particular system configuration is penalized by its relative error rate and rewarded, as usual, by its OTI. The ratio of the error rate, E, to the minimum error rate, *Emin*, is multiplied by OTI. For this FoM, Case 7 provided the highest value, followed by Cases 3 and 4. For this FoM, the relatively low E for the 16-fold tessellation with 20% overlap outweighs the much higher OTIs of the nonoverlapping tessellation cases.

Which of these three FoMs is the most appropriate for the system depends on how well the system is able to detect and respond to errors. If a surfeit of images from which to identify the one or more individuals of interest are available, for example, with video streams, then the first FoM, rewarding the approach for both accuracy and relative throughput, is very appropriate. If, on the other, each image is of high value in and of itself—for example, when only a few frames show the subject(s) of interest—then each image for which a high level of identification confidence is not obtained may be reanalyzed as a nonsegmented image. For this situation, the FoM defined as $(A/OTI) + E$ is very appropriate, since it predicts the amount of time necessary to perform full image analysis in those cases when the original parallelism by component approach (e.g., one of Cases 2–10) fails. Finally, in cases where the primary cost of the system is in responding to errors—for example, when false positives lead to false arrests—the third FoM, OTI $\times$ (*Emin/E*), is a very appropriate FoM to use comparatively.

5.2.3.4 Scaled-Correlation Parallelism and Biometrics

Scaled-correlation parallelism is somewhat related to model down-sampling. In model down-sampling, there is the simultaneous scaling of specific data analysis features and of the data itself. Thus, the "model" for analyzing biometrics data—for example, a signal or image—is made to be largely independent of the signal or image characteristics. Scaled-correlation parallelism, on the other hand, can use the down-sampled model to perform analysis at different scales—for example, at different sampling frequencies or image resolution—to perform higher-confidence identification than can be obtained at only a single scale. It is important to note, however, that scaled-correlation parallelism does not require model down-sampling. The same (or different default, if preferred) set of features can be analyzed at each scale and the results processed as seen in the following example.

Having discussed fingerprint identification in the context of componentization through decomposition, we now look at how fingerprint detection can benefit from scaled-correlation. In order to address this, however, it must be made clear which form of "correlation" is part of the "scaled-correlation" approach to parallelism. In this context, correlation of feature behavior and scale is generally a disadvantage, since it implies that no new interpretation is available at any scale. What is more important here is the correlation of the relative accuracy (or other FoM) for each approach for each of the different scales. This is illustrated by the example in Table 5.16.

Table 5.16 Table distinguishing the feature correlation ("correlation across scale"), which may be helpful at indicating the best resolution to perform the analysis, from relative accuracy correlation, indicated by the delta, Δ = Accuracy(Approach A) – Accuracy(Approach B). The latter is used to determine the applicability of scaled-correlation parallelism to the task. The individual accuracies of both approaches at each resolution are indicted, and the cumulative accuracy of all scales at or above the current resolution is given in parenthesis—for example, "0.867" is the accuracy when the 600 ppi, 400 ppi, and 300 ppi results for Approach A are combined

Fingerprint Identification Results	Mean Feature Correlation with the Results at Highest Resolution ("Correlation across Scale")		Accuracy of the Approach at Each Scale (Leads to Scaled-Correlation Parallelism)		
Resolution (ppi)	Approach A	Approach B	Approach A	Approach B	Δ = Approach A – Approach B
600	1.000	1.000	0.832	0.735	0.097
400	0.584	0.643	0.761 (0.845)	0.744 (0.789)	0.017 (0.056)
300	0.725	0.711	0.736 (0.867)	0.687 (0.813)	0.049 (0.054)
200	0.583	0.634	0.803 (0.912)	0.756 (0.837)	0.047 (0.075)

In Table 5.16, fingerprint identification results are shown for original images captured at 600 ppi, and then processed in parallel at 400, 300, and 200 ppi. Two fingerprint identification algorithms are considered, labeled "Approach A" and "Approach B." The columns for mean feature correlation show that the results at the different resolutions correlate with the results at the highest resolution with values in the range of 0.58–0.73; in other words, not particularly highly correlated. This implies that there may be extra information obtained by including the results from two or more resolutions in parallel. The scaled-correlation parallelism values are given in the last three columns. These results show that the cumulative accuracy of each approach continues to increase as the lower-resolution results are added—from 0.832 to 0.845 to 0.867 to 0.912 for Approach A, and from 0.735 to 0.789 to 0.813 to 0.837 for Approach B. The scaled-correlation parallelism, therefore, merits deployment here, as the combination of results from all four scaled image sizes results in significantly higher accuracy (91.2% or 83.7%) than the results for the highest accuracy "best" set (83.2% or 73.5%). Further, Approach A is consistently better than Approach B.

Thus, scaled-correlation parallelism is useful when higher accuracy results can be obtained by combining the results at multiple scales, rather than simply selecting the optimal scale or configuration (as is consistent with model down-sampling).

5.2.4   Security Printing

The fourth and final "core domain" in which parallelism by component is considered in this chapter is that of security printing. As introduced in earlier chapters, security printing is not a single domain, but rather a broad, interdisciplinary field comprising the important tasks of serialization, inspection, authentication, forensics, and access rights validation. As such, the preparation of security printing tasks for parallelism by component is generally quite straightforward.

The first of two security printing-related approaches benefiting from parallelism by componentization is a natural consequence of the use of VDP to create multiple variable printed marks, or "deterrents," used for one or more the principle security printing tasks—this approach is termed *variable element parallelism*. Figure 5.11 illustrates one such hybrid mark. The hybrid mark in Figure 5.11 comprises three independent variable data codes—a two-dimensional (2D) Data Matrix barcode (black and white barcode in the middle of the image), a larger 3D (color 2D) barcode, and microtext of the character string "30581D4D025DC3400000002F" beneath these two nested barcodes. Each of these three components can be analyzed by a separate pipeline. In this case, each pipeline is custom configured to the particular analysis task, emphasizing the manner in which preparing the security printing tasks for parallelism by itself provides system architecture advantages.

Processing these three elements using hybrid parallelism by componentization will also illustrate how other componentization principles are brought to bear in security printing. The amount of preprocessing required in this case is minimal—the image is simply sectioned into two parts between the microtext and the color barcode. Because of the white space surrounding the 2D barcode—even though it is inside of the color barcode—the 2D barcode is efficiently and accurate read by commercial barcode reading software, which also conveniently ignores the surrounding color barcode as "noise." A separate, custom imaging algorithm is used to extract the intentional data embedded in the color barcode. Both of the barcodes can be read

30581D4D025DC3400000002F

Figure 5.11   Security printing image amenable to componentization through decomposition. The 2D barcode and the white space surrounding it can be extracted and analyzed by a 2D barcode reading pipeline; the surrounding color barcode can be analyzed by a second parallel pipeline; and the microtext below these barcodes can be analyzed by a third parallel pipeline

successfully at relatively low resolution—for example, at 75–150 ppi. This means that the common image shared by these two barcode pipelines can be down-sampled before sending the image (and a copy of it) to the two pipelines in parallel—a form of *model down-sampling*. The microtext image, however, must be retained at higher resolution or it will be unreadable to an optical character recognition (OCR) engine. The down-sampling of two pipelines with the maintenance of full resolution for the third pipeline is thus a form of *structural reframing*. Forcing the issue a bit, using the black and white image plane for 2D barcode reading and the color planes for 3D barcode reading can be loosely considered a form of *componentization through decomposition*—truly forcing the issue, having the same image used for two parallel pipelines can be considered a form of *overlapped parallelism*. Regardless, in this example, since the original microtext-bearing image is much smaller in size than the barcode images, after structural reframing the three parallel pipelines have much more uniform processing times.

The second novel pattern for parallelism that arises from security printing is related to the POD nature of many variable data printers and presses. Suppose we use the security printing hybrid image shown in Figure 5.11, and assume that it is associated with a label. Each of the three variable marks comprising the label—the 2D barcode, the 3D barcode, and the variable microtext—may provide a unique identifier for each label, a process called *mass serialization* of the mark (with the serialization being stochastic, or unpredictable, from one label to the next). Each 2D barcode is different from every other 2D barcode in the set; each 3D barcode is unique; and each microtext string is different from every other microtext string. Thus, the microtext "30581D4D025DC3400000002F" as shown in Figure 5.11, may on the next label be "A62BE7DC23905CC2B290DD21," and so on. It is important to recognize that, during the creation of these marks, the number of variable elements is also the number of mass serialized sets that must be created.

When mass serialized sets of identifiers—suitable for writing to any of the marks in Figure 5.11—are created, they must appropriately make use of random sequence generation (RSG). Inevitably, such an approach results in some repetition of already existing identifiers, a phenomena designated "identifier collision." Every time a new potential identifier is supplied by the RSG algorithm, then, identifier collision must be avoided. In order to do that, each new potential identifier must be compared to the existing database of identifiers, and discarded if they collide. If there are multiple variable elements, then there are multiple databases, each storing an existing set of identifiers for that variable element. It is, therefore, logical to deploy a form of *search parallelism* to this security printing task, wherein each of these identifier sets is searched in parallel for collisions in preparing the identifiers for the next hybrid mark. This leads to a parallelism by component that occurs *before* the next label is printed, so that the identifiers are already validated for the hybrid mark when it is time to print it.

This form of search parallelism provides what is termed the *prelocking of content*. The parallelism by component occurs before the content is actually locked to the specific variable element in the printed material—that is, before it is printed. Nevertheless, its impact on overall design is similar to the processing pipeline parallelism architectures featured in much of this chapter: the throughput is limited by the throughput of the longest data string used for one of the deterrents. In some ways, then, this search parallelism can be considered a specialized form of temporal parallelism.

In this brief consideration of parallelism by component for security printing, the final type of pattern discussed in this chapter but not yet considered is that of scaled-correlation parallelism. The staggered barcodes described in Chapter 4 illustrate the application of scaled-correlation parallelism to security printing, and I consider this to be a "parallelism by task" application more so than a "parallelism by component" task since the same component is used for, effectively, multiple tasks. The distinction, however, is largely semantic, and so the reader is referred to Chapter 4 on that topic.

5.3 Summary

In this chapter, eight specific patterns for the application of parallelism by component were considered. These were explored to exemplify some of the complexities involved in deploying a parallelism-by-component system, and more importantly to add tools to the toolkit of the parallel system architect. These eight patterns are:

1. Structural reframing, in which a procedure is prepared for parallelism by component, and benefits immediately from the restructuring of the analysis (generally through disproportionate improvement in throughput due to the new structuring).
2. Model down-sampling, in which a model for analysis is able to scale with a reduced data set without the loss of accuracy.

3. Componentization through decomposition, in which a data set is parsed into multiple subsets, or planes, of information, allowing independent parallel analysis with improved throughput.
4. Temporal parallelism, in which different elements in a sequence can be analyzed in parallel without loss of analysis accuracy.
5. Overlapped parallelism, in which case a data set—for example, image—is subsegmented and the subsegments can overlap each other. This can provide improved throughput without significantly increased error rate since the overlap in general prevents segmentation from occurring within a feature of interest.
6. Scaled-correlation parallelism, in which analysis at different scales—for example, sampling frequencies or image resolution—are used to provide higher-confidence identification than can be obtained at only a single scale.
7. Variable element parallelism, in which two or more variable data elements are analyzed in parallel. This is pure componentization—each variable data element is, in effect, a separate component.
8. Search parallelism, in which searching is divided into parallel partitions and then recombined.

Combined, these patterns comprise a useful toolkit for the would-be parallel system architect. One of the realities of the new world of computing is that resources, while still not unlimited, are much more prevalent than in the past. For complex systems, we are still a long way from being able to consider exhaustive search as a viable means of system optimization. However, because the complexity of many applications—such as those of document processing, image understanding, biometrics, and security printing outlined in this chapter—has not increased as quickly as Moore's law, processing availability, and storage availability over the past decade, we are in the position of considering the next 1–2 orders of magnitude increase in the movement toward exhaustive search. It is specifically this opportunity that the patterns in this chapter exploit. I argue here that it is in general a better use of this plentiful but not unlimited surfeit of processing and storage capability to consider multiple patterns for the parallel system before using these excess resources to optimize an *a priori* determined architecture. Parallel processing patterns such as those outlined in this chapter lead to a large exploration of the solution space for the overall system, and as such are *efficient* within the context of the excess—but not excessive—new incremental resources.

Perhaps more importantly, this varied set of patterns provides a widespread, introspective search of the solution space. Because such different design approaches as temporal and spatial, overlapping and down-sampling, are considered, the system designer is not likely to be blindsided by missing an obvious approach to improving system accuracy, robustness, performance, or efficiency. Moreover, in finishing with the hybrid patterns—variable element parallelism and search parallelism—the approaches to parallelism by componentization naturally lead us to the consideration of the parallelism-enhancing patterns that comprise the backbone of this book: meta-algorithmics. We now move our attention to that topic.

References

Cheyenne Mountain (1996) Zoo Brochure, http://cmzoo.org/ (accessed March 10, 2012).
Kittler, J. and Illingworth, J. (1986) Minimum error thresholding. *Pattern Recognition*, **19** (1), 41–47.

Nanavati, S., Thieme, M., and Nanavati, R. (2002) *Biometrics: Identity Verification in a Networked World*, John Wiley & Sons, Inc., New York, 300 pp.

Otsu, N. (1979) A threshold selection method from gray level histograms. *Pattern Recognition*, **9** (1), 62–66.

Rosenberg, A.E., Lee, C.H., and Soong, F.K. (1994) *Cepstral Channel Normalization Techniques for HMM-Based Speaker Verification*. Proceedings of the IEEE ICASSP, pp. 1835–1838.

Simske, S.J. and Arnabat, J. (2003) *User-Directed Analysis of Scanned Images*. ACM Symposium on Document Engineering, pp. 212–221.

Simske, S.J. and Baggs, S.C. (2004) *Digital Capture for Automated Scanner Workflows*. ACM Symposium on Document Engineering, pp. 171–177.

Simske, S.J., Li, D., and Aronoff, J.S. (2005) *A Statistical Method for Binary Classification of Images*. ACM Symposium on Document Engineering, pp. 127–129.

Simske, S.J., Sturgill, M.M., and Aronoff, J.S. (2010) Authentic versus counterfeit image classification after re-sampling and compression. *Journal of Imaging Science and Technology*, **54** (060404), 1–5.

Wahl, F.M., Wong, K.Y., and Casey, R.G. (1982) Block segmentation and text extraction in mixed/image documents. *Computer Vision Graphics and Image Processing*, **2**, 375–390.

Zramdini, A. and Ingold, R. (1993) Optical font recognition from projection profiles. *Electronic Publishing*, **6** (3), 249–260.

6

Introduction to Meta-algorithmics

Ideally a book would have no order to it, and the reader would have to discover his own.
—Mark Twain

6.1 Introduction

Mark Twain might suggest that the core set of meta-algorithmic patterns—the focus of this chapter—should be presented randomly, with the reader figuring out her own order to impose on them. However, in fleshing them out it has become apparent that there is actually a natural order to these, not necessarily on the complexity of the patterns, but on the way in which the patterns are constructed from their components. As such, they are presented as first-, second-, and third-order meta-algorithm patterns, building in complexity—and often on the previous patterns—as we progress through them. As described in Chapter 2, meta-algorithms are the pattern-driven means of combining two or more algorithms, classification engines, or other systems. They are powerful tools for any data scientist or architect of intelligent systems. Collectively, meta-algorithmics are called intelligence generators, and I will often allude to them as such, except where I explicitly call out algorithms, systems, or large engines such as optical character recognition (OCR) or automatic speech recognition (ASR) knowledge engines. Meta-algorithmic generators are designed to provide the means of combining two or more sources of knowledge generation even when, or especially when, the combined generators are known only at the level of black box (input and output only).

The first-order patterns are, naturally, the simplest. But they still provide nuanced, and often highly flexible, systems for improving—or optimizing—accuracy, robustness, performance, cost, and/or other factors of interest to the system architect. The first algorithm is the simplest of all—Sequential Try—and is used to illustrate the ready amenability of meta-algorithmic patterns to parallelism. In the case of Sequential Try, the parallelized equivalent is the Try pattern. These are therefore considered together. With the Constrained Substitute pattern, we explore the approach of obtaining "good enough" performance using meta-algorithmics. This is very important when system cost is a consideration, and also has implications for parallel processing. Next, Voting and its fraternal twin, Weighted Voting, are considered together. Voting approaches are eminently suited to parallel processing, and perhaps the most difficult

Meta-algorithmics: Patterns for Robust, Low-Cost, High-Quality Systems, First Edition. Steven J. Simske.
© 2013 John Wiley & Sons, Ltd. Published 2013 by John Wiley & Sons, Ltd.

part for the system architect is in determining the weights. Next, perhaps the most important of all meta-algorithmic patterns, namely Predictive Selection, is considered. This powerful pattern allows the system architect to assign different analysis approaches to different parallel pathways based on a—preferably highly reliable—prediction of how well, relatively, each pathway will perform on the task at hand. The first-order meta-algorithms conclude with another broadly applicable pattern: Tessellation and Recombination. This pattern requires the most domain expertise of any of the first-order meta-algorithmics, but also provides the greatest amount of reorganization of the information, with higher potential for emergent results and behavior, under certain conditions, more like that of a genetic algorithm.

The second-order meta-algorithmics, on the other hand, are distinguished from the first-order meta-algorithmics largely by the two or more stages usually required to perform the meta-algorithmic task. Here, powerful tools such as confusion matrices, output space transformation, expert-guided weighting, and thresholding of decision confidence are used to guide the best selection from a plurality of selections for the output. Second-order patterns begin with the confusion matrix, which is a succinct and convenient means of conveying both the absolute and specific system errors. The confusion matrix allows us to differentiate readily between precision and recall, both important measures of system accuracy, but quite different in terms of their value for meta-algorithmics. Weighted Confusion Matrix patterns allow us to evolve the overall system decisions as more data comes in, providing continual, adaptive machine learning. Next, output space transformation is introduced, which allows us to enforce more dynamic collaborative behavior amongst the individual generators, which in turn allows us to use the same generators and the *same meta-algorithmic pattern* to optimize the system for different factors; for example, accuracy or robustness or cost. The value of this transformation tool therefore resides in the high degree of flexibility it provides for the system designer. This transformation tool is incorporated in the Confusion Matrix with Output Transformation pattern.

Expert-guided decision-making underpins the Tessellation and Recombination with Expert Decisioner pattern. Specifically, domain expertise is internalized into the "recombination" step of the algorithm, dictating how aggregates of the data will be reconstructed from the primitives resulting from the tessellation step. The Predictive Selection with Secondary Engines pattern uses the engine with the highest precision for the predicted class unless it does not meet a certain criteria. If not, then it uses a secondary meta-algorithmic pattern—usually the Weighted Confusion Matrix pattern. The Single Engine with Required Precision pattern, in contrast, uses the best single meta-algorithmic engine if it meets a threshold level of precision. If not, in an analogy to the Sequential Try pattern, it tries the next best single meta-algorithmic pattern: if it has the required precision, it is selected. This continues until an engine with required precision for the particular classification is obtained.

Majority Voting or Weighted Confusion Matrix, and Majority Voting or Best Engine, are two second-order meta-algorithmic patterns based on the Voting pattern. These are relatively simple composite patterns that are particularly useful when a large number of meta-algorithmic engines are deployed, but do not reach a majority consensus. Another such "composite" pattern is the Best Engine with Differential Confidence or Second Best Engine pattern. This pattern comprises a minimized Sequential Try when and only when the engine with the highest precision for its proposed output also has a corresponding confidence value that is below a particular threshold. Lastly, the Best Engine with Absolute Confidence or Weighted Confusion Matrix provides another confidence-dependent composite pattern: here if the confidence is not acceptable, instead of selecting the best engine, the Weighted Confusion Matrix pattern is applied.

Like second-order meta-algorithmics, third-order meta-algorithmics generally combine two or more decision steps. Third-order meta-algorithmics are distinguished from the second-order patterns, however, in the tight coupling between the multiple steps in the algorithm. As such, the analysis tools—feedback, sensitivity analysis, regional optimization, and hybridization being the primary ones—tightly couple not only one step to the next but also connect the downstream steps back to the earlier steps. Nowhere is this more evident than in the first third-order meta-algorithmic pattern: the simple Feedback pattern, in which errors in the reconstructed information are immediately fed back to change the gain—for example, weights—on the final system.

A longer-viewed third-order meta-algorithmic pattern is the Proof by Task Completion pattern, which dynamically changes the weighting of the individual knowledge-generating algorithms, systems, or engines after tasks have successfully completed. This approach allows infinite scalability (new data does not change the complexity or storage needs of the meta-algorithmic pattern), and a variable level of flexibility, depending on how heavily weighted old and new are. This pattern illustrates well the fact that meta-algorithmic system design is both a craft and an art. The basic patterns provide a structural framework for the application of meta-algorithmics, which comprises the science of meta-algorithmic patterns. However, there is a lot of art involved in the form of the application of domain expertise. Applying specific domain expertise allows the designer to finalize the behavior of the system using a combination of experience (based on learned rules, constraints, and/or preferences), statistics collected as part of the processing, and/or forward-looking estimation of current and future system needs.

The Confusion Matrix for Feedback pattern incorporates the relationship between the intelligence generators elucidated by the confusion matrix. The feedback is therefore directed to the most salient *binary decisions* in the problem. The use of expert-driven rules and learned constraints is incorporated into the next third-order pattern, the Expert Feedback pattern. Then, gears are shifted as the Sensitivity Analysis pattern is introduced. This powerful pattern is focused on identifying stable points within the solution space among the top choices, which is mainly targeted at providing an optimally robust—rather than accurate—system. While the confusion matrix can definitely be mined for its stable areas, alternatively the sensitivity to weighting within the plurality of weighting-driven meta-algorithmic patterns may be considered. Finally, stable areas within the correlation matrix for algorithmic combination can be considered as part of this pattern.

The Regional Optimization pattern is another intricate pattern, focused on what could be considered "introspective meta-algorithmics," wherein individual engines are tuned for subclasses of the overall task. This pattern powerfully extends the Predictive Selection pattern in which different first- or second-order meta-algorithmic patterns (not just meta-algorithmic algorithms, systems, or engines) are selected based on which has the highest expected precision for the specific subclass of the overall problem space. Effectively, then, the Regional Optimization pattern could also be termed the Extended Predictive Selection pattern. The final meta-algorithmic pattern defined and elaborated in this chapter is the Generalized Hybridization pattern. This pattern is concerned with optimizing the combination and sequence of first- and second-order meta-algorithmic patterns for a given—generally large—problem space. In this sense, it shares the complexity of the Regional Optimization pattern: it compares more than one meta-algorithmic pattern for an optimally performing system.

For each of these 21 meta-algorithmic patterns, this chapter provides a system diagram; a table of "fast facts" of relevance to the system designer considering more than one meta-algorithmic approach for her task at hand; and a general discussion of the pattern as an introduction. The three subsequent chapters will provide more in-depth examples, analysis, and comparison of and among these 21 patterns.

6.2   First-Order Meta-algorithmics

First-order meta-algorithmics are characterized by their relative simplicity—most involve a single transformation of the multiple sources of data output from the meta-algorithmic intelligence generators to make a decision. But the simplicity belies the power and value of these algorithms. In addition, these patterns are disproportionately deployed. The Pareto principle holds for meta-algorithmics as it does for many other systems—the first-order meta-algorithmic patterns comprise roughly 20% of the patterns in this book, while comprising 80% of the meta-algorithmic deployments used in real systems. We begin our discussion of meta-algorithmics with the Sequential Try pattern.

6.2.1   Sequential Try

The Sequential Try pattern evaluates multiple knowledge generators in a specific order, continuing until a sufficient accuracy or other specification is obtained. It is one of the simpler design patterns to conceptualize—it comprises trying one algorithm at a time in a logically ordered fashion, for example, by highest likelihood of success, minimum licensing cost, best performance/throughput, and so on; measuring the output in terms of a specific quality metric, and continuing until expected quality is achieved. Generally, the quality metric will be a given confidence in the output, such as is typical for most classification and intelligence-generating algorithms, systems, and engines.

The basic procedure is simple, as illustrated in Figure 6.1. With this pattern, the system will try one knowledge generator at a time. The output of the generator is measured, and if satisfactory according to the output quality metric, then it is selected as the final output, with no further generators in the list executed. If, however, the output does not meet the required quality metric, then the next generator in the list is used to analyze the input data. The process is continued until either a satisfactory output is reached, or the list of generators is exhausted.

As illustrated in Figure 6.1, the Sequential Try pattern is 100% serial, and should not be made parallel. This is because it is usually deployed when the later generators are more expensive, slower, require more resources, or are otherwise less preferred than the generators tried earlier. The system facts and considerations for the Sequential Try pattern, affording the meta-algorithmic system architect with important information for deciding on whether to select this or an alternate pattern, are provided in Table 6.1.

Table 6.2 presents data for three meta-algorithmic generators. Table 6.3 illustrates how to choose the order of a Sequential Try set based only on the expected value (i.e., the mean) of the processing time and the probability of success. Note that in each case the overall probability of success, p_{overall}, is given by

$$p_{\text{overall}} = 1.0 - ((1.0 - 0.6) \times (1.0 - 0.4) \times (1.0 - 0.2)).$$

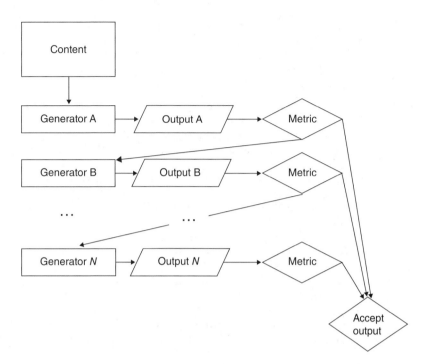

Figure 6.1 Schematic for the Sequential Try (first-order meta-algorithmic) pattern. In this pattern, the algorithms (or systems or engines) are performed in sequence until acceptable output is obtained. This pattern is 100% sequential (serial) and usually cannot be made parallel

Thus, $p_{\text{overall}} = 0.808$ (an 80.8% chance of success) irrespective of the order of the Sequential Try. But, if the order is {3,2,1}—that is, Generator 1 is tried first, Generator 2 is tried next, and Generator 3 is tried last—the mean time for processing is the lowest, at 5.80 s. Table 6.2 illustrates that this is due to the relative p/t ratios of the generators.

The parallel processing analog to the Sequential Try pattern is the Try pattern (Figure 6.2). In some applications—for example, document digitization, video processing, or other analysis of large, relatively uniform media—the Try pattern is used to test, in parallel, a small but representative sample of the overall content for selection of the best intelligence generator. Each of the candidate generators is tested on the representative sample set and the results are evaluated. The generator with the best performance is selected for the overall job.

In the more general case, the Try pattern is used to simultaneously process data with multiple generators and select the generator that provides the best overall output quality. Unlike the first case, this instantiation of the Try pattern is not meant to select a single generator for every sample: it can select a different optimum generator for each sample. This per sample "customization" of generator selection is seemingly inefficient, but since the Try pattern is fully parallel the selection processing time is limited only by the processing required for the slowest generator. As with the Sequential Try pattern, the output of a generator must match or surpass a prescribed metric for this parallel Try pattern. This general incarnation is effectively a "try several, keep the best" approach.

Table 6.1   Meta-algorithmic pattern system facts and considerations for the "Sequential Try" first-order meta-algorithmic pattern

Topic	System Facts and Considerations
Pattern name	Sequential Try
Related patterns	Try (direct parallel form), Single Engine with Required Precision (Sequential Try with precision being the deciding factor), Generalized Hybridization (Sequential Try of multiple meta-algorithmic patterns)
Input considerations	No special training required
Output considerations	Output should be generated that is compatible with the quality metric used to assess pass/fail
Decisioning	If the quality of the output is above the threshold level (required minimum quality), the process completes successfully
Scalability	Readily scalable, especially when the sequential algorithms are arranged, where possible, so that the generator(s) most likely to pass (see "statistical behavior" below) are tried first
Processing	Serial only. Slowest possible arrangement, but memory and storage footprints are only equal to those of the "worst" generator
Parallelism	None
Statistical behavior	For like generators (e.g., having the same cost), the task completes fastest when the generators are ordered based on the ratio of processing time to probability of success ratio (see Tables 6.2 and 6.3)
Repurposability	Due to its simplicity, the Sequential Try pattern is easily repurposed. It is directly repurposable for homologous data sets since they can share the quality metric
Domain expertise	Moderate to substantial domain expertise is required to create the quality metric. The rest of the system is relatively generic
Best used on	Images and other data types for which quality metrics are well known

Table 6.2   Probabilities of success (p), processing time (t), and the ratio of p/t for three generators. This is used to define the optimal ordering for a Sequential Try pattern

Factor	Generator 1	Generator 2	Generator 3
p (success)	0.6	0.4	0.2
t (processing time in s)	5.0	3.0	1.0
p/t ratio (in s^{-1})	0.120	0.133	0.200

Table 6.3   Mean processing time for the six different orderings of the generators presented in Table 6.2. As is shown here, the optimal ordering {3,2,1} corresponds to ordering by p/t ratio

Ordering	Expected Total Processing Time Formula	$E(t)$ (s)
{1,2,3}	$(0.6 \times 5.0 \text{ s}) + (0.16 \times 8.0 \text{ s}) + (0.24 \times 9.0 \text{ s})$	6.44
{1,3,2}	$(0.6 \times 5.0 \text{ s}) + (0.08 \times 6.0 \text{ s}) + (0.32 \times 9.0 \text{ s})$	6.36
{2,1,3}	$(0.4 \times 3.0 \text{ s}) + (0.36 \times 8.0 \text{ s}) + (0.24 \times 9.0 \text{ s})$	6.24
{2,3,1}	$(0.4 \times 3.0 \text{ s}) + (0.12 \times 4.0 \text{ s}) + (0.48 \times 9.0 \text{ s})$	6.00
{3,1,2}	$(0.2 \times 1.0 \text{ s}) + (0.48 \times 6.0 \text{ s}) + (0.32 \times 9.0 \text{ s})$	5.96
{3,2,1}	$(0.2 \times 1.0 \text{ s}) + (0.32 \times 4.0 \text{ s}) + (0.48 \times 9.0 \text{ s})$	5.80

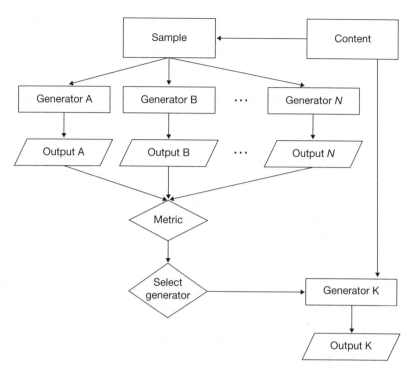

Figure 6.2 Schematic for the Try (first-order meta-algorithmic) pattern. The "Try" pattern is a parallel—that is, simultaneous attempt of all possible "tries"—instantiation of the Sequential Try pattern

The system facts and considerations for the Try pattern are given in Table 6.4. The primary difference from the related Sequential Try pattern is the full parallelism by design of the Try pattern.

6.2.2 Constrained Substitute

The second first-order meta-algorithmic pattern is the Constrained Substitute pattern, which allows the choice of a suitable reduced-expense (in terms of cost, processing time, bandwidth, a combination thereof, or other metric) algorithm, engine, or system to—effectively—replace a higher-expense approach. The Constrained Substitute pattern is applicable when there are a set of competing intelligence generators that are each, independently, capable of performing a specific task. In addition, some of these generators are preferable to others based on constraints such as resources (processing, storage, etc.), licensing and other costs, and/or digital rights.

One of the principal tenets of meta-algorithmic systems is that relatively simple "lightweight" systems, alone or in combination with other systems, can perform a function with similar accuracy, robustness, and so on, as a more expensive and/or extensive "heavyweight" system. The "lightweight" system consumes less system resources (memory or execution cycles) or provides a simplified analysis for a different purpose. At run-time, if the system is not constrained, then the heavyweight system is used. However, if the one or more constraints hold

Table 6.4 Meta-algorithmic pattern system facts and considerations for the "Try" first-order meta-algorithmic pattern

Topic	System Facts and Considerations
Pattern name	Try
Related patterns	Sequential Try (serial form of "Try"), Single Engine with Required Precision (selects best engine based on precision)
Input considerations	No special training required
Output considerations	Output should be generated that is compatible with the quality metric used to assess pass/fail
Decisioning	If the quality of the output of at least one generator is above the threshold level (required minimum quality), the process completes successfully (and the output with highest quality is selected)
Scalability	Readily scalable, requires only an additional parallel processor for each additional generator tried
Processing	Fully parallel. Fastest possible processing, assuming processor availability. Maximum memory and storage requirements
Parallelism	Innate
Statistical behavior	After several iterations, if the same generator continually is chosen, then the Try pattern is discontinued and the single best generator is used thereafter
Repurposability	Due to its simplicity, the Try pattern is easily repurposed. It is directly repurposable for homologous data sets since they can share the quality metric
Domain expertise	Moderate to substantial domain expertise is required to create the quality metric. The rest of the system is relatively generic
Best used on	Images and other data types for which quality metrics are well known

true, then the "lightweight" generator is used to provide the overall system output. Generally, the output of the lightweight generator will have been shown to historically be correlated with one or more of the heavyweight generators, and so reasonably be deployed as a substitute for the original generator.

The heavyweight generators can perform their function in a complex yet accurate manner. In this discussion, they are designated "full generators" (FGs). We also have one or more lightweight generators. A lightweight generator performs a similar function and consumes fewer resources or time. We must have a lightweight generator that performs the same function (e.g., OCR, ASR, or image understanding) as the FG even though they differ in the technique and the utilization of the resources. A lightweight generator, herein termed a partial generator (PG), reduces run-time performance penalties (memory, speed, etc.) of the FG. This can be useful in the cases where you have similar competing FG systems that do one function and a partial algorithm, or PG, that does the same thing, but more quickly, without a licensing fee, and so on. In general, the PG is less accurate than the FG.

To use a PG instead of FG, one has to make sure that the PG correlates with one of the full algorithms and at the same time correlates with the ground truth data for a set of test data. If such correlation fails, then the use of partial algorithms is not possible. There are, therefore, two phases for applying the substitute pattern: the correlation (or "training") phase and the run-time (or "deployment") phase, as illustrated in Figure 6.3.

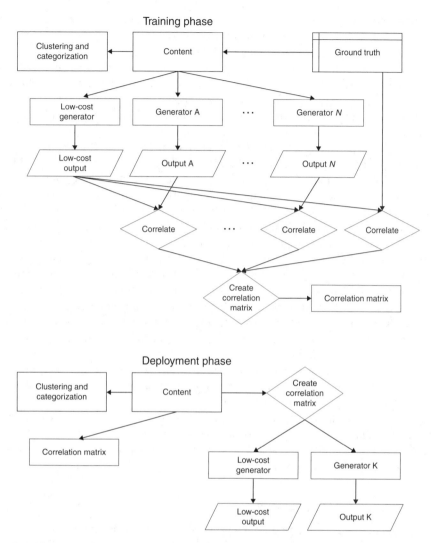

Figure 6.3 Constrained Substitute (first-order meta-algorithmic pattern). The schematic for the correlation or "training" phase is shown at the top, and the schematic for the run-time or "deployment" phase is shown at the bottom

In the *correlation phase*, we determine whether a PG can be used as a substitute for the original FG. The PGs and the FGs are executed against a set of ground truth samples. The output from the application of the PG and the FG is then used to statistically obtain the following correlations:

1. $C(p,f)$ correlation between the results of the FGs and PGs
2. $C(f,GT)$ the correlation between the results of the full algorithm and the ground truth data
3. $C(p,GT)$ the correlation between the results of the partial algorithm and the ground truth data

The output of training gives the answers to the questions: (1) can we use a PG instead of the full competing generators, and (2) which FG is highly correlated with the PG? Figure 6.3 (upper diagram) provides a schematic diagram for the correlation or "training" phase of the Constrained Substitute pattern. The output of training is the clusters or categories of input training—which can be used to define input data categories or clusters, each of which may have a distinct $C(p,f)$, $C(f,GT)$, and $C(p,GT)$ calculation. This concept is compatible with the Predictive Selection pattern to be described in Section 6.2.4, where it will be discussed in greater depth.

In the *deployment phase*, the PG is used instead of the FG to which it is highly correlated. There are two alternatives to consider as output of the pattern. First, one can consider the output of the PG as the output of the system (or the output of the FG). Though this is not true, it is a good approximation if the PG and the FG are highly correlated. Second, one can prorate (transform) the output of the PG to obtain an estimate of the actual output if a transformation function can be used. The transformation function (and the necessary parameters) is dependent on the nature of the generator (might require training).

As an example of FGs and PGs, consider document segmentation algorithms, informally referred to as "zoning engines." The FG is a full-featured zoning generator that creates regions with fully polygonal boundaries. The PG is a Manhattan-layout-based zoning generator that creates regions that are only rectangular in shape. Similarly, for an OCR generator, an FG could be an OCR generator that recognizes the text of various languages and a PG could be an OCR engine that only detects English text. The Constrained Substitute pattern can also be used where the results are presented to a user. In this case, it might be useful to use the PG to present a quick analysis of the results to the user, while the FG is actually performing the intensive work behind the scenes. Once the FG has completed its analysis, its results can be presented to the user, replacing the PG's results.

The Constrained Substitute pattern is optimal for performance and for run-time utilization of resources; however, it may be difficult to implement, can be impractical for some generators, and requires a lot of data to train. Table 6.5 summarizes the key facts and considerations for the Constrained Substitute pattern.

6.2.3 Voting and Weighted Voting

The third of the first-order meta-algorithmic patterns is the Voting pattern, included here with its often more powerful variant, the Weighted Voting pattern. This pattern is the first to include the output of multiple algorithms, services, or systems in the final output, rather than simply selecting the best knowledge generator.

The Voting pattern is one of the simplest of the meta-algorithmic patterns. Voting is usually performed on the most atomic level possible. Each of the meta-algorithmic intelligence generators is run against the content. For OCR, this means voting occurs on the individual characters. For ASR, this means that the voting occurs on the individual phonemes or words.

The vote itself incorporates the output of two or more intelligence generators. Both majority and plurality voting can be addressed. Past work (Lin *et al.*, 2003) has shown that when unweighted voting occurs, in some cases the addition of more algorithms actually prevents a consensus—typically, three engines works well for "strict voting," and even numbers of voting generators perform more poorly. Plurality voting can also provide a good balance of accuracy with less misclassifications than majority voting when the number of generators gets large.

Table 6.5 Meta-algorithmic pattern system facts and considerations for the "Constrained Substitute" first-order meta-algorithmic pattern

Topic	System Facts and Considerations
Pattern name	Constrained Substitute
Related patterns	Proof by Task Completion (can be used to generate a correlation matrix when similar tasks are completed by different generators)
Input considerations	The task may have to be partially classified to know which substitutions can be used
Output considerations	No special requirements except that the substitute generators must produce the same type of output as the heavyweight generators
Decisioning	If the correlation of the PG and FG is high, the PG can substitute for the FG, saving system resources
Scalability	The correlation data scales as N^2, meaning the amount of training increases geometrically as more generators are added to the system
Processing	Same as the generator chosen during run-time. Training processing needs are high
Parallelism	Training phase fully parallel. Run-time phase generally requires only a single generator, so that parallelism does not apply per se
Statistical behavior	Accuracy of substitution improves (both mean and variance of the correlation) as more training data is added
Repurposability	The correlation data may be reused for other predictive meta-algorithmic approaches
Domain expertise	No special domain knowledge required
Best used on	Large, relatively uniform data sets

The simple Voting pattern (Figure 6.4, top), however, has some unsatisfying limitations. Each of its primary incarnations—namely, majority and plurality voting—assigns the same weight to each of the Voting patterns. But this is rarely appropriate. Even when such a uniform weighting seems appropriate, it may only be due to an incomplete understanding of the problem at hand. That is, the problem space likely needs to be further discretized such that the weights can be assigned to subspaces of the original problem space. Performing this subspace analysis leads naturally to the Predictive Selection first-order meta-algorithmic pattern, described in the next section.

Even without Predictive Selection, however, there is often the possibility to improve performance of the system using Weighted Voting. The Weighted Voting form of the Voting pattern (see Figure 6.4, bottom diagram) results in a different relative confidence coefficient for each meta-algorithmic intelligence generator. In a previous article (Lin *et al.*, 2003), we showed that for a simplistic classification problem—wherein there are N_{classes} number of classes, to which the a priori probability of assigning a sample is equal, and wherein there are $N_{\text{classifiers}}$ number of classifiers, each with its own accuracy in classification of p_j, where $j = 1, \ldots, N_{\text{classifiers}}$—the following classifier weights are expected:

$$W_j = \ln \left(\frac{1}{N_{\text{classes}}} \right) + \ln \left(\frac{p_j}{e_j} \right),$$

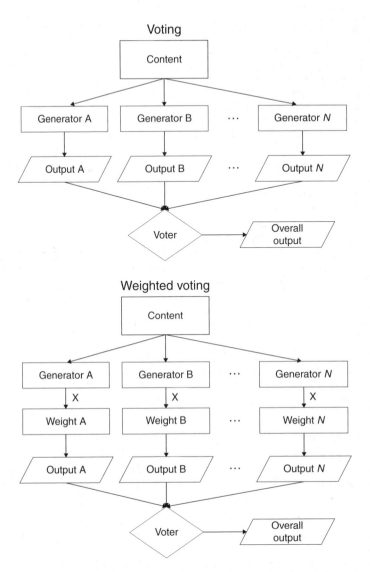

Figure 6.4 Voting and Weighted Voting (first-order meta-algorithmic pattern). The schematic for the Voting pattern is shown at the top, while the schematic for the Weighted Voting pattern is shown at the bottom. The "voter" decision block tallies the votes and decides on the winner of the voting

where the weight of Classifier j is W_j and where the term e_j is given by

$$e_j = \frac{1 - p_j}{N_{\text{classifiers}} - 1}.$$

However, note that in Lin *et al.* (2003), the optimal values are mistakenly given for a system in which there are only two classes (not five as reported erroneously).

Table 6.6 Weighted Voting weights as predicted, separately, by Lin *et al.* (2003), by the inverse of the error rate, and by the square of the accuracy for the 5-class, 3-classifier problem described in the text

Information	Classifier 1	Classifier 2	Classifier 3
Accuracy (*p*)	0.99	0.80	0.90
Weights (Lin *et al.*, 2003)	0.678	0.087	0.236
Weights [1/error]	0.870	0.043	0.087
Weights [accuracy²]	0.404	0.263	0.333
Weights [hybrid]	0.637	0.153	0.210
Weights [1/sqrt(error)]	0.649	0.145	0.205

The weights of a particular classification problem are given in Table 6.6. In this example, the meta-algorithmic intelligence generators are classifiers, and the three classifiers have accuracies $p_1 = 0.99$, $p_2 = 0.8$, and $p_3 = 0.9$. Here, we consider a 5-class problem for which the weights are

$$W_1 = \ln(0.2) + \ln(0.99/(0.01/2)) = -1.609 + 5.288 = 3.679;$$
$$W_2 = \ln(0.2) + \ln(0.8/(0.2/2)) = -1.609 + 2.079 = 0.470;$$
$$W_3 = \ln(0.2) + \ln(0.9/(0.1/2)) = -1.609 + 2.890 = 1.281.$$

Normalizing these weights so that they sum to 1.0, we obtain the $\{W_1, W_2, W_3\}$ values of $\{0.678, 0.087, 0.236\}$ shown in Table 6.6.

Five other weighting schemes are also included in Table 6.6. When the weights are proportional to the inverse of the error, then the weight for Classifier *j* is given by

$$W_j = \frac{1.0/(1.0 - p_j)}{\sum\limits_{i=1}^{N_{classifiers}} 1.0/(1.0 - p_i)}.$$

From this equation, $W_1 = 0.870$, $W_2 = 0.043$, and $W_3 = 0.087$. The weights derived from the inverse-error proportionality approach are already normalized—that is, sum to 1.0—by design.

The next weighting scheme shown in Table 6.6 is one based on proportionality to accuracy squared. The associated weights are described by the following simple equation:

$$W_j = \frac{p_j^2}{\sum\limits_{i=1}^{N_{classifiers}} p_i^2}.$$

Looking at Table 6.6, it is clear that these two methods—proportionality to inverse error and accuracy squared—differ considerably in how they weight the individual classifiers. The inverse-error-based method heavily favors the more accurate classifiers in comparison to the "optimal" weighting of Lin *et al.* (2003), while the accuracy-squared-based method favors the less accurate classifiers in comparison to the "optimal" weighting. This implies that a hybrid method, taking the mean weighting of these two methods, may provide performance closer to the optimum method. This is definitely the case, as shown by the set of weights $\{W_1, W_2, W_3\}$ of $\{0.637, 0.153, 0.210\}$ shown in Table 6.6. The generalized hybrid scheme is given by the following equation:

$$W_j = C_1 \frac{1.0/(1.0 - p_j)}{\sum_{i=1}^{N_{\text{classifiers}}} 1.0/(1.0 - p_i)} + C_2 \frac{p_j^2}{\sum_{i=1}^{N_{\text{classifiers}}} p_i^2},$$

where $C_1 + C_2 = 1.0$. In Table 6.6, $C_1 = C_2 = 0.5$. Clearly, varying these coefficients allows the system designer to tune the output for different considerations—accuracy, robustness, lack of false positives for a given class, and so on.

The final weighting approach given in Table 6.6 is one based on the inverse of the square root of the error, for which the weights are defined:

$$W_j = \frac{1.0/\sqrt{1.0 - p_j}}{\sum_{i=1}^{N_{\text{classifiers}}} 1.0/\sqrt{1.0 - p_i}}.$$

The behavior of this weighting approach is similar to the hybrid method and not greatly dissimilar from that of the optimal method.

So, if there is an "optimal" method, then why would any other weighting approach be selected? One reason is insufficient training data to be confident about the real mean of the value for accuracy. For example, $p_1 = 0.99$ in the example. If the real value of p_1 should actually prove to be 0.98, then the optimal method (and certainly the inverse-error method) of computing the weights would result in $\{W_1, W_2, W_3\}$ values of $\{0.630, 0.099, 0.271\}$, with absolute value differences of $\{0.048, 0.012, 0.035\}$ from the original weighting set. On the other hand, this same change in p_1 only changes the weights for the accuracy-squared approach to $\{0.398, 0.266, 0.336\}$, with absolute value differences of only $\{0.006, 0.003, 0.003\}$. Using this as an example of *sensitivity analysis*, for this change in p_1, it is clear that the optimum approach is eight times as sensitive as the accuracy-squared approach. Thus, the estimates provided by the accuracy-squared method are more robust to changes in the estimation of accuracy (or, concomitantly, error).

Another reason to use an alternate weighting approach is to create a system more robust to changes in the class attributes. Different classes may be more frequently encountered over time, changing both the a priori classification assumptions and the relative accuracy of the classifiers. In addition, the attributes of the classes themselves may drift over time, which can significantly change the relative accuracy of the different classifiers. Thus, methods more robust to these types of changes—such as the hybrid method provided in Table 6.6—may be deployed so that the system accuracy does not significantly change with such changes in input.

Table 6.7 Meta-algorithmic pattern system facts and considerations for the "Voting"/"Weighted Voting" first-order meta-algorithmic pattern

Topic	System Facts and Considerations
Pattern name	Voting (and Weighted Voting)
Related patterns	Majority Voting or Weighted Confusion Matrix, Majority Voting or Best Engine (both of these use a binary voting scheme)
Input considerations	Accuracy of the individual algorithms, systems, or engines needs to be determined with relatively low variance, especially for the optimum (Lin *et al.*, 2003) or inverse-error-based approaches
Output considerations	Output is very simple: either simple counting (Voting pattern) or linear combinations of the generators (Weighted Voting pattern)
Decisioning	The outcome receiving the greatest number of votes is the output. With all forms of voting, though, the outcomes can be ranked in order
Scalability	Eminently scalable, with only linear increases in processing time
Processing	All generators contributing to the vote can be performed in parallel
Parallelism	Generators can be run 100% parallel; voting tally and decision is performed in a single (serial) thread
Statistical behavior	Multiple statistic models can be used for the weighting
Repurposability	Can be used also to test for system robustness using sensitivity analysis approaches
Domain expertise	Very little required, since only output accuracy is required
Best used on	Systems with a large number of intelligence generators, especially systems where the intelligence generators have largely different accuracy

The system facts and considerations for the Voting and Weighted Voting pattern are provided in Table 6.7. Like the Try pattern, the Voting-based pattern is usually associated with parallel processing of all the salient knowledge generators simultaneously.

6.2.4 Predictive Selection

The fourth pattern, Predictive Selection, is quite powerful—at least among the meta-algorithmic patterns that result in the selection of a single intelligence generator from a plurality of them—and usually involves choosing the information generator that has the highest precision in a specific predictor test. Figure 6.5 outlines the two main components of the Predictive Selection pattern: the statistical learning (training) phase of the pattern (top) and the run-time phase of the pattern (bottom). Combined, these two processes allow the Predictive Selection pattern to identify the generator that reports the highest confidence in its output for the specific sample. This depends on the generators' performance on training data belonging to the same category as the specific sample being processed. Since there is a high degree of confidence that the output of the generator will correlate well with the historical data, the generator that reports the most confidence in its output is usually the best choice for run-time input. New input samples must, therefore, be classified by their "category" and the generators are ranked according to their historical (training) performance on each category. The Predictive Selection pattern will therefore select the generator to use based on the type of input and will only execute the one selected generator (algorithm, system, or engine).

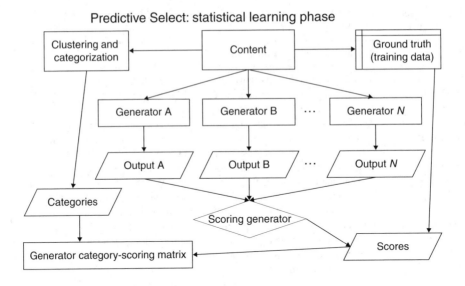

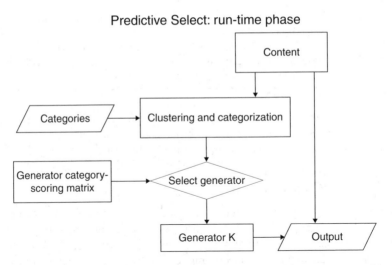

Figure 6.5 Predictive Selection (first-order meta-algorithmic pattern). The statistical learning system diagram is shown at the top, and the run-time system diagram is shown at the bottom

The selection of the best generator for a specific type of input is based on finding an algorithm—usually a simple one—for assigning an input sample to a specific category (subset of the input space), and then selecting the best generator based on which generator provides the highest accuracy for that category. Obviously, some statistical data gathering must occur first. This training set data is used to define classification categories and for each category rank the generators based on their performance. This performance data is collected into a generator-ranking matrix. This matrix allows us to accurately assign the input to a category (or subset), called the *categorization* process. Thus, when a new input is to be processed, its category is first determined, and then the best performing meta-algorithmic generator for that category is selected for execution. This is, in practice, relatively easy to accomplish. The training data is used to determine the precision of each generator for each category. During the run-time phase, the generator with the highest precision for its reported category is selected as correct.

Note that in many cases, the categories are relevant not only as the means of comparing predictive ability of the different generators but also may be the output of the generator. That is, the categories are the same classes that the generators are supposed to accurately assign input to. This does not have to be the case, but it often is. This means that all we need do is compare the precision of the generator for the output, and select the output classification as the one reported by the generator with the highest precision. In other cases, the categories are distinct from the output; for example, the categories may be black and white image or color image, and the classes output by the generators are different image objects. In this case, a best pairing of {category, generator output} is the accepted output.

The Predictive Selection pattern, therefore, is heavily dependent on the existence of sufficient ground truth, or "training," data to define each of the categories, or "classes," which are later used to assign the generator. The categories selected, when they are distinct from the output classification, are generally comprised of one or more—sometimes many more—clusters, depending on the following factors:

1. Central tendency of the data
2. Ratio of clusters/categories
3. Total amount of training data available
4. Differential predictive capabilities of the categories

Central tendency of the data is an important factor because the better the separation between the categories, the better the predictive value of the *generator category-scoring matrix* (as shown in both diagrams in Figure 6.5). The F-score, as usual, is the key metric defining this central tendency, and it is the ratio of the mean-squared error within the clusters to the mean-squared error of the cluster means. The more Gaussian the input clusters, the more reliable the F-scores. The ratio of clusters to categories is another important factor in generating sufficient ground truth for the prediction. In general, the number of clusters/categories should be substantially less than the number of samples/cluster or the predictive behavior of the training data will have very limited statistical power. The third factor, total amount of training data available, seems at first an easy one: the more, the better. However, this is not necessarily true if in collecting more, multiple behaviors are being modeled as a single behavior. As an example, if there is some systematic drift to the behavior of the training data over time or location or with the device(s) used to collect the data, and the differences are not compensated for (e.g., through calibration and/or normalization), then multiple categories may be arbitrarily lumped together. This will increase overlap between what would otherwise be the true distinctions between categories, with obvious impact on the F-score and predictive power of the categorization. The fourth factor to be considered in the definition of the categories is the differential predictive capabilities of the categories. If all of the categories result in similar relative predictive capabilities—that is, the relative ranking of the generators is similar for each of the categories—then the predictive value of the categorization is of limited value. In such a case, it would be better to explore a new metric for prediction.

Given these considerations, it is clear that the overall utility of this pattern is entirely dependent on the quality of the prediction—that is, the definition of the categories. This definition is associated with the *statistical learning phase*. As outlined above, this is the phase of the pattern required to assess which generators work better for different types of input data. A set of ground truth samples is required as input to the training phase. The output of this phase is a data structure that we can call the generator-ranking, or generator category-scoring, matrix. A "category" is a set of inputs that possess some common characteristics (metrics associated with the input data type). The ranking provided is the set of generators in order of their correlation, or accuracy, with the ground-truthed data of the training set. Figure 6.5 (top diagram) illustrates the schematic diagram for the statistical learning (training) phase. A set of ground truth samples with at least one sample (but generally far more, as discussed above) in each category is assumed available. These input samples are then processed by a classification algorithm to assign each to a specific category. These samples are also processed by each of the generators. The output from each generator is then assessed (compared) to the ground truth data using a scoring algorithm to determine scores (or correlation factors) for each sample. The scores and the categories are then processed to produce the algorithm-ranking matrix.

During the *run-time (deployment) phase*, the input samples will be processed to determine the type (category) they best match. The best performing—for example, highest accuracy—generator for that category will then be selected by looking up the input category in the generator category-scoring matrix. The output of this deployment phase, therefore, is simply the output of the selected generator.

The Predictive Selection pattern (Table 6.8) is certainly not new to machine learning. If the categories for prediction are different from the classes for the output of the selection process, then the approach is similar to that of decision-tree-based classification. That is, the categories comprise the first decision and the classes comprise the second decision. If the categories for prediction are the same as the classes for the output of the selection process, however, the approach is more closely related to the confusion matrix meta-algorithmic approaches, as will be described in Section 6.3 and Chapter 8.

6.2.5 Tessellation and Recombination

The fifth and final first-order meta-algorithmic pattern, Tessellation and Recombination, is shown to be especially useful for creating correct results even when none of the individual generators produces a correct result—a process called emergence. This process is perhaps best introduced through analogy. Suppose three card players have poker hands (five cards each). Each of them offers to give up one or more cards from their hand and pools their discards with the other two players. Then, each player selects from the discard pile the cards they most want, or just selects them randomly, potentially changing their hand for the better. A player with two aces, two kings, and a queen, for example, may discard the queen and select a king from the common discard, changing the player's hand from "two pairs" to a "full house"; in other words, substantially improving the output.

There are, therefore, two primary parts of the pattern: the tessellation reduces the input space to atomic units (separates cards from the five card hand, in our analogy), the smallest identifiable units from the combined output of the set of generators. The output from each generator is analyzed and decomposed (split) into a set of basic (atomic) primitives. As a result,

Table 6.8 Meta-algorithmic pattern system facts and considerations for the "Predictive Selection" first-order meta-algorithmic pattern

Topic	System Facts and Considerations
Pattern name	Predictive Selection
Related patterns	Confusion Matrix and all Confusion Matrix-incorporating patterns; Predictive Selection with Secondary Engines and other hybridized Predictive Selection patterns; Regional Optimization (Extended Predictive Selection) and other extended (e.g., recursive) Predictive-Selection-based patterns
Input considerations	During run-time phase, categories and algorithm category-scoring information must be loaded along with the content
Output considerations	Output is straightforward since only a single generator is selected
Decisioning	As opposed to the "Voting" and "Tessellation and Recombination" patterns, the "Predictive Select" pattern is aimed at using only one generator's output for each sample to be analyzed
Scalability	Scalability is linear with the number of generators during the training phase, and limited only by the performance of the generator chosen during the run-time phase. Under most conditions, no additional ground truthing is needed when new generators are introduced
Processing	Since only one generator is used during the run-time phase, the processing time is minimized, but it is also more likely to give errors in the output
Parallelism	Whatever parallelism can be garnered must come from the parallelism integral to the generator selected. In terms of processing the algorithm, system, or engine, the process has no innate parallelism
Statistical behavior	The scoring algorithm selects the best generator to use. Optimum scoring algorithms provide different relative rankings among the generators for different categories
Repurposability	The Predictive Selection algorithm is the front end to a number of composite, or hybrid, meta-algorithm patterns, and so is highly repurposable
Domain expertise	Considerable domain expertise is often required to design the predictive scoring metric(s) and approach
Best used on	Problems involving a plurality of quite different generators, one or more of which is useful on a wide variety of inputs

we obtain a set of data primitives larger than the set produced by an individual algorithm. For instance, if two segmentation algorithms give two different assessments for a region, the primitive (atomic) regions in this case will be: the common (overlapped) subregions, additional subregions from segmenter 1, and additional subregions from segmenter 2. Hence, we end up with a number of regions greater than that produced by either individual segmentation algorithm.

The second part of the pattern is the recombination, where a merging algorithm considers the fully tessellated primitive output data and merges primitive outputs into larger-grained output. This would be selecting the desired king from the discard pile and creating a better five card hand, the larger-grained output in our analogy. One challenge in implementing the Tessellation and Recombination pattern is in defining the atomic elements produced by the tessellation, and the clusters produced by the recombination. These intermediate data heavily depend on

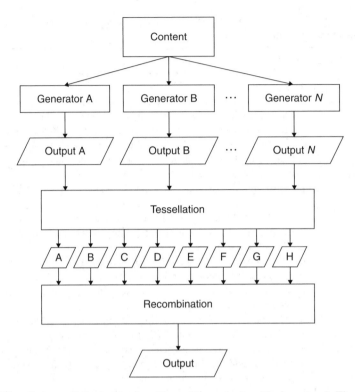

Figure 6.6 Tessellation and Recombination (first-order meta-algorithmic pattern). The differences in the outputs (i.e., the atomic elements created) of the individual algorithms (or other generator) are used to create the tessellations

the type of the function provided by the algorithms. Therefore, it is most likely that domain knowledge will be needed for implementing this pattern. The technique by which the atomic elements are merged is encapsulated in the recombination algorithm. Some of those atoms represent the common output (agreed upon) by all algorithms. Other atoms could be produced by one algorithm and an alternative atom is produced by another algorithm. For instance, in a document analysis system, a segmented region could be detected as "photo" by one algorithm and "drawing" by another algorithm. Specific examples are provided in Chapter 7.

As in the Voting pattern, each of the algorithms in the set is run against the content. The recombination step can involve alignment when algorithms produce different output (deletion, insertion, substitution). Also similar to the Voting pattern, the Tessellation and Recombination pattern has high processing requirements at run-time, since all of the generators will be executed. Unlike the Voting pattern, however, the Tessellation and Recombination pattern requires the additional processing associated with forming and merging atomic elements after the tessellation (Figure 6.6).

The system facts and considerations for the Tessellation and Recombination pattern are given in Table 6.9. It should be noted that the Tessellation and Recombination pattern provides no innate parallelism, and requires the highest relative processing of any first-order meta-algorithmic pattern. Adding to this, there is a large amount of domain knowledge

Table 6.9 Meta-algorithmic pattern system facts and considerations for the "Tessellation and Recombination" first-order meta-algorithmic pattern

Topic	System Facts and Considerations
Pattern name	Tessellation and Recombination
Related patterns	Tessellation and Recombination with Expert Decisioner (expert rules guide the recombination approach)
Input considerations	No special requirements
Output considerations	The rules for recombination for the basic pattern are generally based on voting (otherwise, please see the addition of the expert decisioner in Section 6.3.3)
Decisioning	Domain knowledge is generally required for the recombination portion of the algorithm
Scalability	In theory, the addition of another generator could result in significant additional tessellation. In practice, there are a diminishing number of new elements created as more generators are added due to the generally high degree of correlation for the outputs of the plurality of generators
Processing	Since each generator is used in creating the tessellation, processing costs increase as the number of generators increases
Parallelism	Parallelism occurs at the generator level. To the extent that the generators themselves are crafted to be processed in parallel, further parallel processing can be used
Statistical behavior	The recombination can be guided by domain-specific statistics
Repurposability	In general, the rules for tessellation and especially recombination are specific to the domain, and not highly repurposable
Domain expertise	The tessellation generally requires little domain expertise. The recombination, however, often requires a considerable amount of domain expertise
Best used on	Complicated problem spaces, especially those in which the number of possible outputs is similar to or greater than the number of generators. Like the Confusion Matrix pattern, it can also be used to create results that none of the individual generators list as their highest ranked result—a process called emergence

required for the tessellation and recombination, making this the most complicated of the first-order patterns. And, with this, we move on to the second-order meta-algorithmic patterns.

6.3   Second-Order Meta-algorithmics

More complicated patterns comprise the *second-order meta-algorithms*. A new set of analysis tools—namely, output space transformation, confusion matrices, and expert decisioners—is required for these second-order meta-algorithmic patterns. In addition, second-order patterns can represent conditional combinations of two simpler, first-order, patterns.

6.3.1   *Confusion Matrix and Weighted Confusion Matrix*

The first of the second-order meta-algorithmic patterns is the Confusion Matrix pattern (Simske, Wright, and Sturgill, 2006), which is in some ways a combination—and

Table 6.10 Sample confusion matrix with three classes. 94% of the samples belonging to class A are classified correctly; 85% of the samples belonging to class B are classified correctly; and 88% of the samples belonging to class C are classified correctly. The off-diagonal elements indicate the misclassifications, of which roughly half are misclassified as belonging to class A

Normalized Confusion Matrix		Classifier Output (Computed Classification) Prediction		
		A	B	C
True class of the samples (input)	A	0.94	0.03	0.03
	B	0.08	0.85	0.07
	C	0.08	0.04	0.88

extension—of the Voting and the Predictive Selection patterns. Its variant, the Weighted Confusion Matrix pattern, is more generally applicable to the combinatorial second-order patterns, where the different generators have different confidence values (weights).

In spite of its name, this pattern uses individual instances (events) that, when accumulated, generate a confusion matrix. A confusion matrix is a two-dimensional matrix wherein each column contains the samples of the classifier output (computed classification) and wherein each row contains the samples in the true class. The confusion matrix is also known as the error matrix or the contingency table (Stehman, 1997).

Table 6.10 provides an example of simple confusion matrix. The matrix is normalized so that each row sums to 1.0, corresponding with 100% of the occurrences of the specific class of the true samples. The confusion matrix is very useful for diagnosing the types of errors that occur. In Table 6.10, for example, the overall classification accuracy is 89% (the mean of the diagonals of the matrix), and 8% of true samples of both classes B and C are identified—that is, misclassified—as belonging to class A.

The Confusion Matrix pattern, and many of the subsequent complex meta-algorithmic patterns, are based, at least in part, on the output probabilities matrix (OPM), in which each of the classifiers (a generator used to produce class membership as its output) reports its estimate of confidence (usually, but not always, as a probability) for classification for each possible class. A simple OPM example, where there are three classes of input and four classifiers, and a single test sample from class C is analyzed, is given in Table 6.11. In this table, the confidence values, or weights, are indeed probabilities. For some generators, however, only a relative ranking of outputs is provided, and so the confidence values must be manufactured to sum to 1.0, and will not usually directly relate to probabilities.

In the example of Table 6.11, class C is the correct classification. However, Classifier 1 identifies class B as the classification for the sample considered, since the confidence weighting for the sample (0.51) is higher than for class A (at 0.08) or class C (at 0.41). Similarly, Classifiers 2, 3, and 4 identify class A, class A, and class B, respectively, as the classification for the sample considered. In the example, the overall classification made by the combination of all the classifiers is for class C, since the sum of partial probabilities for class C (1.52) is greater than for class A or class B. This is an example of an emergent result, wherein the output of the OPM—the enabling technology of the Confusion Matrix pattern—provides the correct result (class C) even though none of the individual classifiers were correct.

Table 6.11 Example of the output probabilities matrix (OPM) used for the decision in the "Confusion Matrix" second-order meta-algorithmic pattern, operating on a single sample (which actually is from class C). While Classifiers 1 and 4 identify class B as the correct class, Classifiers 2 and 3 identify class A as the correct class. Interestingly, the combination of the four classifiers identifies Class C as the correct class (the sum for class C is the highest, at 1.52)

		Classifier				
Output Probabilities Matrix		1	2	3	4	Sum
Classifier confidence	A	0.08	0.48	0.44	0.11	1.11
(usually	B	0.51	0.13	0.24	0.49	1.37
probability) output	C	0.41	0.39	0.32	0.40	1.52
for class						

The output of the individual sample OPM (specifically, the "sum" column, furthest right in Table 6.11) can be used as one input in the generation of a full confusion matrix. Since the correct class is class C, the sums are entered as follows into the "class C" row of a confusion matrix such as that shown in Table 6.12.

Two more samples—one each from class A and class B—are analyzed using the OPM in Tables 6.13 and 6.14. In Table 6.13, the sample is correctly classified as belonging to class A, while in Table 6.14 the sample is incorrectly identified as belonging to class C. The output of these three samples—Tables 6.11, 6.13, and 6.14—are collected in the confusion matrix of Table 6.15.

The information in Table 6.15 shows how a confusion matrix is populated by multiple OPM. The accuracy (mean of the diagonal elements) is 41%, even though two of the three classifications were correct. Table 6.16 is the digital summarization of Table 6.15. Here the accuracy is 67% but none of the probabilities are retained. Either approach—that of Table 6.15 or 6.16—is a legitimate means of converting the OPM into a confusion matrix.

Having provided this introduction to the OPM and showing its relationship with the confusion matrix, we turn to its use in meta-algorithmics. Figure 6.7 outlines how it is deployed. When new content to be classified is obtained, each classifier produces a ranked set of output with associated probability values normalized to 1.0. These probability values represent the classifier output (computed classification) values provided in Tables 6.11, 6.13, and 6.14. Ideally, they represent the relative confidence the classifier has in each class. If not, they can

Table 6.12 Confusion matrix with one data set, the output of Table 6.11, entered. The entries—1.11, 1.37, and 1.52—are normalized to sum to 1.0 and entered into the row corresponding to the correct classification (class C)

		Classifier Output (Computed Classification)		
Confusion Matrix		A	B	C
True class of the	A	0.00	0.00	0.00
samples (input)	B	0.00	0.00	0.00
	C	0.28	0.34	0.38

Table 6.13 Example of the output probabilities matrix (OPM) applied to a sample that comes from class A. The "sum" correctly identifies the sample as belonging to class A

		Classifier				
Output Probabilities Matrix		1	2	3	4	Sum
Classifier confidence	A	0.55	0.29	0.33	0.66	1.83
output for class	B	0.31	0.43	0.22	0.14	1.10
	C	0.14	0.28	0.45	0.20	1.07

Table 6.14 Example of the output probabilities matrix (OPM) applied to a sample that comes from class B. The "sum" incorrectly identifies the sample as belonging to class C

		Classifier				
Output Probabilities Matrix		1	2	3	4	Sum
Classifier confidence	A	0.11	0.23	0.17	0.20	0.71
output for class	B	0.45	0.30	0.48	0.40	1.63
	C	0.44	0.47	0.35	0.40	1.66

Table 6.15 Confusion matrix with the three data sets, the output of Tables 6.11, 6.13, and 6.14, entered. The entries are all normalized to 1.0 across the rows as in Table 6.12, and associated with the correct classification

		Classifier Output (Computed Classification)		
Confusion Matrix		A	B	C
True class of the	A	0.45	0.28	0.27
samples (input)	B	0.18	0.41	0.41
	C	0.28	0.34	0.38

Table 6.16 Confusion matrix with the three data sets, the output of Tables 6.11, 6.13, and 6.14, entered as correct or incorrect classifications only. No normalization is required here since only one sample came from each class. Classes A and C are 100% accurate while class B has 0% accuracy for this sparsely populated matrix

		Classifier Output (Computed Classification)		
Confusion Matrix		A	B	C
True class of the	A	1	0	0
samples (input)	B	0	0	1
	C	0	0	1

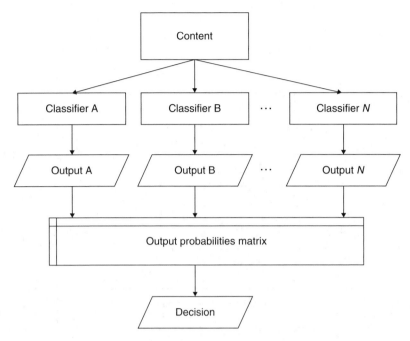

Figure 6.7   Confusion Matrix (second-order meta-algorithmic pattern)

be transformed to do so: please see Section 6.3.2 on the Confusion Matrix with Output Space Transformation (Probability Space Transformation) pattern. The output of all the classifiers is collected in the OPM. Each of the columns should sum to 1.0 (the output of the classifier should be normalized). The rows are then summed, and the row with the largest sum is the *decision* of the meta-algorithmic pattern. This is in some ways similar to weighted voting.

Figure 6.8 illustrates the Weighted Confusion Matrix pattern. Here, the classifiers are weighted relative to one another (the weights need not sum to 1.0—though in practice this is often adopted). The weights are then multiplied by the elements in the OPM to produce the weighted output probabilities matrix (WOPM), as shown in Table 6.17. Because the weights will favor more accurate classifiers—see, for example, the Voting and Weighted Voting pattern above for ways in which such weights are derived—if sufficient ground-truthed (training) data is available, better overall accuracy is expected. An example of this is provided in Table 6.17, in which a previously incorrect classification is corrected by weighting.

The system facts and considerations for the Confusion Matrix pattern are given in Table 6.18.

6.3.2   Confusion Matrix with Output Space Transformation (Probability Space Transformation)

The second of the second-order meta-algorithmic patterns is the first combinatorial pattern: it combines the Confusion Matrix pattern with an Output Space Transformation. With this

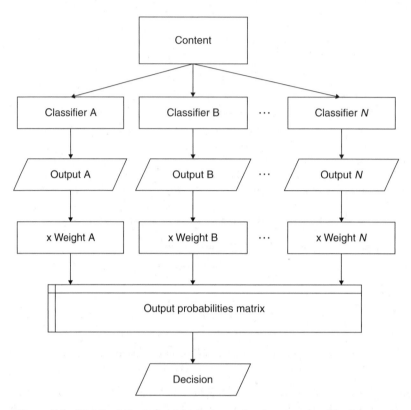

Figure 6.8   Weighted Confusion Matrix (second-order meta-algorithmic pattern)

pattern, the multiple intelligence generators (classifiers in the case of the Confusion Matrix pattern) can not only work on the same input data and create compatible output data but also produce well-behaved output data. Because of its association with the OPM, this pattern is also known as Probability Space Transformation.

Figure 6.9 provides the schematic for this pattern, which can be combined with either the Confusion Matrix or the Weighted Confusion Matrix pattern (see Section 6.3.1). The premise

Table 6.17   Example of the weighted output probabilities matrix (WOPM) applied to the same sample, from class B, given in Table 6.14. The "sum" now correctly identifies the sample as belonging to class B

		Classifier (Weight)				
Weighted Output Probabilities Matrix		1 (0.15)	2 (0.2)	3 (0.4)	4 (0.25)	Sum
Classifier confidence output	A	0.02	0.05	0.07	0.04	0.18
for class	B	0.07	0.06	0.19	0.10	0.42
	C	0.07	0.09	0.14	0.10	0.40

Table 6.18 Meta-algorithmic pattern system facts and considerations for the "Confusion Matrix" second-order meta-algorithmic pattern

Topic	System Facts and Considerations
Pattern name	Confusion Matrix and its associated Weighted Confusion Matrix
Related patterns	Confusion Matrix with Output Space Transformation (Probability Space Transformation), Majority Voting or Weighted Confusion Matrix, Best Engine with Absolute Confidence or Weighted Confusion Matrix, Confusion Matrix for Feedback
Input considerations	The individual classifiers must provide, at minimum, ranked preferences for their classification. The system architect may need to provide her own set of probabilities if only a ranking is provided—for example, if N rankings are provided, the rankings can be given values $Np, (N-1)p, \ldots, 2p, p$ where the sum of this is 1.0 (e.g., if $N = 4, p = 0.1$)
Output considerations	The decision process is unambiguous, making the decision output straightforward
Decisioning	The maximum sum of scores identifies the class
Scalability	Adding new classifiers or classes requires classifier-specific training—complexity increases with the addition of either, meaning that in general scaling is expensive both in terms of training and run-time processing costs
Processing	Since each classifier is used for the generation of the output probabilities matrix, the processing costs are high
Parallelism	The classifiers may be processed in parallel
Statistical behavior	The combination of probability matrix and weighting use provides significant flexibility to the design. Further flexibility can be added, if needed, using the related Confusion Matrix with Output Space Transformation pattern
Repurposability	The pattern is instantly repurposable if one or more of the classifiers must be removed (e.g., for cost reasons)
Domain expertise	Because of the unambiguous nature of the decision process, significant domain expertise is generally not required
Best used on	Classification systems of all types

of the Probability Space Transformation pattern is to modify the probability curves of one or more classifiers. This is performed to coordinate the behavior of the probability differences between consecutively ranked classes among the multiple engines, as introduced in Simske, Wright, and Sturgill (2006). Under certain circumstances such a transformation of the output probability space, p, can result in significantly improved classification accuracy. One simple mapping is given by

$$p \to p^{\alpha}.$$

This signifies the mapping of an output probability p, where $0.0 \leq p \leq 1.0$, to p^{α}, where $\alpha > 0.0$. When there are N classifiers, no more than $N - 1$ classifiers need to be transformed, since $N - 1$ is the number of degrees of freedom in coordinating N outputs. This was validated in Simske, Wright, and Sturgill (2006), where the value for α was varied over the range 0 to 10^6,

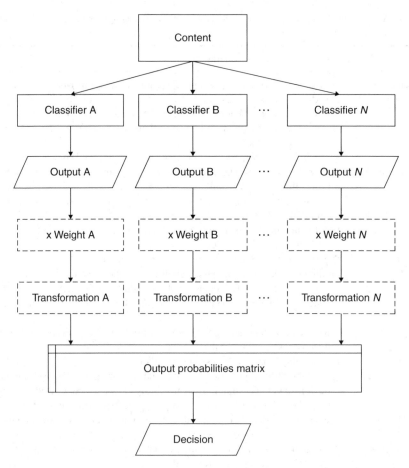

Figure 6.9 Confusion Matrix with Output Space Transformation (Probability Space Transformation) second-order meta-algorithmic pattern. The boxes with dashed outlines may not be present. The "× Weight" boxes are present only if the Weighted Confusion Matrix pattern is used prior to any output space transformation. The "transformation" boxes are present wherever the output is transformed

and the optimum value for the Nth classifier was indeed $\alpha = 1.0$, which effects no change in p. The general procedure for determining the values of α is as follows:

1. Rank all engines from best to worst as $1, \ldots, N$.
2. Optimize α_1.
3. For all engines $k = 1, \ldots, N - 1$, iteratively hold all $\alpha_1, \ldots, \alpha_k - 1$ steady, and optimize α_k.
4. Leave $\alpha_N = 1.0$.
5. Perform pattern as shown in Figure 6.9.

This approach may appear to have similarity to boosting; however, the goal is not to relatively weight one classifier over another, but simply to change the relative output behavior of the classifiers so that they can work better together. Thus, many other output transformation functions are possible. In general, the transformation should be monotonic, so that polynomials, logistic curves, and staircase functions are other good candidates for the transformation. These curves are used in other domains, such as perceptrons and boosting, but for another purpose—learning—than they are used for here. Again, the goal of this pattern is to better explore the relationship between multiple classifiers by transforming the output curves of $N - 1$ of them to allow them to work together better—not to work together better individually. It should also be noted that there are different ways to determine the value of α; for example, ranking the engines in order of their mean correlation with the other engines, ranking the engines in several random orders (trial and error) and selecting the one with the best overall accuracy on the training data, and so on.

The system facts and considerations for the Confusion Matrix with Output Space Transformation pattern are given in Table 6.19.

6.3.3   Tessellation and Recombination with Expert Decisioner

Several second-order meta-algorithmic patterns are based on the combination of a first-order meta-algorithmic pattern with another decision-making process when the system confidence in the first-order pattern is low. Tessellation and Recombination with Expert Decisioner, Predictive Selection with Secondary Engines, and Single Engine with Required Precision are three such patterns. All three are built on first-order meta-algorithmics: Tessellation and Recombination, Predictive Selection, and Sequential Try, respectively.

The Tessellation and Recombination with Expert Decisioner diagram is shown in Figure 6.10. Most of the diagram shares its architecture in common with the earlier Tessellation and Recombination pattern (see Section 6.2.5). The additional elements in the schematic are outlined with dashed borders, and are the two locations where the expert decisioner elements may be added. The first of these converts the tessellated elements "A, B, C, D, E, F, G, H" into partially recombined elements "I, C, J, K" guided by the expert decisioner. A simple example of such an expert decisioner is as follows. Suppose two OCR engines provide the following two outputs, representing the words in a particular sentence:

1. Acom monkey stone material
2. A common key stonemate rial.

The tessellation of these two sentence representations involves creating the smallest atomic units possible, accounting for any and all breaks between words in the set of sentences:

A com mon key stone mate rial.

The expert decisioner may be, in this case, an English language dictionary that looks for all terms in the tessellated set that are in the dictionary. This set, where proper nouns are capitalized, is:

A com common monkey key keystone stone Stonemate mate material

Table 6.19 Meta-algorithmic pattern system facts and considerations for the "Confusion Matrix with Output Space Transformation (Probability Space Transformation)" second-order meta-algorithmic pattern

Topic	System Facts and Considerations
Pattern name	Confusion Matrix with Output Space Transformation (Probability Space Transformation)
Related patterns	Confusion Matrix and Weighted Confusion Matrix, Majority Voting or Weighted Confusion Matrix, Best Engine with Absolute Confidence or Weighted Confusion Matrix, Confusion Matrix for Feedback
Input considerations	The individual classifiers must provide, at minimum, ranked preferences for their classification, and the system architect may need to provide his own set of probabilities if only a ranking is provided, as with the Confusion Matrix pattern
Output considerations	The key to this pattern is transforming the probabilities of one or more classifier outputs to provide better overall system accuracy (training phase only). Run-time output as for the Confusion Matrix pattern
Decisioning	The maximum sum of scores identifies the class. While the maximum sum of scores changes with probability space transformation, optimizing this transformation only affects the training phase
Scalability	Adding new classifiers or classes requires classifier-specific training, and reconsideration of probability space transformation. This is order N^2 and thus scales poorly
Processing	Since each classifier is used for the generation of the output probabilities matrix, the processing costs are high. The transformation costs during run-time are trivial
Parallelism	The classifiers may be processed in parallel
Statistical behavior	The base Confusion Matrix approach has high flexibility due to both probability matrix and weighting generation. The ability to transform probability spaces adds even further flexibility
Repurposability	The pattern is somewhat repurposable if one or more of the classifiers must be removed (e.g., for cost reasons), although the probability space transformations may need to be reoptimized
Domain expertise	Because of the unambiguous nature of the decision process, significant domain expertise is generally not required. The final output transformation approach and coefficients are determined based on maximizing accuracy on the training data
Best used on	Classification systems of all types

The expert decisioner then provides an overall weight for each of the possible recombined sentences, which are captured in Table 6.20.

As shown in Table 6.20, there are two 100% scores, "A common key stone material" and "A common keystone material." The traditional "recombination" then proceeds. Effectively, this block decides between "key stone" and "keystone," which is an easy decision since both OCR engines provide a break (i.e., a "space" character) between "key" and "stone," making the final output:

A common key stone material

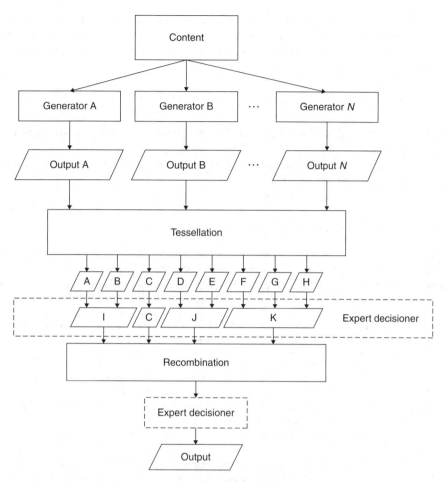

Figure 6.10 Tessellation and Recombination with Expert Decisioner pattern. The tessellation proceeds as for the first-order Tessellation and Recombination pattern (Figure 6.6), but instead of simple recombination (e.g., voting) the expert decisioner either performs some form of recombination before or after the traditional "recombination" block. See text for detail

Table 6.20 Possible recombined sentences from the tessellated string "A com mon key stone mate rial"

Recombined Sentence Candidate	Weight
A com monkey stone material	95.0
A com monkey Stonemate rial	70.0
A com monkey stone mate rial	78.3
A common key stone material	100.0
A common key Stonemate rial	70.0
A common key stone mate rial	83.3
A common keystone material	100.0
A common keystone mate rial	80.0

In Figure 6.10, it is shown that the "expert decisioner" may also occur after the traditional "recombination" operation. If this approach is used, then the expert decisioner will only operate on the following tessellated set, since the "recombination" block has eliminated "keystone" as a possible tessellated element:

A com common monkey key stone Stonemate mate material

Following the approach of Table 6.20, then, the only 100.0 score is obtained for "A common key stone material."

The system facts and considerations for the Tessellation and Recombination with Expert Decisioner pattern are given in Table 6.21.

6.3.4 Predictive Selection with Secondary Engines

The second of the combinational second-order meta-algorithmic patterns is the Predictive Selection with Secondary Engines pattern. For this pattern, the training—or "statistical learning"—phase is identical to the Predictive Selection pattern (Section 6.2.4), and so is not

Table 6.21 Meta-algorithmic pattern system facts and considerations for the "Tessellation and Recombination with Expert Decisioner" second-order meta-algorithmic pattern

Topic	System Facts and Considerations
Pattern name	Tessellation and Recombination with Expert Decisioner
Related patterns	Tessellation and Recombination, Expert Feedback (which uses expert rules to dictate learning through feedback)
Input considerations	No special requirements
Output considerations	The rules for recombination can be guided by expertise, for example, intelligent look-up tables—such as dictionaries for text or speech—and object lists for images
Decisioning	Expert domain knowledge is required for the recombination
Scalability	Same as for Tessellation and Recombination (Section 6.2.5)
Processing	Processing costs increase as the number of generators increases, since each is used during tessellation
Parallelism	Parallelism can occur at the generator level. It is therefore up to the individual generator to provide parallelism (the pattern does not)
Statistical behavior	The recombination can be guided by domain-specific statistics, as shown in Table 6.20
Repurposability	Because the recombination is generally domain-specific, the pattern is not highly repurposable
Domain expertise	The tessellation generally requires little domain expertise. The recombination, especially the expert decisioner, requires a considerable amount of domain expertise
Best used on	Complicated problem spaces, especially those in which the number of possible outputs is similar to or greater than the number of generators. Mature problem spaces are highly amenable to expert decisioning

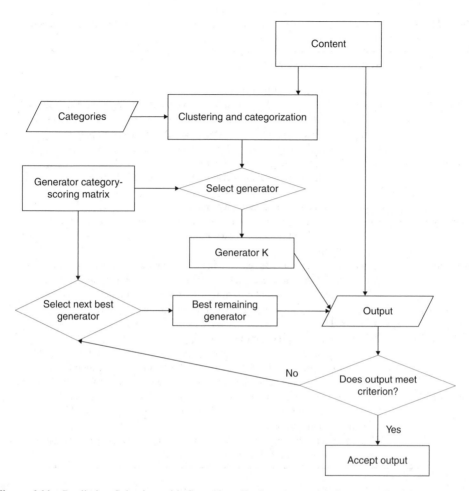

Figure 6.11 Predictive Selection with Secondary Engines (second-order meta-algorithmic pattern). The statistical learning system diagram is not shown, since it is identical to that of the Predictive Selection pattern (Figure 6.5). If the algorithm (originally Algorithm K in the first pass, and thereafter the "best remaining algorithm") does not result in acceptable output (passing a specific criterion), then the predicted best remaining algorithm is selected until either acceptable output is attained or the maximum number of iterations have been used up

repeated in Figure 6.11. In Figure 6.11, the run-time diagram is a straightforward extension of Figure 6.5: if the first generator selected is shown to provide output with unacceptable quality, then the first generator is marked as unacceptable and the best remaining generator is selected. This process continues until (a) a suitable generator is found, (b) the number of allowed generator substitutions has occurred, or (c) there are no generators remaining. In case (c), the output will generally be that of the original best generator, although system failure will be reported.

Most of the system facts and considerations (Table 6.22) for this pattern are similar to those of the Predictive Selection pattern. As for the Predictive Selection pattern, when the meta-algorithmic generators are classifier algorithms, the best classifier algorithm is the one

Table 6.22 Meta-algorithmic pattern system facts and considerations for the "Predictive Selection with Secondary Engines" second-order meta-algorithmic pattern

Topic	System Facts and Considerations
Pattern name	Predictive Selection with Secondary Engines
Related patterns	Confusion Matrix and all Confusion-Matrix-incorporating patterns, Predictive Selection pattern, Regional Optimization (Extended Predictive Selection), and other extended (e.g., recursive) Predictive-Selection-based patterns
Input considerations	During run-time phase, categories and algorithm category-scoring information must be loaded along with the content
Output considerations	Output is compared to a quality criterion for a pass/fail decision
Decisioning	A single best generator is identified in the first pass
Scalability	Under most conditions, no additional ground truthing is needed when new generators are introduced, allowing the pattern to scale to a very large set of generators
Processing	If only one generator—the one predicted to provide the best output accuracy—is used, and it produces acceptable output quality, the processing time is minimized. If the output criterion fails, though, additional processing—for example, using the generator with the next highest predicted accuracy—may be required
Parallelism	The system is serial (no parallelism)
Statistical behavior	The scoring algorithm selects the best generator to use. Optimum scoring algorithms provide different relative rankings among the generators for different categories, allowing the next best generator to be readily identified
Repurposability	The Predictive Selection portion of the pattern is highly repurposable
Domain expertise	Considerable domain expertise is often required to design the predictive scoring metric(s) and approach
Best used on	Systems where multiple generators are available, and where output quality can be readily (and hopefully efficiently) computed

for which the classification *precision* (as opposed to recall or accuracy) is highest for its reported class.

The difference between this pattern and the earlier Predictive Selection pattern is that some form of check on the output is provided after the best remaining classification has been assigned to the sample. In many ways, this is a look ahead to the Feedback (Section 6.4.1) and related patterns, except that the conditions on quality are less strict for this pattern in that they do not require comparison of the output to the input. For this pattern, the unacceptable quality can be based only on the output.

6.3.5   Single Engine with Required Precision

The Single Engine with Required Precision pattern is a second-order meta-algorithmic pattern that elaborates the first-order Sequential Try (Section 6.2.1). It is also somewhat similar to the just-introduced Predictive Selection with Secondary Engines pattern. In this relatively simple pattern (Figure 6.12), the generators are considered in order of a certain criterion. Example criteria are (a) overall accuracy, (b) cost, (c) mean processing time, (d) availability, and (e) training maturity/robustness.

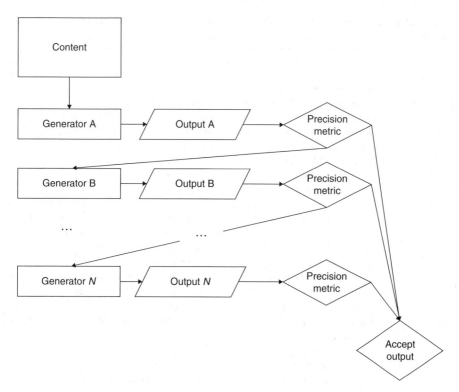

Figure 6.12   Single Engine with Required Precision (second-order meta-algorithmic pattern). The statistical learning system diagram is not shown, since it is identical to that of the Predictive Selection pattern (Figure 6.5) where the generator category-scoring matrix is the precision matrix. Generators are ordered based on a criterion and are checked in order until one reports an output for which the associated precision on training data meets an expected minimum value

Considering the generators in order, the precision of each generator for its computed output on the training data is compared to the system requirement. If the output classification reported is associated with a sufficiently high precision (predicted accuracy) on the training data, then this output is accepted and the process terminates. If not, the next ordered generator—for example, the next most cost-sensitive, next most accurate, next best throughput, and so on—is considered. Unlike the Predictive Selection with Secondary Engines pattern, the output criterion is precision of the generator on its training data (not the generator-ranking criterion). The system facts and considerations for this pattern are given in Table 6.23.

6.3.6   Majority Voting or Weighted Confusion Matrix

This pattern is the first of two variations on the first-order Voting pattern. This is another relatively simple second-order pattern. It is especially useful when a certain level of confidence in the output is required. Since majority voting is required, in general this pattern works best when there are an odd number of generators. The pattern first determines if a majority of the generators agree on their decision. If so, that decision is the output of the pattern and there is

Table 6.23 Meta-algorithmic pattern system facts and considerations for the "Single Engine with Required Precision" second-order meta-algorithmic pattern

Topic	System Facts and Considerations
Pattern name	Single Engine with Required Precision
Related patterns	Predictive Selection and related patterns, Sequential Try
Input considerations	(a) Overall accuracy, (b) cost, (c) mean processing time, (d) availability, or (e) training maturity/robustness are used to order the generators. Precision of each generator for each category of input must be computed
Output considerations	No special considerations since precision is computed before the generator is run
Decisioning	Precision is the decision criterion
Scalability	Readily scalable, since the generators are run sequentially until precision criterion is met. However, new generators must be trained to provide relative ranking
Processing	Processing costs are potentially high, if the highest ranked generator does not succeed, as each successive generator must be run in sequence
Parallelism	The algorithm is sequential, there is no parallelism
Statistical behavior	On simple input samples, the highest ranked engine will usually be the only one run (assuming all are sufficient to pass). It is good practice, then, to use the least expensive and/or minimal processing generators higher in the rank
Repurposability	The precision matrix generated during ground truthing is fully repurposable to other Predictive Selection and related patterns
Domain expertise	No special domain expertise is required
Best used on	Processes where a large number of different generators (with widely different costs, accuracy, etc.) are available; processes where ground truthing is relatively easy

no additional processing. If not, we use the Weighted Confusion Matrix approach outlined in Section 6.3.1 (Figure 6.13).

An interesting aspect of this combinational pattern is that the majority voting step can, under many conditions, be completed in less than half the time required to process all of the generators. This is possible if the generators are ordered based on processing time (with the least processing time first). Suppose, for example, that 20% of the generators are run in parallel. The fastest 20% to completion can be run in the first set, the 20% next fastest to completion in the next set, and so on. If there is an overwhelming consensus after the 60% fastest to completion are processed (i.e., more than 83.3% of these generators— equating to more than 50% of the entire set of generators—vote for the same output), then the 40% slowest to completion need not be processed, and the output will have been determined by this faster-to-completion set. Often, the 60% fastest to completion will take less processing time than the 40% slowest to completion, a sort of processing Pareto rule. Regardless, the system facts and considerations for this pattern are given in Table 6.24.

6.3.7 Majority Voting or Best Engine

This pattern is the second of two variations on the first-order Voting pattern. It is especially applicable to qualitative (pass/fail) problems. The first step again involves majority voting, such that this pattern should generally employ an odd number of generators. The pattern first

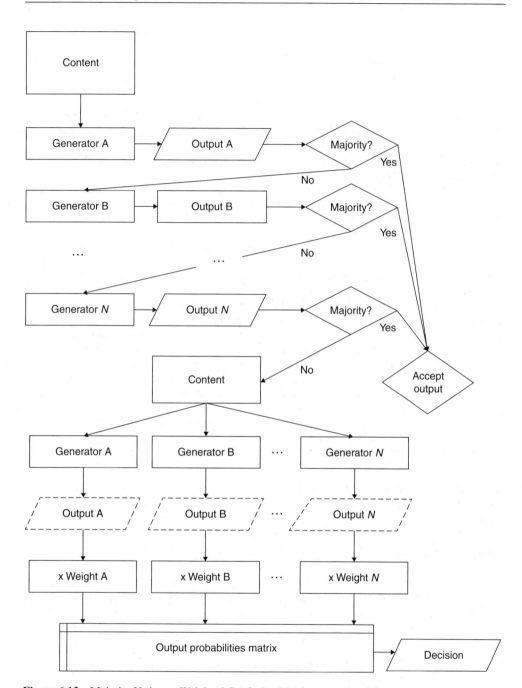

Figure 6.13 Majority Voting or Weighted Confusion Matrix (second-order meta-algorithmic pattern). If, and once, a majority is achieved, the decision output of the majority is accepted. Otherwise, the Weighted Confusion Matrix pattern is followed. The output blocks of the generators may need to be recalculated (dashed boxes) if the output in the Majority Voting portion of the pattern has not provided a weighted output probabilities matrix (WOPM). The output of the WOPM is the decision of the combinational pattern

Table 6.24 Meta-algorithmic pattern system facts and considerations for the "Majority Voting or Weighted Confusion Matrix" second-order meta-algorithmic pattern

Topic	System Facts and Considerations
Pattern name	Majority Voting or Weighted Confusion Matrix
Related patterns	Confusion Matrix, Weighted Confusion Matrix, Majority Voting or Best Engine, Best Engine with Absolute Confidence or Weighted Confusion Matrix, Confusion Matrix for Feedback
Input considerations	No special considerations, except that generally this works better with a relatively large amount of input, since the voting output is so discrete
Output considerations	Majority Voting provides only a single output, whereas Weighted Confusion Matrix provides, and uses for decisioning, the weighted output probabilities matrix (WOPM). Generators that provide a WOPM automatically are therefore preferable
Decisioning	Majority Voting shows nonparametric consensus, while the Weighted Confusion Matrix shows parametric consensus
Scalability	Readily scalable, although an odd number of generators is preferable
Processing	Performing the Majority Voting first can provide reduced processing if not all generators are processed in parallel. See text for details
Parallelism	Processing of all generators can be performed in parallel
Statistical behavior	The Majority Voting subpattern eschews statistics for raw vote counts. If no majority is achieved, then the Weighted Confusion Matrix and its already-described statistical behavior govern the output
Repurposability	The Majority Voting approach can be used as a front end to many hybrid systems
Domain expertise	Majority Voting and (Weighted) Confusion Matrix operations require no special domain expertise
Best used on	Systems with binary or other simplified output, systems with a large number of available generators

determines if a majority of the generators agrees on their decision, using the same approaches described in Section 6.3.6. If a majority is achieved, their decision is the output of the pattern and there is no additional processing. If not, we use the output of the Best Engine (i.e., the engine with the overall best accuracy on the training set) (Figure 6.14).

This pattern is well suited to systems in which a large set of generators, each of which can be processed very quickly, is available. Since both parts of the pattern require simple voting for only one output among many, each generator may be replaced by a simplified form, akin to a "constrained substitute" (see Section 6.2.2), of itself. As usual, the introduction of this pattern concludes with the system facts and considerations (Table 6.25).

6.3.8 Best Engine with Differential Confidence or Second Best Engine

How do we know that we have made the right choice at the end of any meta-algorithmic pattern? This is a fundamental question, especially inasmuch as meta-algorithmic patterns are designed with being able to treat generators, especially engines (i.e., generators that produce both data and a classification), as black boxes. In so doing, the focus of the meta-algorithmic patterns is on the input and output. One crucial element of the output, as introduced with

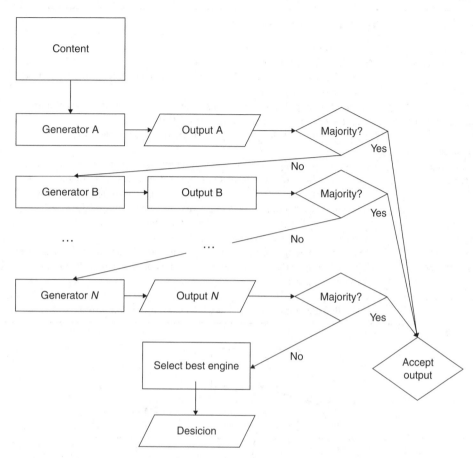

Figure 6.14 Majority Voting or Best Engine (second-order meta-algorithmic pattern). If a majority is achieved, the decision output of the majority is immediately accepted. Otherwise, the output of the best engine is selected

the OPM (see Section 6.3.1), is the confidence the engine reports for its output choice. The confidence is the maximum value in the column corresponding to a given engine (or classifier) in the OPM. This confidence value is computed for each engine in the overall system. If the best engine is highly confident in its output classification, it is concomitantly more likely to have the highest confidence amongst all of the engines. This fact is the basis of the combinatorial pattern described in this section; namely, the "Best Engine with Differential Confidence or Second Best Engine" pattern (Figure 6.15).

In this pattern, the counterintuitive notion of selecting the Second Best Engine by default is employed. Here, the Best Engine is used whenever it has a higher confidence, C_1, in its output than the maximum confidence of any of the other engines, designated C_{others}. How much higher is determined by the value of the threshold, shown in the decision diamond in Figure 6.15. This threshold value is determined during the training phase, preferably on a set of (validation) samples distinct from the samples used to compute the engine accuracies. When

Table 6.25 Meta-algorithmic pattern system facts and considerations for the "Majority Voting or Best Engine" second-order meta-algorithmic pattern

Topic	System Facts and Considerations
Pattern name	Majority Voting or Best Engine
Related patterns	Majority Voting or Weighted Confusion Matrix, Best Engine with Differential Confidence or Second Best Engine, Best Engine with Absolute Confidence or Weighted Confusion Matrix
Input considerations	Majority Voting generally this works better with a relatively large amount of input, since the voting output is so discrete
Output considerations	Majority Voting and Best Single Engine provide a single decision
Decisioning	Both Majority Voting and Best Single Engine provide nonparametric consensus
Scalability	Readily scalable, although an odd number of generators is preferable
Processing	Performing the Majority Voting first can provide reduced processing if not all generators are processed in parallel. See Section 6.3.6 for details
Parallelism	Processing of all generators can be performed in parallel
Statistical behavior	Simple decisions are involved in this pattern
Repurposability	The Majority Voting approach can be used as a front end to many hybrid systems
Domain expertise	Majority Voting and selecting the Best Engine operations require no special domain expertise
Best used on	Systems with binary or other simplified output, systems with a large number of available generators, systems in which consensus is preferred

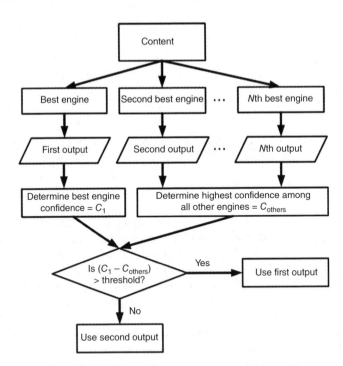

Figure 6.15 Best Engine with Differential Confidence or Second Best Engine (second-order meta-algorithmic pattern). The Best Engine is the engine with the overall highest accuracy (among all classes, or among all engines for the classification they report). The rest of the engines are ranked using the same metric. If the best engine is differentially more confident (by a minimum of the threshold value) in its output than the rest of the engines, it is chosen. If not, the Second Best Engine is chosen

Table 6.26 Results for different values of $(C_1 - C_{others})$ in a test of document classification, related to the work in Simske, Wright, and Sturgill (2006). Note that the overall accuracy of the Best Engine was 0.877; the overall accuracy of the Second Best Engine was 0.858; and the overall accuracy of the Best Engine with Differential Confidence or Second Best Engine, when the threshold is 0.10 (column 4) is 0.882. This is a decrease in error rate from 0.123 to 0.118, or 4.1%

$(C_1 - C_{others})$ Range	[1] Best Engine Accuracy	[2] Second Best Engine Accuracy	Selection	Percentage of Samples in this Range
≥ 0.30	0.999	0.843	[1]	0.002
0.25–0.30	0.998	0.845	[1]	0.034
0.20–0.25	0.985	0.877	[1]	0.089
0.15–0.20	0.943	0.908	[1]	0.133
0.10–0.15	0.928	0.909	[1]	0.219
0.05–0.10	0.912	0.915	[2]	0.231
<0.05	0.869	0.887	[2]	0.246

$(C_1 - C_{others}) \leq$ threshold, however, we choose the Second Best Engine. Since threshold is generally greater than 0.0, this means on a number of occasions, we will choose the Second Best Engine. This approach works if the Best Engine is more accurate when $(C_1 - C_{others}) >$ threshold and/or the Second Best Engine performs better than the Best Engine when $(C_1 - C_{others}) \leq$ threshold. This is often the case, as shown for data from a document classification task in Table 6.26.

As Table 6.26 illustrates, this is a meta-algorithmic pattern that requires direct evaluation of the OPM before deciding (a) whether to use it, and (b) what threshold to use.

This pattern is well suited to slight variations, based on the nature of the data and the willingness/ability of the system architect to thoroughly analyze the training data before recommending deployment settings. In fact, for the same data set as used for Table 6.26 (Simske, Wright, and Sturgill, 2006), we had initially investigated the related "Best Engine with Absolute Confidence or Second Best Engine" pattern (diagram not provided, but obvious from context), for which we did not obtain as consistently predictive results as when using the approach in Table 6.26 (the meta-algorithmic accuracy, of 0.874, was actually less than that of the Best Engine). There is actually a family of five alternatives to the "Best Engine with Differential Confidence or Second Best Engine" approach that should be investigated by the meta-algorithmic system architect before settling on the specifics of this pattern. They are:

1. Best Engine with Absolute Confidence or Second Best Engine
2. Second Best Engine with Differential Confidence or Best Engine
3. Second Best Engine with Absolute Confidence or Best Engine
4. For $N = 1, \ldots, N_{engines}$, Nth Best Engine with Absolute Confidence
5. For $N = 1, \ldots, N_{engines}$, Nth Best Engine with Differential Confidence

For (1), we replace the equation "$(C_1 - C_{others}) >$ threshold" in Figure 6.15 with the simpler equation "$C_1 >$ threshold." For (2) and (3) we change out the role of the Best Engine in the root pattern (Best Engine with Differential Confidence or Second Best Engine) with that of the Second Best Engine. The premise is that if the Second Best Engine is highly confident in its output—either in a relative sense (2) or an absolute sense (3)—it should be seriously considered to be used instead of the Best Engine, unless these two engines are very highly correlated. This is because these two engines disagree on some decisions in which the Second Best Engine is the correct one, and this is most likely to occur in general when the Second Best Engine is most confident in its output.

Not surprisingly, this approach is often quite successful, especially when the Best and Second Best Engine have similar overall accuracies. Having considered, then, the extension of high confidence to all engines, we arrive at the two variants (4) and (5), in which every engine—from best to worst in terms of accuracy—is compared to a threshold (possibly a different threshold for each engine!) for its absolute (4) or relative (5) confidence. Once an engine exceeds its appropriate threshold, its output is accepted and the pattern is complete. This approach is often useful when the different engines provide a wide range of confidence values—meaning that when they report high confidence, it is generally meaningful.

The system facts and considerations for this pattern are given in Table 6.27.

Table 6.27 Meta-algorithmic pattern system facts and considerations for the "Best Engine with Differential Confidence or Second Best Engine" second-order meta-algorithmic pattern

Topic	System Facts and Considerations
Pattern name	Best Engine with Differential Confidence or Second Best Engine
Related patterns	Best Engine with Absolute Confidence or Weighted Confusion Matrix, Generalized Hybridization
Input considerations	A large amount of training data must be generated and analyzed in order to make the reported confidence values meaningful
Output considerations	All engines in the system must provide confidence values, preferably in the form of an output probabilities matrix, or OPM
Decisioning	The pattern—and its five main variants outlined—makes a threshold-based decision. If enough training data is available, this is the only variable in the design, and it can be determined automatically using data such as shown in Table 6.26
Scalability	Scaling is linear with the number of engines, since each engine is responsible for reporting its own confidence values
Processing	This pattern—and its variants—require a lot of processing time, since each engine must have reported its confidence values before the pattern can proceed
Parallelism	The individual engines can be processed in parallel
Statistical behavior	The confidence values should be closely associated with probability. This will be further discussed in Chapter 8
Repurposability	The information gathered for this pattern can be used for the five variants described, and the confidences can be used for many other patterns
Domain expertise	No special domain expertise is required, and the setting of the threshold can be easily automated
Best used on	Systems with a large number of engines, especially where two or more of the engines are highly accurate

6.3.9   Best Engine with Absolute Confidence or Weighted Confusion Matrix

The final second-order meta-algorithmic pattern is also a combinational pattern in which the Best Engine is selected with a certain level of confidence or else a secondary pattern is selected. The secondary pattern in this case is itself a second-order meta-algorithmic: the Weighted Confusion Matrix pattern. The diagram of the pattern is shown in Figure 6.16.

As with the previous section, this is a meta-algorithmic pattern that requires direct evaluation of the OPM to decide between selecting the output of the Best Engine and the output of the WOPM, which makes a decision based on the output of all the engines. This pattern is also

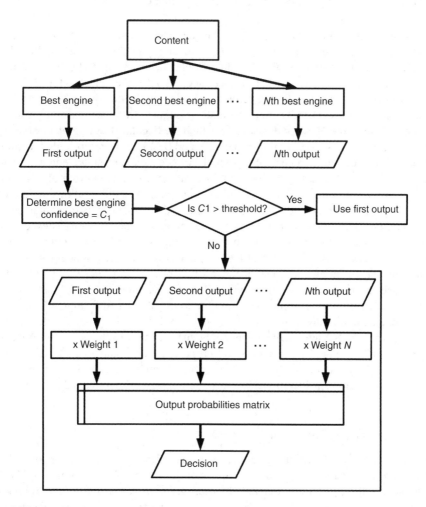

Figure 6.16   Best Engine with Absolute Confidence or Weighted Confusion Matrix (second-order meta-algorithmic pattern). The Best Engine is the engine with the overall highest accuracy (among all classes, or among all engines for the classification they report). The rest of the engines are ranked using the same metric. If the best engine reports a confidence value for its selected output above a given threshold, it is chosen. If not, the Weighted Confusion Matrix (Section 6.3.1) is used to choose the output

well suited to slight variations. There is actually a family of five alternatives to the "Best Engine with Absolute Confidence or Weighted Confusion Matrix" approach that should be investigated by the meta-algorithmic system architect before settling on the specifics of this pattern. They are:

1. Best Engine with Differential Confidence or Weighted Confusion Matrix
2. Second Best Engine with Absolute Confidence or Weighted Confusion Matrix
3. Second Best Engine with Differential Confidence or Weighted Confusion Matrix
4. For $N = 1, \ldots, N_{\text{engines}}$, Nth Best Engine with Differential Confidence
5. For $N = 1, \ldots, N_{\text{engines}}$, Nth Best Engine with Absolute Confidence

For (1), we replace the equation "$C_1 >$ threshold" in Figure 6.16 with the equation "$(C_1 - C_{\text{others}}) >$ threshold." For (2) and (3) we change out the role of the Best Engine in the root pattern (Best Engine with Absolute Confidence or Weighted Confusion Matrix) with that of the Second Best Engine. Here, as in Section 6.3.8, the rationale for (2) and (3) is that if the Second Best Engine is highly confident in its output—either in a relative sense (2) or an absolute sense (3)—it is used in place of the Best Engine. The same reasoning holds here: the two engines disagree on some decisions in which the Second Best Engine has the correct output, and this is most likely as a rule when the Second Best Engine has its highest confidence in its output. This rationale can be applied to high confidence in any single engine, providing two more variants on this pattern (4) and (5), in which every engine—from best to worst in terms of accuracy—is compared to a threshold (again, possibly a different threshold for each engine). Once any engine exceeds any given threshold, the output of this individual engine is accepted as the system decision and the pattern terminates. If no threshold is exceeded, the pattern terminates with the Weighted Confusion Matrix as described in Section 6.3.1. This family of six approaches is generally useful when there is a multiplicity of engines with widely different accuracies with an additional characteristic: for one large subset of input, the most accurate or second most accurate engine will suffice; for the other large subset, the WOPM is generally the most accurate output generator. One example is an ASR engine being used to handle customer calls. Many of the callers are native speakers of a language, for which the best or second best ASR engine will suffice. For nonnative speakers, however, custom engines (e.g., ASR engines good at handling Eastern European, South Asian, or East Asian accents) will contribute strongly to the WOPM and therein significantly increase the system accuracy.

The system facts and considerations for this pattern are given in Table 6.28.

6.4 Third-Order Meta-algorithmics

The second-order meta-algorithmic patterns, as we have just seen, added complexity largely by opening up the system chosen to options and in most cases combinations of two. The third-order meta-algorithmic patterns—generally focused on feedback from the output to input—add a further level of complexity. For one thing, it is often harder to predict the behavior of the patterns. For another, they often require two or more variables to be set. However, these patterns also add a higher degree of flexibility and tunability, since they provide multiple subpatterns joined together. We begin this section with the Feedback pattern, which is the first one to involve the feedback of error signals to improve the real-time output behavior of a system.

Table 6.28   Meta-algorithmic pattern system facts and considerations for the "Best Engine with Absolute Confidence or Weighted Confusion Matrix" second-order meta-algorithmic pattern

Topic	System Facts and Considerations
Pattern name	Best Engine with Absolute Confidence or Weighted Confusion Matrix
Related patterns	Best Engine with Differential Confidence or Second Best Engine, Weighted Confusion Matrix, all combinational and hybridized patterns involving a weighted confusion matrix
Input considerations	A significant amount of training data is required in order to make the reported confidence values meaningful
Output considerations	All engines in the system must provide confidence values, preferably in the form of a weighted output probabilities matrix, or WOPM
Decisioning	The pattern—and its five main variants outlined—makes a threshold-based decision upfront. If no confidence value exceeds its threshold, then a Weighted Confusion Matrix approach is performed
Scalability	Scaling is linear with the number of engines, since each engine is responsible for reporting its own confidence values. Computing weights for each engine requires negligible additional computation
Processing	This pattern—and its variants—require a lot of processing time, since each engine must have reported its confidence values before the pattern can proceed. However, the confidence values can be used for both of the two parts of the pattern
Parallelism	The individual engines can be processed in parallel
Statistical behavior	The confidence values should be closely associated with probability
Repurposability	The information gathered for this pattern can be used for the five variants described, and the confidences can be used for many other patterns (e.g., this pattern and the pattern in Section 6.3.8 require the same basic data sets)
Domain expertise	No special domain expertise is required
Best used on	Systems for which there are a large set of engines, with widely different accuracies. For one large subset of input, the most accurate or second most accurate engine will suffice. For the other, the WOPM is generally accurate

6.4.1   Feedback

The first third-order meta-algorithmic pattern is, in fact, the simple Feedback pattern. This pattern is important inasmuch as it allows two algorithms—a specific algorithm and its inverse—to collaborate in the production of highly accurate output. The Feedback system has one algorithm implementing a specific function. This algorithm must be designed to accept error signals that allow the algorithm to adjust its operation based on errors in the output.

The Feedback pattern relies on the design of an inverse transformation of the algorithm; apply the inverse algorithm to the output to regenerate a second (regenerated) copy of the original input. The regenerated output is then compared to the original input document to check for similarities and differences. An error signal is then fed back to the algorithm (or system of algorithms) to correct the output. This process can be continued iteratively until the error signal is reduced to an acceptable level. If the error signal cannot be reduced to a sufficiently small value, then a different error correction method may be attempted, as available.

Since this algorithm is a break from the first- and second-order meta-algorithmic patterns, it will be beneficial to provide an example of it in operation. Suppose the system is an OCR system, focused on digitizing paper documents. The text output from the OCR system can be used—along with the font and connected component information—to create a raster image representing the document. The original document image is then compared to the regenerated document image. The comparison, performed after aligning the binarized representation of the two images and subtracting one from the other, may show that the OCR algorithm incorrectly recognized a character/word since the difference between the original and the regenerated image exceeds a specific threshold. Or, the difference image (taken from subtracting one image from the other) can be used to indicate different salient OCR errors have occurred, including in order of the magnitude of the differences: (a) insertion of a character, (b) deletion of a character, (c) transposition of two characters, and (d) substitution of one character for another.

A simplified version of the Feedback pattern uses the error signal to accept or reject the output data. In this case, one algorithm is used in the forward path and its inverse is used in the feedback path. Figure 6.17 illustrates this incarnation. Two additional algorithms are required: one algorithm is required for regenerating the original via an inverse transformation and another algorithm is required for comparison. It is likely that for many algorithms the comparator will not be a simple implementation. For example, for an image-based Feedback patterns, alignment of the original and regenerated images is a must, and both affine (translation, rotation, scaling) and nonaffine (warping, bending, blurring) image correction may be required as part of the alignment process.

The Feedback pattern is especially important when there is only one generator available. It is the first meta-algorithmic pattern introduced that can operate when there is only one (forward) generator available. This "forward generator" therefore performs a specific functionality (with

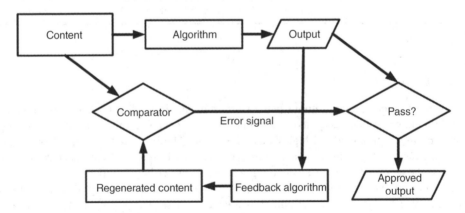

Figure 6.17 Feedback (third-order meta-algorithmic pattern). The (forward) algorithm provides output that is then the input to the (inverse) feedback algorithm. The feedback algorithm is used to regenerate the input ("regenerated content"). Comparing the regenerated content to the original content results in an error signal. If the error signal is acceptably small, then the output is approved

no competing/collaborating generators), making it important for us to continually (real-time) monitor and measure the quality of its output. If this sounds like quality control, that is no accident, as the Feedback pattern was designed specifically to be useful in quality control of the output data. For example, we have used the Feedback pattern to check the quality of the output of a document-remastering engine (Yacoub and Simske, 2002). In that application, the generator analyzes an input document raster image and produces an information-rich document (e.g., PDF format). To validate that the PDF output does not contain any flaws related to, for example, missing content or misclassification of parts of the document during document image processing, we employed a feedback algorithm that transformed the output PDF into a raster image format—the same format as the original input image format, in this case TIFF—and compared the original input image and the regenerated image, in accordance with Figure 6.17. The error message is then fed to the (forward) algorithm in an iterative process to successfully improve the document (or, iteratively, sections of the document).

Another, related, example of the Feedback pattern is when it is applied to improve the quality of the OCR output by regenerating a raster image from the text identified in the OCR output—along with its location, or bounding box, information—and then comparing the input image with the regenerated image for inconsistencies or misclassifications (as with the previous example, this requires excellent registration of the two images).

The Feedback pattern differs from the first- and second-order meta-algorithmic patterns in that it does not rely on the comparison of the output of two or more collaborating (or competing) generators, or the combination of the output of multiple generators; instead, it is self-contained and combines the generator with its inverse and a comparator operator and/or metrics to determine detectable flaws in the output data. The Feedback pattern, of course, can only be used for detection and correction of errors in the output of a generator if the generator accepts—and can act upon—feedback. It should be clear, then, from this section that the application of the Feedback pattern often requires one or more of the following: (a) a large amount of processing (as illustrated by the regenerative image creation in the examples); (b) a high degree of domain expertise (as is often needed to reverse complicated signal or image transformations); and (c) a high degree of expertise in a related or even unrelated domain, illustrated by the need for excellent image registration skills when performing inverse-OCR above.

As usual, this section concludes with the table summarizing the system facts and considerations for the pattern (Table 6.29).

6.4.2   Proof by Task Completion

The next third-order meta-algorithmic pattern is the Proof by Task Completion pattern, which dynamically changes the relative weighting of the individual generators, or the collaborative deployment of the individual generators, in response to the successful completion of intended generator tasks. The "proof" is evidence that the current system configuration is the right one for its deployment. The Proof by Task Completion pattern allows the overall system architecture to be switched between the various system deployment patterns (or, more simply, various system deployment settings) in order to provide a system that is continually redeployed to meet user performance, accuracy, licensing, and other requirements. The pattern can also be used to create a system that is tuned to focus on the end user's desired task. Thus, this

Table 6.29 Meta-algorithmic pattern system facts and considerations for the "Feedback" third-order meta-algorithmic pattern

Topic	System Facts and Considerations
Pattern name	Feedback
Related patterns	Proof by Task Completion, Confusion Matrix for Feedback, Expert Feedback
Input considerations	If possible, the feedback information provided as part of the pattern should be in the form of some (or even all) of the input to the system. In the case of OCR above, for example, scanned raster images are the normal input of the OCR engine, and so feedback in the form of images is readily accommodated by the OCR engine
Output considerations	Output should be translatable (or transformable) into the same form as the input. This transformation is performed by the feedback algorithm, which regenerates content in the form of the input
Decisioning	Decisioning in the pattern is governed by the error signal from the comparison of the original and regenerated content. This makes the comparator a crucial design element in the pattern
Scalability	There is no innate scalability to the Feedback pattern, since each input instance requires its own sequential (even if iterative, still sequential) set of operations
Processing	Processing costs are often quite high for the Feedback pattern, since inverse algorithms are often not optimized as readily as forward algorithms
Parallelism	There is no innate parallelism to the Feedback algorithm, since all operations must occur in a given sequence
Statistical behavior	The comparator embodies all the statistical behavior of the pattern
Repurposability	Some of the feedback algorithms may be repurposable to other tasks; for example, image registration was used in both examples in this section
Domain expertise	Relatively high domain expertise is not uncommon with the Feedback pattern
Best used on	Problems in which there are not a lot of meta-algorithmic generators available, especially in cases where there is only one generator available. The Feedback pattern can be held in "reserve" (since it is generally slower than other patterns) for when licensing costs or other "right to operate" costs associated with other generators make deployment of a less costly and/or in-house developed generator necessary

pattern supports the automatic, adaptive redeployment of the entire information-creating to self-optimize for its desired, real-time changing, "cost function." This cost function is based all or in part on one or more of the following seven factors:

1. *Throughput/performance*

How many documents need to be processed in a given amount of time, given the resources of the back end system? One important consideration here is, of course, amenability of the system components to parallelism. Patterns that are more suited to parallelism—for example, Weighted Voting and Tessellation and Recombination—may be chosen over patterns that are serial only (like Sequential Try and many combinational patterns). If the system is bottlenecked, for example, the best engine can be used in place of any meta-algorithmic pattern, removing the more expensive—in terms of performance, at least—combinational patterns from consideration.

2. *Given a certain amount of processing time, which meta-algorithmic pattern is most likely to converge on the correct result?*

This consideration, designated the "residual error," gives an indication of how well a meta-algorithmic pattern succeeds in identifying the correct output as one of a certain number of top candidates. An equivalent metric is how highly ranked the correct output is. As shown in Simske, Wright, and Sturgill (2006), several meta-algorithmic patterns are successful in increasing the accuracy of a document classification system. The maximum improvement in accuracy was 13%. However, these same meta-algorithmic patterns were even more successful in reducing residual error, with values as high as 55%.

3. *Severity of errors made in generating the output*

One of the advantages of meta-algorithmics—when sufficient processing is available—is that a "committee of experts" is less likely to make truly egregious errors than an individual algorithm, system, or engine. This is the flip side of the "residual error" described in factor 2 above, but it is an important one. In many systems, certain types of errors result in significant (and disproportionately expensive) costs. Suppose, for example, that in a large-scale ASR system, any errors in translation of voice to electronic data result in the need to forgo the ASR and call in a human operator. This will result in a huge incremental cost, perhaps 100 times as much as the ASR connection. So, a meta-algorithmic pattern that best prevents complete misunderstanding of the speech (say, e.g., the Try method) is quite powerful.

Interestingly, the severity of potential errors can be predicted *using a meta-algorithmic pattern*. The Predictive Selection pattern can be deployed to collect indicators of certain error types, and use the occurrence of these predictors to reconfigure the deployment of the meta-algorithmics (to avoid the predicted errors). Taking the ASR example, if a set of voice features derived from the *mel-frequency cepstral coefficients* are correlated with an accent (say, Scouse, or the English Merseyside accent) that the current ASR engine or meta-algorithmic combination of engines has proven to handle poorly, then the system can be reconfigured (or substituted for in entirety) with an ASR meta-algorithmic capable of handling this particular accent.

Such a predictive selection approach can also be used to estimate the system resources to be required downstream for a given class of input. It can, therefore, (a) predict the importance of the content (from characteristics of the content); (b) assign relative resources (processing time, number of candidate outputs to investigate during processing, etc.); and (c) adjust the meta-algorithmic patterns in accord with the available resources.

4. *Licensing concerns (trying to minimize licensing costs for generators requiring a by-use, or "click charge," license)*

This is a different type of cost function than the others, because it will naturally be biased toward patterns that allow the exclusion of one or more of the generators. If certain engines encumber licensing fees, patterns can be selected to steer around the more expensive licensed generators. Assuming the licensed generators are more accurate, then, we have observed that in many cases several nonlicensed generators will be required, along with the correct meta-algorithmic pattern, to provide the same accuracy and/or system robustness. This means that in order to reduce system licensing costs, usually greater system performance costs will be incurred.

5. *Differential training (ground truthing) on the multiple generators*

Special-purpose generators may be added to the system, in which case the meta-algorithmic pattern to be used may differ depending on the content type being analyzed. This selection is related to the "Predictive Select" pattern, and allows the choosing of an optimal deployment pattern for a given workflow, or a "shorthand for meta-algorithmic pattern selection" based on a cue from the content. One such decision-making process is as follows: (a) is the content of a special class?; (b) if yes, then apply the pattern specific to this special class; and (c) if no, then apply the appropriate pattern, selected after considering the other cost function attributes.

6. *Automatic updating: when additional training (ground truthing) cases are added, the system is retrained and if indicated redeployed*

This is an important aspect of the system adaptability—as new training-augmenting content is added after workflows using the different meta-algorithmic patterns are successfully completed, the various patterns are reassessed against this augmented ground truth. From the previous training data and the augmenting "completed task" data, the relative efficacy of each pattern under the various conditions is updated. This can change the patterns chosen even in a system in which the cost function is static, since the overall cost changes as the ground truth is augmented.

7. *Responsiveness to the different needs of the overall system*

One example of this cost function element is when the back end system is spending too much time performing the post-meta-algorithmic system processes, more thorough—that is, requiring more processing—meta-algorithmic patterns can be deployed on the front end, and vice versa.

The relative importance of the seven different cost function considerations outlined above will vary between systems and their associated workflows, and their disparate needs and preferences are as much a matter of domain expertise as meta-algorithmic patterning expertise. The Proof by Task Completion system (Figure 6.18) is thus responsive to both changes in end user expectations (in the form of the cost function) and the changing nature of the training (ground-truthed) and successfully processed content. The salient facts about this pattern are accumulated in Table 6.30.

6.4.3 Confusion Matrix for Feedback

The Confusion Matrix for Feedback pattern incorporates the relationship between the intelligence generators elucidated by the confusion matrix. The feedback is therefore directed to the most salient *binary decisions* in the problem. In general, these decisions will be classifications since the classic form of the confusion matrix (see Section 6.3.1) is a two-dimensional matrix with each column containing the computed classification and each row containing the samples in the true class. The Confusion Matrix for Feedback pattern is diagrammed in Figure 6.19.

In-depth illustration of the application of this pattern, along with the Confusion Matrix mathematics involved, is given in Chapters 8 and 9. But a simple, application-independent illustration will be given here to show how this pattern is deployed. Suppose we have a simple

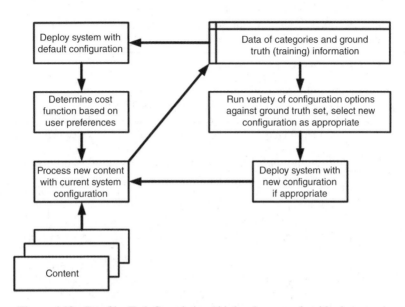

Figure 6.18   Proof by Task Completion (third-order meta-algorithmic pattern)

3-class system with a confusion matrix defined by

$$\text{From} \begin{array}{c} \\ A \\ B \\ C \end{array} \begin{array}{ccc} A & B & C \\ \left[\begin{array}{ccc} 0.89 & 0.07 & 0.04 \\ 0.15 & 0.76 & 0.09 \\ 0.09 & 0.07 & 0.84 \end{array}\right]. \end{array}$$

Here the overall accuracy—when each class has the same number of training samples—is the mean of the elements in the main diagonal, or 0.83. Class A is the most likely class to be assigned to a sample (37.7%), followed by class C (32.3%) and class B (30.0%). This pattern, however, relies on the comparison of the binary confusion matrices associated with each pair of classes. The three pairs that can be separated from the confusion matrix above are

$$\text{The AB confusion matrix is}\quad \text{From} \begin{array}{c} \\ A \\ B \end{array} \begin{array}{cc} A & B \\ \left[\begin{array}{cc} 0.89 & 0.07 \\ 0.15 & 0.76 \end{array}\right]; \end{array}$$

$$\text{The AC confusion matrix is}\quad \text{From} \begin{array}{c} \\ A \\ C \end{array} \begin{array}{cc} A & C \\ \left[\begin{array}{cc} 0.89 & 0.04 \\ 0.09 & 0.84 \end{array}\right]; \end{array}$$

$$\text{The BC confusion matrix is}\quad \text{From} \begin{array}{c} \\ B \\ C \end{array} \begin{array}{cc} B & C \\ \left[\begin{array}{cc} 0.76 & 0.09 \\ 0.07 & 0.84 \end{array}\right]. \end{array}$$

Table 6.30 Meta-algorithmic pattern system facts and considerations for the "Proof by Task Completion" third-order meta-algorithmic pattern

Topic	System Facts and Considerations
Pattern name	Proof by Task Completion
Related patterns	Feedback, Confusion Matrix for Feedback, Expert Feedback
Input considerations	Statistics collected from ground truthing, augmented by the successful completion of tasks, is used to continually update the configuration of the system for any/all of input content
Output considerations	For this pattern, a good portion of the output is downstream from the meta-algorithmic generator. When the overall task or workflow of which the generator is a part completes, that output is fed back to the system and augments the statistics the system uses to decide on its configuration
Decisioning	The decision on what current meta-algorithmic configuration to use is based on a cost function derived from any or all of seven different considerations as described in the text
Scalability	This pattern is infinitely scalable, since new data does not change the complexity or storage needs of the meta-algorithmic pattern. In fact, scaling only helps this pattern, since the more data, the better central tendencies in the statistics enabling the pattern
Processing	The amount of processing incurred by the Proof by Task Completion pattern is dependent on the current pattern chosen as a consequence of the ever-growing set of training and training-augmenting content
Parallelism	There are is no specific parallelism associated with this pattern, although of course multiple generators, each creating output that can later be used to update the configuration, can be run in parallel. The actual meta-algorithmic pattern salient to the current configuration can, of course, use any internal parallelism it has
Statistical behavior	The statistical behavior is relatively simple. When a document successfully completes its task, it is used to augment the training set for the particular output
Repurposability	The system is repurposable inasmuch as it provides, over time, increasingly valuable training data for the generator(s) associated with the meta-algorithmic pattern(s)
Domain expertise	Some domain expertise is required to grade (or relatively weight) the successfully completed output
Best used on	Systems intended for a long lifetime, systems in which the cost of an error is high, and systems in which ground truthing (training) is undesirable due to difficulty, expense, time constraints, and so on.

The AB confusion matrix accuracy is $(0.89 + 0.76)/(0.89 + 0.07 + 0.15 + 0.76) = 0.882$. The AC confusion matrix accuracy is 0.930, and the BC confusion matrix accuracy is 0.909. Thus, the Confusion Matrix for Feedback pattern dictates that we, as system architects concerned with optimizing overall system accuracy, are encouraged most to look for a generator that can distinguish between classes A and B with good accuracy. The overall accuracy of the system will then be dependent on the accuracy of the generator for distinguishing A versus B and the accuracy of the generator for distinguishing between A and B combined versus C. The

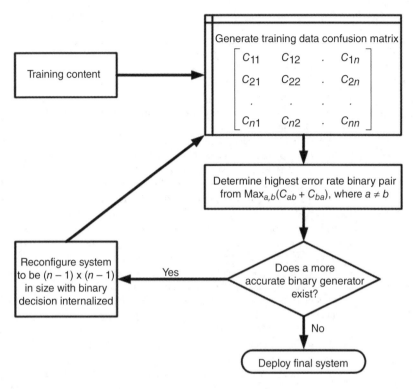

Figure 6.19 Confusion Matrix for Feedback (third-order meta-algorithmic pattern). The training content is collected for the entire set of N classes, and the sum of off-diagonal pairings $(C_{ab} + C_{ba})$, where $c \neq b$, are used to determine which pairs of classes to target with binary generators. As long as further improvement in behavior can be obtained, iterative (including recursive) binary generators can be added to the system

latter is derived from the original confusion matrix and is represented by the following binary confusion matrix:

$$\text{The } (A+B)\,C \text{ confusion matrix is } \quad \text{From } \begin{array}{c} \\ A+B \\ C \end{array} \begin{array}{cc} A+B & C \\ \begin{bmatrix} 1.87 & 0.13 \\ 0.16 & 0.84 \end{bmatrix} \end{array}.$$

Here the accuracy is $(1.87 + 0.84)/(1.87 + 0.13 + 0.16 + 0.84) = 0.903$. However, since 62.3% of the samples are in the correctly labeled $(A + B)$ class—that is, 1.87 is 62.3% of the sum of all the elements in the $(A + B)\,C$ confusion matrix—the overall system accuracy is $0.84/3.00 + 0.623 \times a(A + B)$, where $a(A + B)$ is the accuracy of the $(A + B)$ generator. Suppose that a domain expert can be found to produce a generator that can distinguish class A from class B with 95% accuracy. Then the overall system accuracy is $0.28 + 0.623 \times 0.95 = 0.872$. This new overall system accuracy is thus 87.2%, representing a reduction in error rate from 17% to 12.8% (a reduction of 24.7%).

Table 6.31   Meta-algorithmic pattern system facts and considerations for the "Confusion Matrix for Feedback" third-order meta-algorithmic pattern

Topic	System Facts and Considerations
Pattern name	Confusion Matrix for Feedback
Related patterns	Feedback, Proof by Task Completion, Expert Feedback
Input considerations	For this algorithm, the more training data, the better, since the differences in error rates are important
Output considerations	No special output considerations. Because the pattern results in re-architecting of the system, the results can easily be combined with another meta-algorithmic pattern in a hybridized (larger) meta-algorithmic system
Decisioning	The primary decision is whether a binary generator (usually a binary classifier) exists that can improve the system behavior for a given pair of data classes
Scalability	Readily scalable, although the number of generator pairs scales geometrically, not linearly, as more generators are to the system
Processing	Processing costs can be quite heavy if the binary generators require similar processing to the original all-generator processing. In theory, then, up to $N - 1$ times the processing may be required for an N-generator system
Parallelism	Since the flow of the system depends on the output of individual paired generator decisions, it is not in general parallelizable
Statistical behavior	Dictated by the off-diagonal confusion matrix elements
Repurposability	The binary generators can be reused in other meta-algorithmic systems
Domain expertise	Domain expertise is a big benefit for this pattern, but the domain expertise need not be for the entire set of input—only the subset of the input for where specific pairs of generators are selected (i.e., pairs of two classes). In this way, multiple subdomain experts can contribute to the overall system
Best used on	Meta-algorithmic systems where two or more sets of generators behave more similarly to each other than to the rest of the generators

For systems with more than three initial classes, the approach above can be repeated on the reduced-dimension confusion matrix. Important facts and considerations about the Confusion Matrix for Feedback pattern are given in Table 6.31.

6.4.4   Expert Feedback

The previous pattern, Confusion Matrix for Feedback, introduced the concept of deconstructing the overall problem space to allow targeted improvement of the overall system behavior. Classification accuracy is the chief target of confusion matrix manipulations. The Expert Feedback pattern, introduced in this section, is reliant on rules and learned constraints that are not derived from the confusion matrix but instead are associated with *elements* in the confusion matrix.

Figure 6.20 provides the two diagrams necessary for the Expert Feedback pattern. The *training phase* is used to generate the confusion matrix, and more importantly to provide reasonably large sets of off-diagonal instances (erroneous classifications). Any single generator or meta-algorithmic pattern for combining generators can be used to derive this confusion

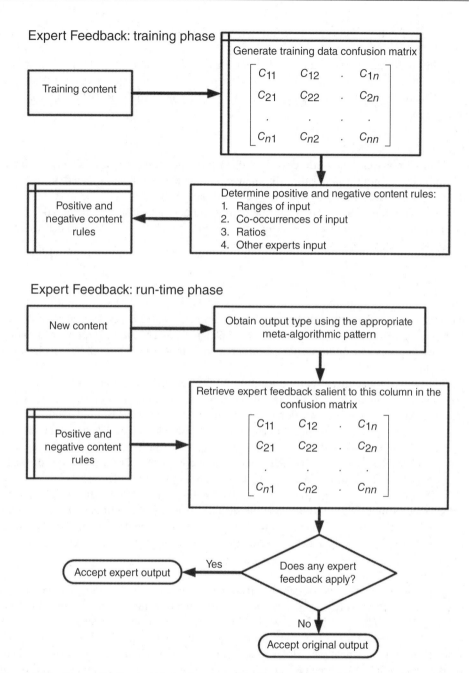

Figure 6.20 Expert Feedback (third-order meta-algorithmic pattern). There are two phases to the pattern. During the training phase, a confusion matrix is generated. Training data should be sufficient to allow the generation of expert rules for the off-diagonal elements $\mathbf{C}_{ij}$, where $i \neq j$. During deployment (run-time phase), the column of the output (assigned classification) is known, and each element in the column is checked for rules that are salient to the output. If a salient rule is identified, the classification output is changed accordingly. If there are one or more conflicting expert rules, the conflict can be resolved through any number of ways See text for details

matrix—it is of course the same generator/pattern of generators to be used in the run-time phase.

After the confusion matrix is generated, the off-diagonal content is probed for the existence of both positive and negative rules, which will provide a form of expert system rules to be applied to increase the overall accuracy of the system. In general, these rules belong to one of three categories:

1. *Input ranges*

An input range is usually described by the minimum and maximum value of an input signal. If the training data is sufficient, then this range describes the expected behavior of the input associated with one or more classes of data. However, since the maximum and minimum in a range are often noisy (the odds of an anomalous sample being added to a given class increase with the size of the training set), it makes more sense to describe the range statistically. The description of the range is

$$\{\text{mean} - H \times \text{std}, \text{mean} + H \times \text{std}\},$$

where std is the standard deviation and H is a parameter that is dependent on a number of factors—including the number of different input signals measured, the number of training samples, the error rate of the system, and each input sample's closeness of fit to a Gaussian distribution.

This type of content rule can be used on all classes of input simultaneously. If there are many input signals, then a simple tally of how many in-range and out-of-range signals occur for each possible classification is accumulated. These data can be used to override the class computed by the meta-algorithmic pattern. For example, if the classification indicated by the meta-algorithmic pattern is that of a class that has several out-of-range input signals, then the classification can be changed to the highest confidence class without out-of-range signals. This can be considered a "voting" approach to using the range values.

In addition to a voting approach, a "weighted voting" approach assigns weights to the out-of-range values based on their relative distance from the mean of each range. This is expressed as a multiple of H, the controlling factor of the range definition. The class with the lowest weighted voting may be used to replace the meta-algorithmic pattern's computed class.

2. *Co-occurrences of input*

Co-occurrences offer the possibility of higher confidence in overriding the computed classification. Effectively, a co-occurrence requires two (presumably rare) events to occur simultaneously. They can be positive (if both occur they absolutely identify a particular class) or negative (if both occur they absolutely preclude a particular class). An example from document classification follows. If a document contains both the text "Loan application" and the text "34 Riverbed Road" occurring on page 1, then the document is positively identified as a loan application for the bank at 34 Riverbed Road. If, however, the document contains both the text "1162 Yerba Buena Boulevard" and the text "art exhibit" on page 1, then it is negatively identified as a loan application for the bank at 34 Riverbed Road.

Co-occurrences, as can be seen above, are a simple way of using specific events known to be associated with a particular class to enforce classification when computed classification fails to provide the right outcome. Importantly, these types of rules can be used a posteriori; that is, after a common feature within a certain class has been identified as being a strong indicator of class membership. This means co-occurrences can be discovered through active search (data mining) or by the incorporation of new data extraction technologies (e.g., for images, by the deployment of improved image recognition technologies). Note that a co-occurrence is deemed valuable even if it is not 100% accurate—it merely needs to improve the overall system accuracy when applied.

3. *Ratios*

Ratios are often highly valuable expert rules. Ratios can build on co-occurrences, especially the extension of co-occurrences to subclasses of information within the larger classification structure of an intelligent system input space. For example, consider surveillance images. The ratio of changed pixels to unchanged pixels can be used to classify the type of downstream tracking algorithms to use. Lower ratios can use simple, efficient tracking approaches such as color histogram back-projection, while higher ratios may require more sophisticated and more processing intensive approaches such as k-shortest paths optimization, model-based tracking, and particle filter methods.

These rules can benefit from expert evaluation of the off-diagonal elements of the confusion matrix. However, in many instances the process can be automated. Events exhibiting notably different prevalence in the off-diagonals compared to the rest of the columns in the confusion matrix are in general indicative of a misclassification. These types of "anomalous" events are rather straightforward to identify with data mining approaches.

In Figure 6.20, the fourth type of content rule listed is "other expert input," which is meant to encapsulate other, usually less data-driven, means of overriding the computed classification. Examples include meta-data considerations (size of the content, data file characteristics, who created the content, where was it created, when was it captured, etc.). Another type of information that can be mined is termed "generator meta-data," which is information provided by the individual generators. One example is the confidence information provided, which can be used in numerous meta-algorithmic patterns, including the Confusion Matrix pattern (Section 6.3.1). The confidence values, collected in the OPM, can additionally be used in a "negative" sense to override computed classification. For example, suppose that when an OPM entry for Generator A is less than a certain threshold value, the accuracy of the overall system is higher when it accepts the second choice for computed classification in place of the first choice. In such a case, then, the low confidence value in the OPM leads to overriding the computed classification, and improvement in system accuracy.

During the run-time, these positive and negative content rules are associated with each computed classification. If any expert feedback applies, then it may be used to override the computed classification. If there is conflicting expert feedback, then the conflict may be resolved in any number of ways, depending on the preferences of the system architect and the policies (and policy conflict resolution rules) associated with the content.

The system benefits from augmented training data, akin to that used to augment the system-governing statistics in the Proof by Task Completion pattern (Section 6.4.2). For the Expert Feedback pattern, it is helpful to augment the off-diagonal (i.e., error) elements of the confusion matrix. For a system with very low generator error rate, this means a substantial investment in training. However, in general, this investment in training is highly valuable. Without it, the content rules generated may be highly noisy and therefore of little actual predictive value.

Table 6.32   Meta-algorithmic pattern system facts and considerations for the "Expert Feedback" third-order meta-algorithmic pattern

Topic	System Facts and Considerations
Pattern name	Expert Feedback
Related patterns	Tessellation and Recombination with Expert Decisioner, Feedback, Proof by Task Completion, Confusion Matrix for Feedback
Input considerations	Ample training data is very important for the generation of positive and negative rules, especially inasmuch as they correspond to the (hopefully) rarer error cases
Output considerations	No special output considerations are required for the meta-algorithmic system itself. The content rules, however, may need to be handled with a policy engine in case of conflicting rules
Decisioning	The decisioning proceeds as for an expert system: once rules are generated they are applied during run-time as salient (assuming conflicting rules do not override them)
Scalability	The system is not easily scalable, as the rules must be regenerated every time the confusion matrix is changed
Processing	The run-time processing is largely dictated by the meta-algorithmic pattern used to provide the original output. Applying the rules requires only a few lines of code at run-time
Parallelism	There is no innate parallelism in the pattern
Statistical behavior	The rules, where possible, should be based on statistical descriptions of the ranges and ratios. Co-occurrences are simply yes/no decisions
Repurposability	In general, the rules are not repurposable
Domain expertise	Significant domain expertise is a benefit when setting up the data mining to find the rules. Thereafter, rule finding can be automated
Best used on	Systems for which a large amount of training data or augmented training data is available

Additionally, having sufficient content to prevent "noisy" rule generation will also help offset the trend for increasing fragility and tight coupling of the system, which is possible unless the rules are expected to remain unchanged over time. More training data helps ensure that the rules have more statistical power. Important aspects of the Expert Feedback pattern are tallied in Table 6.32.

6.4.5   Sensitivity Analysis

When the number of generators increases, several different trends are generally observed. The first is that the generators themselves become more highly correlated in the mean—since there is less overall output space to explore as each successive generator is added. The second is that clusters of generators that are more closely correlated with each other than with the rest of the generators emerge. The third is that often subsets of the generators are more effectively combined with different meta-algorithmic patterns. Each of these trends argues for the removal (or recombination) of one or more generators. The Sensitivity Analysis pattern is focused on the reduction in the number of generators through one of these three mechanisms: (1) correlation,

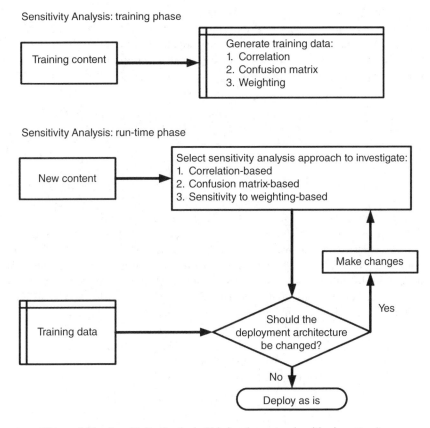

Sensitivity Analysis: training phase

Sensitivity Analysis: run-time phase

Figure 6.21   Sensitivity Analysis (third-order meta-algorithmic pattern)

(2) confusion matrix behavior, and (3) selection among multiple meta-algorithmic pattern options (Figure 6.21).

Correlation is often useful: meta-algorithmic patterns such as Voting make use of the consensus that correlation among generators provides. However, highly correlated generators do not, as a rule, explore the entire input space much more effectively than one of the (correlated) generators by itself. Thus, highly correlated generators can be effectively combined along the lines shown in Section 6.4.3 for the Confusion Matrix for Feedback pattern in order to reduce the dimensionality of the meta-algorithmic system. Reducing the number of generators through the combining of highly correlated generators is thus one incarnation of the Sensitivity Analysis pattern.

As a different means of reducing the number of generators, another Sensitivity Analysis approach can be used to identify one or more stable points within the solution space. This approach targets an optimally robust—rather than accurate—system, and as such often reduces the amount of training data that is required. This aspect of sensitivity analysis investigation is performed using the *confusion matrix*. The off-diagonal elements of the confusion matrix, $\mathbf{C}$, are investigated. When all of the $\mathbf{C}_{ij} \approx \mathbf{C}_{ji}$, where $i \neq j$, this is an indication of generator interaction stability. So, one means of reducing the number of generators from N to M is to select the M most balanced generators; that is, the ones minimizing the quantity $|\mathbf{C}_{ij} - \mathbf{C}_{ji}|$.

The confusion-matrix-based approach can be evaluated quantitatively from an overall perspective. Measurements of entropy are valuable for determining the overall balance of the confusion matrix. The confusion matrix entropy, e, is defined by

$$e = \frac{-\sum_{i=1}^{N}\sum_{j=1+1}^{N}\left[\left(\frac{C_{ij}}{C_{ij}+C_{ji}}\right) \times \ln\left(\frac{C_{ij}}{C_{ij}+C_{ji}}\right) + \left(\frac{C_{ji}}{C_{ij}+C_{ji}}\right) \times \ln\left(\frac{C_{ji}}{C_{ij}+C_{ji}}\right)\right]}{N(N-1)},$$

where the confusion matrix elements $\mathbf{C}_{ij}$ are as described above. The equation explicitly illustrates the pairings of $\mathbf{C}_{ij}$ and $\mathbf{C}_{ji}$, where $i \neq j$. The maximum value for the entropy is not surprisingly the natural log (ln) of 2.0, or 0.6932, achieved when each and every $\mathbf{C}_{ij}$ and $\mathbf{C}_{ji}$ is equivalent. Using the confusion matrix $\mathbf{CM}$ shown below, it is left as an exercise for the reader to show that the paired entropy for classes (A,B) is 0.6255; for classes (A,C) it is 0.6173; and for classes (B,C) it is 0.6853. These results show that we are more likely to have sufficient training data for distinguishing classes B and C than for distinguishing either of these classes from class A. We can thus potentially reduce the order of the classification to differentiating class A from the combined set of classes B and C upfront.

$$\mathbf{CM} = \text{From} \begin{array}{c} \\ A \\ B \\ C \end{array} \overset{\begin{array}{ccc} A & B & C \end{array}}{\begin{bmatrix} 0.89 & 0.07 & 0.04 \\ 0.15 & 0.76 & 0.09 \\ 0.09 & 0.07 & 0.84 \end{bmatrix}}.$$

It is also left for the reader to show that the overall entropy of the example confusion matrix $\mathbf{CM}$ is 0.6427, substantially lower than the maximum entropy of 0.6932.

The third type of sensitivity analysis considered here is the *sensitivity to weighting* of one or more candidate weighting-driven meta-algorithmic patterns. Weighted Voting and Weighted Confusion Matrix patterns (Sections 6.2.3 and 6.3.1) are two such patterns, and Table 6.6 tabulates some of the more common and/or useful voting schemes. Once default weights are calculated, the sensitivity analysis approach—as illustrated by example in Table 6.33—independently alters the weights of each individual classifier and the effect on overall system accuracy is noted. In Table 6.33, in every circumstance where the relative weighting of Classifier 3 goes up (the "1−," "2−," and "3+" rows), the overall accuracy increases (by 1%, 3%, and 4%, respectively), while in every circumstance in which the relative weighting of Classifier 3 goes down (the "1+," "2+," and "3−" rows), the overall accuracy drops (by 1%, 2%, and 3%, respectively). Next, comparing Classifiers 1 and 2, when the weighting for Classifier 1 goes up relative to the weight for Classifier 2 (the "1+" and "2−" rows), the accuracy goes down by 1% or up by 3%, respectively; and when the weighting for Classifier 1 goes down relative to the weighting for Classifier 2 (the "1−" and "2+" rows) the accuracy goes up by 1% or down by 2%, respectively. Overall, this is a mean 1% improvement when Classifier 1 goes up relative to Classifier 2, mildly indicating that Classifier 1 should have its weighting relatively improved in comparison to Classifier 2.

Table 6.33 Sample Sensitivity Analysis pattern, third configuration (sensitivity to weighting) output. Here, the weighting on each classifier was increased by 25% (the "1+," "2+," and "3+" rows) or decreased by 25% (the "1–," "2–," and "3–" rows). Weights are shown nonnormalized except for the default weights. The data indicate that the weight for Classifier 3 should be increased relative to Classifiers 1 and 2, and also that the weight of Classifier 1 should be increased relative to that of Classifier 2

Configuration	Individual Classifier Weights and Overall Meta-algorithmic Pattern Accuracy			
	Classifier 1	Classifier 2	Classifier 3	Overall Classification
Default weights	0.35	0.25	0.40	0.83
1+ Weight	0.44	0.25	0.40	0.82
1– Weight	0.26	0.25	0.40	0.84
2+ Weight	0.35	0.31	0.40	0.81
2– Weight	0.35	0.19	0.40	0.86
3+ Weight	0.35	0.25	0.50	0.87
3– Weight	0.35	0.25	0.30	0.80

Table 6.34 Meta-algorithmic pattern system facts and considerations for the "Sensitivity Analysis" third-order meta-algorithmic pattern

Topic	System Facts and Considerations
Pattern name	Sensitivity Analysis
Related patterns	Confusion Matrix for Feedback, Weighted Voting
Input considerations	One or more of the following should be computed: correlation matrix, confusion matrix, weightings for the generators
Output considerations	The primary output of this pattern is a recommendation of which generators to eliminate or combine
Decisioning	Decisioning is based on one or more of the following: (a) high degree of correlation, (b) anomalous behavior in the confusion matrix, and (c) high sensitivity of the weighting value
Scalability	All sensitivity analysis approaches are scalable, but require retraining of the system each time a new generator is added
Processing	As this pattern is generally associated with iterative redefining of the system architecture, processing costs are high and interrupted by system evaluation
Parallelism	Aside from whatever inherent parallelism is in the original meta-algorithmic patterns, this pattern is iterative and sequential
Statistical behavior	Correlation is itself a statistical approach. The Sensitivity Analysis pattern brings a certain statistical rigor to optimizing the system architecture based on the confusion matrix and generator weighting, as well
Repurposability	The results of this pattern are not, generally, repurposable. However, the pattern "repurposes" the output of other patterns, such as the Weighted Confusion Matrix and Weighted Voting patterns. It can also repurpose correlation statistics that arise from testing new generators when they are considered for inclusion in other meta-algorithmic approaches
Domain expertise	The approaches rely on no specific domain expertise
Best used on	Systems with a large amount of available training data and/or a large number of generators

Complex combined knowledge can be obtained by considering all of the single dimension sensitivities simultaneously. From the above set of results, it is recommended to further increase the weight for Classifier 3 and leave the weight for Classifier 1 nearly steady. This approach was tested by successively changing the overall weights from (0.35, 0.25, 0.40) to (0.35, 0.20, 0.45), and so on, to (0.35, 0.00, 0.65). Among these, an optimal weighting of (0.35, 0.00, 0.65) was found for Classifiers (1, 2, 3) with an overall classification accuracy of 0.89.

Important considerations when deploying or considering deploying this pattern are tallied in Table 6.34.

6.4.6 Regional Optimization (Extended Predictive Selection)

The next pattern is also concerned with "introspective meta-algorithmics," in which the meta-algorithmic pattern may be modified based on the statistics of the input. In the Regional Optimization pattern, however, individual generators are tuned for subclasses of the overall task. As a consequence, different meta-algorithmic configurations (combinations of generators and maybe even different patterns altogether) will be deployed for different subsets of the input. This is related to, and based in part upon, the Predictive Selection (Section 6.2.4) pattern. However, for this pattern, the classes of input data are not based on human-defined classes, but are instead defined so as to be separably relevant to two or more meta-algorithmic patterns.

The Regional Optimization pattern is shown in Figure 6.22. The training and run-time phases are straightforward. Training content is collected for every generator during the training phase. Training data is assigned to different classes and the appropriate meta-algorithmic patterns are

Regional Optimization: training phase

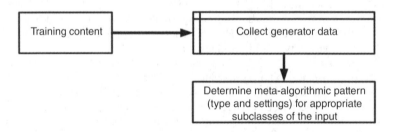

Regional Optimization: run-time phase

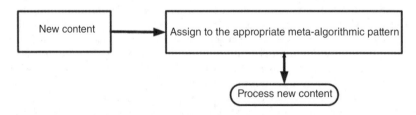

Figure 6.22   Regional Optimization (third-order meta-algorithmic pattern)

configured for each class. This effectively replaces the selection of individual generators (as in the Predictive Selection pattern) with *individual meta-algorithmic patterns.*

How are the individual meta-algorithmic patterns selected? If the previous sections in this chapter have not already driven home the plurality of choice the system architect has in selecting from meta-algorithmic patterns, then the multiple examples for several domains illustrated in the next three chapters will. Needless to say, system architects will usually be able to find two or more meta-algorithmic patterns that perform differentially well on subsets of the input. The Regional Optimization pattern provides for a different meta-algorithmic pattern to be selected on the basis of the input data characteristics. It is important that the input data be accurately classified so these errors do not propagate to the downstream meta-algorithmic patterns.

Alternatively, the Regional Optimization pattern allows the selection of a more germane meta-algorithmic pattern based on the computed classification of an initial meta-algorithmic pattern. In this case, a meta-algorithmic pattern is used, effectively, as the generator

Table 6.35 "Regional Optimization (Extended Predictive Selection)" third-order meta-algorithmic pattern system facts and considerations

Topic	System Facts and Considerations
Pattern name	Regional Optimization (Extended Predictive Selection)
Related patterns	Predictive Selection
Input considerations	The output of a predictive selection generator category-scoring matrix element (which can itself be a meta-algorithmic pattern) is the input that selects a particular meta-algorithmic pattern
Output considerations	The output of the pattern is no different from that of, for example, the Predictive Selection (first-order) meta-algorithmic pattern
Decisioning	The "Predictive Select" portion of this pattern actually chooses a downstream meta-algorithmic pattern, not simply one generator
Scalability	Scalability is linear with the number of generators, since the meta-algorithmic patterns are based only on the generators added to the system
Processing	Sequential processing of two meta-algorithmic patterns means the expected processing time is twice that of most other meta-algorithmic patterns
Parallelism	The only parallelism associated with this pattern is integral to the incorporated meta-algorithmic patterns. The pattern itself is serial. However, when a larger data set is partitioned and a different meta-algorithmic pattern is found optimal for each partition, the door is opened for a form of "meta-algorithmic parallel processing"
Statistical behavior	The scoring algorithm selects the best generator to use. Optimum scoring algorithms provide different relative rankings among the generators for different categories
Repurposability	Not generally repurposable, since tightly coupled to the problem/input space
Domain expertise	Some domain expertise is required in order to simultaneously plan for the types of meta-algorithmic patterns to use with different classes of input, and for assigning of data to classes suitable for different meta-algorithmic patterns
Best used on	Systems with wide ranges of input, or input belonging to two or more separable subclasses of content. Especially useful for the processing of mixed task workflows where the classes of input are readily and accurately ascertained

category-scoring matrix element in the pattern of Figure 6.5; with the downstream meta-algorithmic pattern replacing the run-time selected "Generator K."

The run-time phase is even more straightforward. Input is assigned to the appropriate class and thence the relevant meta-algorithmic pattern, and output defined using that particular meta-algorithmic pattern. Salient system considerations are collected in Table 6.35.

As described in Section 5.2.1, performing image processing tasks on sections of images can produce different results than performing image processing tasks on the intact images. This means that the same image can be processed by the same algorithm or intelligent system and yield a plurality of results. Taken one step further, this implies that the optimal intelligent system for analyzing each partition may vary. In effect, then, this pattern is a form of *meta-algorithmic parallel processing*. This type of approach, in another light, can also be viewed as a form of sensitivity analysis—if the generator output differs for a partition of the image in comparison to the intact image (or other partitions), then it is likely that the generator has less than full confidence in the output, and perhaps that a different pattern may be appropriate.

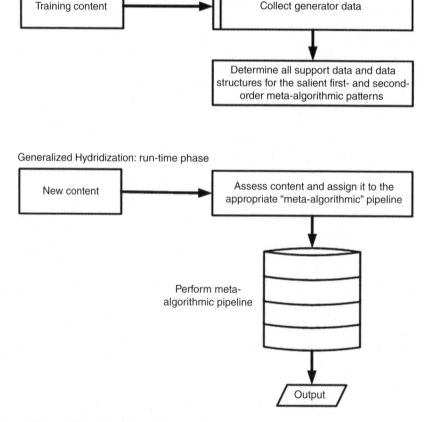

Figure 6.23 Generalized Hybridization (third-order meta-algorithmic pattern)

6.4.7   *Generalized Hybridization*

The Generalized Hybridization pattern is used to optimize the combination and sequence of first- and second-order meta-algorithmic patterns for a given—generally large—problem space. Like the Regional Optimization pattern, it compares more than one meta-algorithmic pattern for an optimally performing system. Prior to the training phase, the system architect decides which meta-algorithmic patterns are to be considered for the deployed system. During the training phase (Figure 6.23), then, the correct data and data structures to support the set of meta-algorithmic patterns are generated. During deployment, some or all of the content is assigned to the one or more meta-algorithmic pattern pipelines.

Examples of hybridized meta-algorithmics include many of the second-order patterns: Predictive Selection with Secondary Engines, Majority Voting or Weighted Confusion Matrix, Majority Voting or Best Engine, Best Engine with Differential Confidence or Second Best

Table 6.36   "Generalized Hybridization" third-order meta-algorithmic pattern system facts and considerations

Topic	System Facts and Considerations
Pattern name	Generalized Hybridization
Related patterns	Regional Optimization, Predictive Selection with Secondary Engines, Single Engine with Required Precision, Majority Voting or Weighted Confusion Matrix, Majority Voting or Best Engine, Best Engine with Differential Confidence or Second Best Engine, Best Engine with Absolute Confidence or Weighted Confusion Matrix
Input considerations	Input data and data structures for all potential first- and second-order patterns to be considered must be generated as part of the input
Output considerations	Output is provided at each stage of the hybridized pattern, affording multiple opportunities for voting strategies in the pipelines
Decisioning	Decisioning is based on whether a logical OR or a logical AND is chosen for the hybridization. With the logical OR, a certain quality metric must be achieved. With the logical AND, voting schemes in their variety are used
Scalability	A high degree of flexibility and robustness to input is possible with more complicated hybridized patterns, which offers some advantages for scalability to changes in input
Processing	Processing costs are high for this pattern, typically the highest of any meta-algorithmic pattern described
Parallelism	Individual meta-algorithmic patterns within logical AND patterns can be run in parallel
Statistical behavior	Logical OR approaches are based on comparison with a threshold, while logical AND patterns are based on voting or weighted voting
Repurposability	This pattern tends to provide a deployment architecture tightly coupled to the input, meaning its architecture is not generally repurposable
Domain expertise	A high degree of domain expertise may be required upfront to decide which meta-algorithmic patterns should be chosen, and when logical OR and logical AND should be used
Best used on	Highly complex systems in which the interactions among the generators is not accurately captured by a simple meta-algorithmic pattern

Engine, and Best Engine with Absolute Confidence or Weighted Confusion Matrix. For each of these patterns, the hybridization is based on a logical OR: either the initial output is acceptable, or another pattern is tested for acceptability. However, the logical AND may also be used to create hybridized meta-algorithmic patterns. When multiple patterns are used in parallel, there must be an unambiguous means of combining their output. Weighted voting is a flexible and efficient means of combining the output of two or more patterns (each pattern itself potentially hybridized). In hybridized patterns, internal voting may also be supported, wherein the output of each stage in the (logical OR) hybridized pattern is voted upon. Later stages in such a pattern may have higher relative weighting.

Important considerations when deploying the Generalized Hybridization pattern are collected in Table 6.36.

6.5 Summary

From a wide perspective, this is the crucial chapter of this book. In it, the three different orders of meta-algorithmic patterns have been introduced. A large—but not exhaustive—set of patterns are provided for each of these three levels. Several of the powerful tools for meta-algorithmics—confusion matrices, weighting, output space transformation, sensitivity analysis, and feedback—are explained. Having provided the patterns—which like all patterns are "templates" for domain-specific implementation—we next turn to application of these patterns to a wide, but again not exhaustive, set of relevant intelligent systems. Chapter 7 will focus on the application of first-order meta-algorithmics to both a core and extended set of domains, followed by Chapter 8 and its focus on second-order meta-algorithmics for the same core domains and a new set of extended domains. Chapter 9 completes the meta-algorithmic application overview by illustrating the use of third-order meta-algorithmics on the same core set and a third extended set of domains. After these chapters, Chapter 10 illustrates some ways in which meta-algorithmics uniquely change the field of machine intelligence. At the conclusion of Chapter 10, the principles of meta-algorithmics are summarized.

References

Lin, X., Yacoub, S., Burns, J., and Simske, S. (2003) Performance analysis of pattern classifier combination by plurality voting. *Pattern Recognition Letters*, **24**, 1959–1969.

Simske, S.J., Wright, D.W., and Sturgill, M. (2006) *Meta-algorithmic Systems for Document Classification*. ACM DocEng, pp. 98–106.

Stehman, S.V. (1997) Selecting and interpreting measures of thematic classification accuracy. *Remote Sensing of Environment*, **62** (1), 77–89.

Yacoub, S. and Simske, S. (2002) Meta-algorithmic patterns. Hewlett-Packard Technical Report HPL-2002-106, 31 pp.

7

First-Order Meta-algorithmics and Their Applications

Just because we don't understand doesn't mean that the explanation doesn't exist.

—Madeleine L'Engle

7.1 Introduction

Chapter 6 introduced the design patterns that form the basis of meta-algorithmic systems. This chapter provides applications of the first five patterns, collectively called the "first-order meta-algorithmics." These patterns include the simplest possible means of combining two or more generators: using serial application (Sequential Try and Try pattern), using simple substitution (Constrained Substitute) or using voting (Voting and Weighted Voting pattern). The other two first-order patterns (Predictive Select and Tessellation and Recombination), however, are more elaborate, and require at least some domain expertise since one (Predictive Select) requires de-aggregating the input space into a useful set of classes, and the other (Tessellation and Recombination) requires de-aggregating the data itself during the tessellation operation. The Tessellation and Recombination approach also usually requires domain expertise, which will be discussed in this chapter for the document understanding domain.

The first-order meta-algorithmics are pivotal, as they are the building blocks of many second- and third-order meta-algorithmics. This chapter provides first-order meta-algorithmic system architecture examples for the following types of intelligent systems: the four primary domains of (1) document understanding, (2) image understanding, (3) biometrics, and (4) security printing; and three additional domains of (1) medical signal processing, (2) medical imaging, and (3) natural language processing (NLP).

7.2 First-Order Meta-algorithmics and the "Black Box"

One of the primary reasons for adopting a meta-algorithmic approach to intelligent system architecture is the huge convenience offered by being able to treat a (usually very complicated) large algorithm, system, or intelligence engine (collectively called generators) as a "black box."

Meta-algorithmics: Patterns for Robust, Low-Cost, High-Quality Systems, First Edition. Steven J. Simske.
© 2013 John Wiley & Sons, Ltd. Published 2013 by John Wiley & Sons, Ltd.

In so doing, the architect needs to be concerned only with the input and output of the system. This is not a substitute for domain knowledge; rather, with meta-algorithmics the application of domain knowledge is focused on the treatment of the input and output. This is usually a less complicated—though not superficial—requirement than having to address the inner complexities of the generator. The reason is that the complexities of the generator are often dissociated from the complexities of the domain. For image understanding, for example, the inner workings of the imaging engine may require advanced knowledge of neural networks, genetic algorithms, image thresholding, texture analysis, differential equations, and shape modeling. The meta-algorithmics used to coordinate the output of two or more segmentation approaches, however, require a completely different set of skills: data analysis, statistical inference, decision sequencing, and so on. This is a powerful consideration. A domain expert, instead of dealing with the nuances of a more esoteric set of technologies otherwise necessary to improve her systems, can instead use her own domain expertise to improve the robustness, accuracy, cost, or other system deployment metrics. The meta-algorithmic system designer is therefore able to treat the individual generators as "black boxes."

In the remainder of this chapter, a nonexhaustive set of examples are provided, intended to illustrate through example the manner in which first-order meta-algorithmic patterns can be deployed on real-world intelligent system design. The primary domains of this book are discussed first (Section 7.3). Then, three of the secondary domains are selected for further illustration: (1) medical signal processing, (2) medical imaging, and (3) NLP (Section 7.4). One or more examples of deploying each of the first-order meta-algorithmic patterns are thus provided.

7.3   Primary Domains

In this book, the primary domains for intelligent systems are document understanding, image understanding, biometrics, and security printing. Document understanding, and in particular zoning analysis, is used to illustrate the application of the Tessellation and Recombination pattern. Image understanding—in particular scene recognition—is used to demonstrate the deployment of the Predictive Selection pattern. The Constrained Substitute pattern is employed for biometric identification of a fingerprint or iris. Finally, the Sequential Try pattern is applied to several security printing applications.

7.3.1   Document Understanding

One of the most important tasks in document understanding is zoning analysis. Zoning engines perform segmentation (definition of region boundaries), then classification (typing of the regions—e.g., as text, image, business graphic, etc.), and then aggregation (clustering of appropriate sets of regions, e.g., image + caption as a "figure," multiple rows of text as a "text column," etc.). Examples of zoning engines are given in several references (Wahl, Wong, and Casey, 1982; Revankar and Fan, 1997; Shi and Malik, 2000). The choice as to what types of zones to identify is generally left to the domain expert and to the types of content in the documents; for example, "equations," "business graphics," "tables," "figures," "text columns," "paragraphs," "articles," and so on, are possible as dictated by the semantics of the information acquired. Regardless, typical zoning analysis engines find regions corresponding, minimally, to text and "nontext" areas on the original document. Subsequent treatment of text (by optical character recognition, or OCR) and nontext (by image compression, bitmap-to-vector conversion, image analysis, etc.) is dependent on the accurate zoning of the document. Representation and storage of the processed file in new formats (PDF, XML, HTML, RTF, DOC, etc.) is also dependent on the overall accuracy of the zoning engine.

Tessellation and Recombination is an ideal first-order meta-algorithmic pattern for combining the segmentation outputs of two or more zoning engines. The output of the zoning engines will typically be a set of normalized probabilities used to identify the relative confidence of the engine for classifying a given zone. For example, a zoning analysis engine (Engine A) may provide the following statistics for a particular logo region (small drawing):

Drawing $p = 0.54$,
Text $p = 0.33$,
Junk $p = 0.10$,
Table $p = 0.02$,
Photo $p = 0.01$.

Now, suppose Engine B provides the following statistics for the same region:

Drawing $p = 0.19$,
Text $p = 0.70$,
Junk $p = 0.07$,
Table $p = 0.02$,
Photo $p = 0.02$.

Suppose now that Engine B was assigned, after evaluation of training data, a confidence rating of 0.3 relative to Engine A's 1.0. Then Engine B's overall (statistical output × confidence value) statistics for this region are:

Drawing $p = 0.06$,
Text $p = 0.21$,
Junk $p = 0.02$,
Table $p = 0.01$,
Photo $p = 0.01$.

The combination of these two—that is, Engine A + 0.3 × Engine B—is, therefore, as follows:

Drawing $p = 0.60$,
Text $p = 0.54$,
Junk $p = 0.12$,
Table $p = 0.03$,
Photo $p = 0.02$.

The classification from this data is "drawing." It is left as an exercise for the reader to show that the classification would have been as "text" had the confidence values for Engines A and B been simply summed. Note also that after the tessellation, the recombination can be driven using an even more sophisticated "recombination weighting" (in place of the blanket 1.0 weighting for Engine A and 0.3 weighting for Engine B in the above example) that uses differential weighting of engines by each region type (e.g., a 0.3 for drawing, a 0.15 for text, etc., in Engine B above). This need for "weighting" during recombination is not relevant to all domains, but is highly relevant to zoning analysis.

As discussed in Section 6.2.5, the recombination portion of the pattern can be thought of as statistically driven aggregation. The primitives, or atomic units, created by the tessellation process can be aggregated if the statistics from the multiple engines so dictate. However, emergent zoning classifications are also possible. This stems from the fact that recombination requires domain expertise. Thus, there are many possible "goodness of recombination" measurements. For zoning analysis, a fully Manhattan layout has a convex hull exactly equal to each region size, resulting in a very clean (usually entirely rectangular) format. Thus, selecting a set of zones that is optimally Manhattan is a good choice for optimization. Another metric of goodness is defining the zones to best match a layout template. A third approach is to compare the variance in the differential background between the different tessellations, and determine whether to classify them as true background (favoring keeping more, smaller zones) or to classify them as foreground (favoring keeping less, larger zones). This trade-off is illustrated in Figure 7.1, where the darker shaded, rectangular regions are the differential background zones in comparing the two-zone Segmentation 1 and the four-zone Segmentation 2. If these two zones have substantially higher histogram entropy, variance, color gamut, or hue than the white areas that are common background for both segmentations, then they are probably not background areas, and so Segmentation 1 is a more credible output. If, however, these darker shaded rectangles have similar histogram and hue characteristics to the white ("known

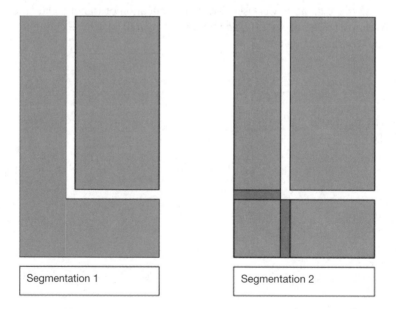

Figure 7.1 Figure showing two different segmentation outputs (larger, shaded polygons) and the white space between the regions that is in common (white) or only visible in Segmentation 2 (polygons with darker shading). See text for details

background") areas, then Segmentation 2 is more credible. Note that Segmentation 2 is also more credible due to its lower convex hull/area ratio, 1.0, when compared to that of Segmentation 1 (approximately 1.7).

One of the important factors in this design pattern is the selection of the zoning engines to include in the design. These are generally selected based on one or more of the following: (1) overall engine accuracy, with the higher-weighted engines generally preferred; (2) performance, if, for example, real-time response is required; (3) digital rights, meaning under some conditions the user may not be allowed to use one or more of the engines; and (4) user intent, meaning, for example, if the user only intends to use the extracted text, then the systems with highest weighting for text identification should be chosen.

Once the set of engines has been chosen, each engine is run separately on the document image and the regions obtained. Each engine will define a large set of data relating to the region type, which will then be converted to a standardized set of region information that comprises the minimum data required to combine the engines effectively and also to allow any region size metrics to be applied. Thus, region bounding box (bbox), polygonal boundaries ($x$vertices, $y$vertices), and size and statistics on region type are required.

The details of the recombination are provided here. First, all regions from all engines are mapped to a special "image-sized map" and each pixel that is part of a region from a particular engine is marked with a unique ID for that engine. For example, if there are M engines used, the unique identifiers for Engines A, B, C, ..., M are $[1, 2, 4, ..., 2^{(M-1)}]$. Now regions are formed (as connected components from this image-sized map) based on their numerical uniqueness, not on their existence in any engine. Suppose we employed two segmentation engines, A and B. Where both engines identify a region, the map is marked "3" (as $1 + 2$); where only Engine A identifies a region, the map is marked "1"; and where only Engine B identifies a region, the map is marked "2." Regions are formed by sequentially searching through the map for "1"s, then "2"s, and then "3"s. In other words, up to three sets of regions are formed in this zoning analysis. When M segmentation engines are used, up to $2^M - 1$ sets of regions, corresponding to different degrees of region identification overlap, are formed.

After regions are classified—for example, as "text" or "drawing" as described above—the regions of the same type abutting one another are merged (recombined). Figure 7.2 illustrates the process for the image originally presented as Figure 5.1. In Figure 7.2a, the zoning output for Engine A is illustrated. In Figure 7.2b, the zoning output for Engine B is presented. Finally, in Figure 7.2c, the recombination of the regions is provided. One improvement in the output of Figure 7.2c is that the large text "ALWAYS NEW... ALWAYS AN ADVENTURE!" is now correctly identified as color line art.

The application of the Tessellation and Recombination first-order meta-algorithmic pattern (originally shown in Figure 6.6) to this two-engine document zoning analysis example provided the means to generate new region segmentation and classification. This is readily observed when comparing Figure 7.2c (the output of the Tessellation and Recombination) to the output of the two individual zoning analysis engines, Figures 7.2a and b.

This example illustrates that domain knowledge is required for implementation of the Tessellation and Recombination pattern (Figure 7.3). In zoning analysis recombination, both the weighting and the decision-making on differently labeled regions require domain expertise. As one example, during the tessellation, often the regions formed do not line up exactly, resulting in rather narrow regions (similar to the darker shaded regions in Figure 7.1, except typically on the outer borders of regions) that may end up being classified differently from the larger regions they outline. These regions may need to be merged with abutting regions even though they are classified differently, in order to preserve a reasonable zoning output for the document. In addition, the merging of like-classified abutting regions also requires zoning-analysis domain-specific expertise. For example, in some instances text regions will be merged into columns; in other instances, text regions abutting each other will be left unjoined since they compose two or more columns.

The example therefore clearly illustrates how the meta-algorithmic pattern allows the system architect to use document analysis expertise, rather than knowledge of the internal workings of the individual engines, to improve the output of the overall system.

7.3.2   Image Understanding

Scene recognition is an important type of image understanding task. Recognition includes the identification of the location (scene identification) and, distinctly, the identification of specific elements within the scene (object recognition). In this example, we are concerned with the

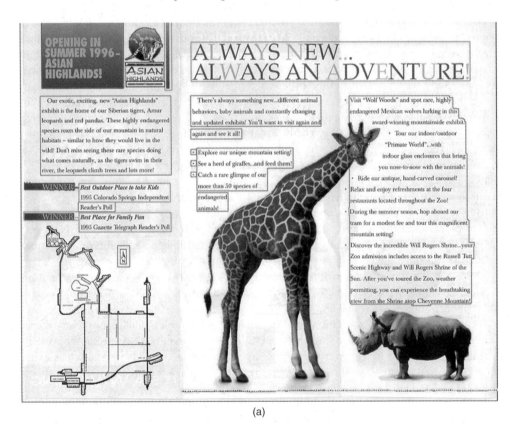

(a)

Figure 7.2 Tessellation and Recombination for a sample image. (a) Output of Engine A. (b) Output of Engine B. (c) Recombination of the output of both engines. Lightest outlines indicate text regions, second lightest outlines indicate photo regions, darkest lines indicate color line art regions, and second darkest lines indicate binary line art regions. Reproduced by permission of Cheyenne Mountain Zoo

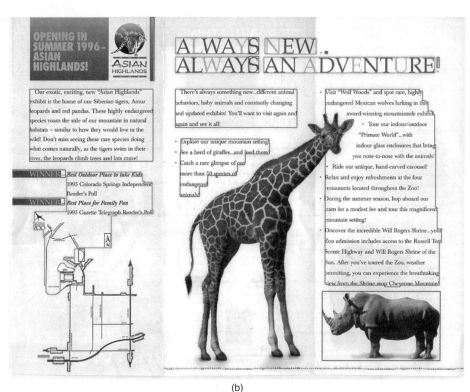

(b)

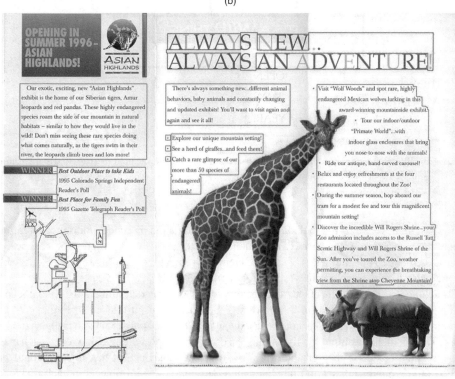

(c)

Figure 7.2 *(Continued)*

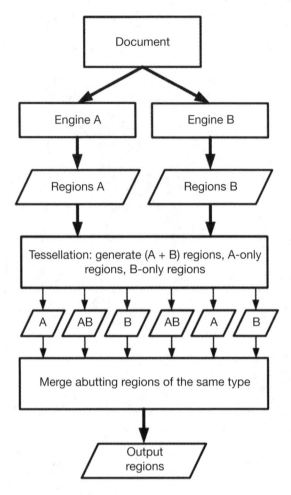

Figure 7.3   Tessellation and Recombination first-order meta-algorithmic pattern as applied to the zoning analysis example described in the text

former, wherein we are trying to deduce a location based on the presence or absence of certain image characteristics. Interestingly, one such characteristic is the presence or absence of specific elements. Thus, the set(s) of specific elements (object recognition) that are identifiable is a predictor for the selected class (scene recognition).

The choice of which objects to use as predictors can be determined in several ways. In Figure 7.4, which is based on the Predictive Select pattern introduced in Figure 6.5, the manner of choosing the objects is not specified, but is left to the "clustering and categorization" block. This block has as input the list of classes along with representative images of each class that combined comprise the "ground truth," or training data. Each of the object recognizers analyzes the ground truth set and tabulates the types and counts of objects in each image. A representative set of results is shown in Table 7.1.

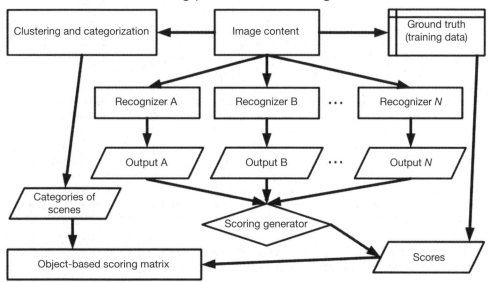

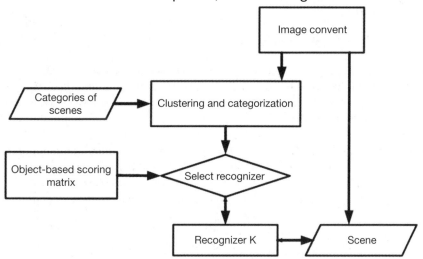

Figure 7.4 Predictive Selection first-order meta-algorithmic pattern as applied to the scene recognition example described in the text

Table 7.1   Objects identified (per 100 images) in four classes of images by three different recognizers

| | Recognizers | | |
Image Type	Recognizer A	Recognizer B	Recognizer C
Indoors, home	134 Chairs, 23 desks, 8 trees, 13 buildings	119 Chairs, 18 desks, 11 trees, 17 buildings	127 Chairs, 14 desks, 15 trees, 16 buildings
Indoors, business	166 Chairs, 143 desks, 14 trees, 37 buildings	177 Chairs, 135 desks, 11 trees, 21 buildings	182 Chairs, 99 desks, 23 trees, 45 buildings
Outdoors, urban	35 Chairs, 24 desks, 278 trees, 366 buildings	15 Chairs, 9 desks, 233 trees, 263 buildings	23 Chairs, 34 desks, 219 trees, 302 buildings
Outdoors, rural	11 Chairs, 7 desks, 456 trees, 103 buildings	6 Chairs, 12 desks, 413 trees, 89 buildings	3 Chairs, 7 desks, 388 trees, 76 buildings

The data provided in Table 7.1 are insufficient to determine the predictions to be used. However, some strategies are immediately visible from the data:

1. If the number of trees and buildings in the image is greater than some threshold, the image is likely outdoors.
2. For outdoor images, if the number of trees is relatively higher than the number of buildings, then the image is likely rural.
3. For indoor images, if the number of chairs is relatively higher than the number of desks, then the image is likely in a home.

These three strategies are readily incorporated into simple formulas, effectively functional as expert system rules, as illustrated in Table 7.2. These rules are used in a simple decision tree classification approach along with the use of Predictive Select at each of the three binary decisions. The three decisions are:

1. Is it indoors or outdoors? This is decided by the first binary classification step. The predictive "object rule" for this decision is the sum of the trees and building objects found.
2. For those considered indoors, is it in a home or a business? This is the second and last binary classification step for the indoors images. The "object rule" for this decision is the number of chairs minus the number of desks found.
3. For those considered outdoors, is it urban or rural? This is the second and last binary classification step for the outdoors images. The "object rule" for this decision is the number of trees minus the number of buildings found.

Table 7.2 summarizes the accuracy of each recognizer for each of the three object rules. Each object rule range is divided into four subranges, and the accuracy of each of the recognizers determined for each subrange. In the table, the highest accuracy recognizer for each subrange is underlined. The percentage of ground truth (or training) images in each subrange is given in parentheses ("pct =") in the table.

Table 7.2 Accuracy of each recognizer for different object rule subranges. The classification accuracy is either for "outdoors" versus "indoors," "rural" versus "urban," or "home" versus "business," and errors are classifications as "indoors," "urban" or "business"

| Object Rule | Recognizers | | |
	Recognizer A	Recognizer B	Recognizer C
Indoors vs. outdoors			
1. Trees + Buildings > 5 (pct = 0.204)	*0.993*	0.974	0.981
2. Trees + Buildings = 5 (pct = 0.266)	*0.988*	0.967	0.976
3. Trees + Buildings = 4 (pct = 0.319)	*0.967*	0.943	0.946
4. Trees + Buildings ≤ 3 (pct = 0.211)	0.898	0.901	*0.904*
Rural vs. urban			
1. Trees − Buildings > 3 (pct = 0.093)	*0.945*	0.937	0.938
2. Trees − Buildings = 3 (*pct* = 0.123)	*0.933*	0.929	0.928
3. Trees − Buildings = 2 (*pct* = 0.144)	*0.887*	0.876	0.873
4. Trees − Buildings ≤ 1 (pct = 0.640)	0.767	0.788	*0.834*
Home vs. business			
1. Chairs − Desks > 2 (pct = 0.042)	0.904	0.913	*0.921*
2. Chairs − Desks = 2 (pct = 0.227)	0.809	0.834	*0.844*
3. Chairs − Desks = 1 (pct = 0.585)	0.756	0.767	*0.804*
4. Chairs − Desks < 1 (pct = 0.146)	0.766	*0.819*	0.754

In Table 7.3, the accuracy of each of the three recognizers is provided for each of the three classes. The reader can compute the values in Table 7.3 by summing the "pct" values multiplied by the accuracy values in Table 7.2. The overall accuracy of Recognizers A, B, and C are 0.769, 0.771, and 0.796, respectively.

The Predictive Select pattern uses the subclass accuracies to select one of the recognizers. For example, suppose that for a new image to be analyzed, Categorizer A finds 4 trees and 1 building; Categorizer B finds 4 trees and 2 buildings; and Categorizer C finds 3 trees and 2

Table 7.3 Accuracy of each recognizer for different object rule ranges. The overall accuracy is the mean of the accuracies for the complete "outdoors" (indoors vs. outdoors, followed by rural vs. urban) and the complete "indoors" (indoors vs. outdoors, followed by home vs. business)

| Object Rule | Recognizers | | |
	Recognizer A	Recognizer B	Recognizer C
1. Indoors vs. outdoors	0.963	0.947	0.952
2. Rural vs. urban	0.821	0.832	0.861
3. Home vs. business	0.776	0.796	0.811
4. (1) × (2) = outdoors accuracy	0.791	0.788	0.820
5. (1) × (3) = indoors accuracy	0.747	0.754	0.772
6. ((4) + (5))/2 = overall accuracy	0.769	0.771	0.796

buildings. According to Table 7.2, then, the subrange accuracy for Categorizer A is 0.988; for Categorizer B it is 0.974; and for Categorizer C it is 0.976. Thus, we select Categorizer A for this image for the "indoors versus outdoors" decision. We may select a different categorizer for the next decision based on the number and type of objects each recognizer finds in the same manner as the "object-based scoring matrix" generated from the data in Table 7.2. Thus, for one image, we may accept the output of Recognizer B followed by the output of Recognizer A; for the next, we may accept the output of Recognizer C followed by the output of Recognizer B, and so on.

In order to estimate how much this approach will improve the overall system accuracy when deployed, we can look at the predicted accuracies under each of the "object rules" with the simplified assumption that all three recognizers will report the same subclass for the same image. For example, we assume that when Recognizer A finds exactly 5 "trees + buildings," so do the other two recognizers. This is not the case, of course—for example, Recognizer A may find 4 trees and 1 building when Recognizer B finds 3 trees and 1 building—but it does provide a (usually) fair way to predict the system behavior given only the data in Table 7.2. For this estimate of the overall predictive select system accuracy, we simply use the underlined accuracies in each row, multiplied by the appropriate "pct," and sum. This "optimized" predictive selection system is estimated to achieve an overall accuracy of 0.813, which corresponds to an error rate of 18.7%, a rate 8.3% less than the error rate (20.4%) of the best overall Recognizer, C. This subrange-based approach is also predicted to be better than using the best individual categorizers for the three classes of categorization—that is, Recognizer A for the "indoors versus outdoors" decision and Recognizer C for the other two decisions, with an accuracy of 0.805 and thus an error rate of 19.5%.

The observed accuracy for the overall system on a data set equal in size to the training set (500 images) was very similar to that predicted—0.815. This example demonstrates how predictive selection can be used on multiple categorizations in a decision tree classification system to improve the overall system accuracy. The approach outlined is generally effective since Predictive Select (Section 6.2.4) results in the selection of a single intelligence generator from a multiplicity of generators. In the example provided in Tables 7.2 and 7.3, accuracy is reported rather than the preferred precision metric. However, choosing the information generator with the highest accuracy is still a good approach (since we do not know the precision or recall), as accuracy, a, is derived from precision, p, and recall, r, as

$$a = \frac{2pr}{p+r}.$$

It is obvious that if precision and recall are highly correlated, then so are accuracy and precision. Therefore, the selection of the best generator based on which generator provides the highest accuracy for the subclass identified by the generator is generally valid.

7.3.3   Biometrics

Biometrics comprises an interesting class of machine intelligence problems, which have a broad range of applications and thus a broad class of possible constraints. Biometric authentication, for example, may need to be established on devices with limited processing and/or limited connectivity. Authentication may be performed once during sign-on; intermittently as the security policy dictates when the user changes the tasks he wishes to perform; or continuously in, for example, a multi-user environment.

Because of variable and often unpredictable biometric system requirements, biometric authentication systems can benefit from the Constrained Substitute first-order meta-algorithmic pattern, originally described in Section 6.2.2. This pattern provides a suitable reduced-expense replacement for a higher-expense generator, where expense is measured in terms of bandwidth, cost, processing time, or a combination thereof. For deployment, this pattern requires a set of competing intelligence generators that are each, independently, capable of performing a specific task. As discussed in Section 6.2.2, the substitute generators are named partial generators, and they can substitute for a full generator under various conditions. If the correlation between the results of a partial generator and the full generators is sufficiently high, then the substitution can take place.

The decision for substitution can be crafted to address the constraints of biometrics. Figure 7.5 illustrates how the Constrained Substitute pattern (Figure 6.3) is modified to accommodate static, intermittent, and real-time biometrics. During training, the full biometric algorithm, which provides the highest authentication accuracy and statistical confidence, is correlated with each of the potential substitutes, which in this case are low-cost biometric algorithms. The results can be assigned to different clusters or categorizations, as shown. Additionally, the output of the low-cost biometric algorithms are compared to ground truth in order to provide a relative ranking among each of the low-cost algorithms for different clusters/categorizations.

During run-time, the content is assigned to a specific category and the different biometric algorithms are rated for their ability to perform in place of the full biometric algorithm. As an example here, suppose we wish to establish and maintain biometric authentication during a conversation. We have three low-cost biometric algorithms, labeled A, B, and C. In addition, we have a full biometric algorithm. We wish to determine under what conditions one of the low-cost biometric algorithms may substitute for this full biometric algorithm.

The first condition is concerned with the probability of a false identification. The data for this condition is shown in Table 7.4. Three different conditions are presented. One, the high security case, requires high biometric identification confidence, with less than one chance in a billion of a false identification; that is, $p < 10^{-9}$. The next, the security within a potentially noisy environment case, requires relatively high biometric identification confidence, $p < 10^{-6}$. Finally, the lowest security case, $p < 10^{-3}$, may be used when the system is used mainly for tracking and not for access control. The biometric confidences are established by analyzed fixed length strings of speech data. A different number of fixed length strings are required by the different biometric algorithms. The full algorithm (FA), for example, may have a more extensive database for comparison, better analysis algorithms, better signal filtering, and so on, and thus require more processing time. The algorithm processing times vary from 110 to 250 ms, as listed in Table 7.4. The processing time is the amount required to analyze 2.0 s of voice data. The data presented are mean processing times; in general, variance is a relatively small percentage of the means for speech-related processing.

In Table 7.4, the low-cost biometric algorithm C cannot provide an equivalent level of authentication security—measured in terms of statistical confidence—to the confidence that an FA can provide after an equivalent amount of processing time. However, low-cost biometric algorithm A can provide $p < 10^{-9}$ and $p < 10^{-3}$ security after 880 and 220 ms of processing, improving on FA. Algorithm B can provide $p < 10^{-9}$ security after 850 ms of processing, improving on FA. So, if matching or improving on the performance time of the FA is the constraint, we may effectively substitute for FA under two of the three security conditions.

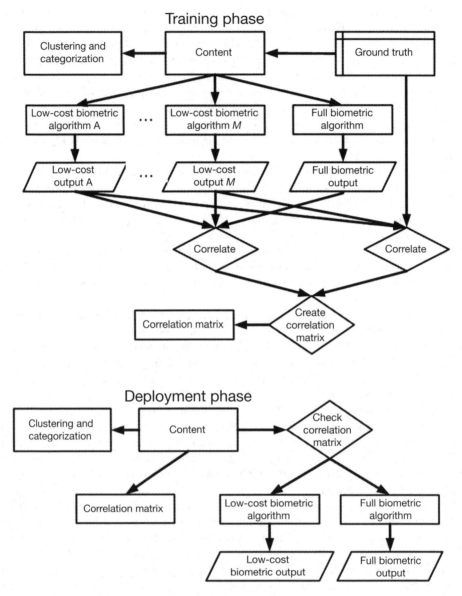

Figure 7.5 Constrained Substitute first-order meta-algorithmic pattern as applied to the scene biometric example described in the text

A more realistic set of constraints is addressed when we consider the cost of operating each algorithm along with the cost of bandwidth. The high-level equation describing this model is

$$\text{Cost} = (C_A + C_p t_p + C_B t_B) n_p,$$

where n_p is the number of times to run the algorithm, C_A is the cost (in \$) to run the algorithm once, C_B is the cost of the bandwidth for a given time period, t_B is the time to send the

Table 7.4 Relative processing time required for three different probabilities of false biometric identification, for three low-cost biometric algorithms (A, B, and C) and the full biometric algorithm (FA). Italicized table entries are those for which a low-cost biometric algorithm can meet or exceed the confidence of an FA after the same amount of processing time

Datum	Biometric Algorithm			
	Low-Cost A	Low-Cost B	Low-Cost C	Full Algorithm
Processing time, t_p (ms)	110	170	140	250
# Times to process 2.0 s of voice data to obtain $p < 10^{-9}$	8	5	8	4
# Times to process 2.0 s of voice data to obtain $p < 10^{-6}$	5	4	5	2
# Times to process 2.0 s of voice data to obtain $p < 10^{-3}$	2	3	3	1
Processing time for $p < 10^{-9}$ (ms)	*880*	*850*	1120	1000
Processing time for $p < 10^{-6}$ (ms)	550	680	700	500
Processing time for $p < 10^{-3}$ (ms)	220	510	420	250

information for one run of the algorithm, C_p is the cost of the processing for a given time period, and t_p is the processing time. Note that the governing constraint on the system is that

$$\left(t_p + t_B\right) n_p < t_{max},$$

where t_{max} is the maximum amount of time allowed for processing and data transmission for the n_p number of times the algorithm is to be run. Most importantly, this equation is used to determine n_p from t_{max} using

$$n_p = (\text{int}) \left[\frac{t_{max}}{t_p + t_B} \right].$$

Thus, n_p is the integer part of $t_{max}/(t_p + t_B)$. Using this set of equations, we can now determine which constrained substitute, if any, can perform in place of the FA. Table 7.5 collects the data for one incarnation, where $C_p = \$0.001/s$, $C_B = \$0.01/s$, t_p are the values from Table 7.4, t_B is $\$0.04/s$, and t_{max} is set at 1.0 s. The values for C_A, which cover development, testing, support, and licensing costs, are $\$0.0001$, $\$0.0002$, $\$0.0001$, and $\$0.0005$ per use, respectively, for algorithms A, B, C, and FA. In 1.0 s or less, the algorithms can be performed three to six times, affording up to $p < 10^{-6}$ statistical security. Thus, functionally, each of the low-cost algorithms can provide a constrained substitution for the functionality of the FA when a 1.0 s maximum task time is allowed. However, only two of the so-called low-cost algorithms substitute for the FA at a lower overall cost. Low-cost algorithm A reduced the overall cost by 10.7%, and low-cost algorithm B reduces the overall cost by 7.2%.

Suppose t_{max} is set so that the FA achieves the highest level of security; that is, $p < 10^{-9}$. In order to achieve this, a $t_{max} \geq 1.16$ s is required. In 1.16 s, the low-cost algorithms A, B, and C process 7, 5, and 6 sets of 2.0 s of audio, respectively, which means that only low-cost algorithm B can also provide authentication false positive probability of less than $p < 10^{-9}$ in the same time frame as the FA. However, the overall system cost for the low-cost algorithm B is $\$0.004\,88$ per event, higher than the overall system cost for the FA, at $\$0.004\,60$ per event.

Table 7.5 Deployment fact comparison for three low-cost biometric algorithms (A, B, C) with a full algorithm (FA). See text for details

Datum	Speech Analysis Biometric Algorithm			
	Low-Cost A	Low-Cost B	Low-Cost C	Full Algorithm
C_A (\$)	0.000 1	0.000 2	0.000 1	0.000 5
C_p (\$/s)	0.001	0.001	0.001	0.001
C_B (\$/s))	0.01	0.01	0.01	0.01
t_p (s)	0.110	0.170	0.140	0.250
t_B (s)	0.040	0.040	0.040	0.040
t_{max} (s)	1.0	1.0	1.0	1.0
n_p	6	4	5	3
$C_p t_p$ (\$)	0.000 11	0.000 17	0.000 14	0.000 25
$C_B t_B$ (\$)	0.000 40	0.000 40	0.000 40	0.000 40
$C_A + C_p t_p + C_B t_B$ (\$)	0.000 61	0.000 77	0.000 64	0.001 15
Cost (\$)	0.003 66	0.003 08	0.003 20	0.003 45
Security approaches supportable	$p < 10^{-6}$, $p < 10^{-3}$	$p < 10^{-6}$, $p < 10^{-3}$	$p < 10^{-6}$, $p < 10^{-3}$	$p < 10^{-6}$, $p < 10^{-3}$

This example illustrates, therefore, a situation in which the FA provides a very competitive overall system cost in spite of its greater processing time and algorithm cost. In this case, the relatively high bandwidth costs are the reason. For mobile, distributed computing, however, such relative costs for the different parts of the system are realistic.

The same high bandwidth costs provide the rationale for an entirely different use of the Constrained Substitution pattern for this task. Namely, if one of the low-cost algorithms could be made to process information locally, its bandwidth cost could be eliminated, although the effective processing cost would likely rise. Nevertheless, since the bandwidth costs are more than half the overall costs for the low-cost algorithms, this type of system redesign would likely be effective.

7.3.4   Security Printing

The final primary domain, security printing, can use the Sequential Try or Try first-order meta-algorithmic pattern (introduced in Section 6.2.1) to solve several different machine intelligence problems. The first is based on the chroma-enhancing tile (CET) approach described in Section 4.2.4. The CET approach implements a scaled data embedding approach whereby different-sized clusters of the barcode tiles—1 × 1, 2 × 2, 4 × 4, and so on—encode related sets of information. The larger the clusters, the lesser the density of information. However, the larger clusters are directly related to the smaller clusters. The Sequential Try pattern can be deployed when the correct sequence is known; for example, it may be included as another printed mark on the same package, label, or document. In this case, the first "try" is the algorithm that attempts to read all of the tiles in the color barcode. If successful, and the tile sequence matches the correct sequence for the printed object, then the analysis is complete. If unsuccessful, then the next try is for the 2 × 2 tile clusters. This continues until a successful read occurs or no larger tile clusters for which to attempt decoding exist.

Table 7.6 Agreement matrix, depicting the percentage of decoded strings that map correctly for clusters comprised different numbers of tiles. An agreement of 0.167 is equivalent to random guessing, since there are six colors used in this color barcode—red, green, blue, cyan, magenta, and yellow

Cluster Size	1×1	2×2	4×4	8×8
1×1	1.000	0.345	0.284	0.247
2×2	0.345	1.000	0.997	0.995
4×4	0.284	0.997	1.000	1.000
8×8	0.247	0.995	1.000	1.000

However, the correct sequence for the color barcode may not be known at the time of decoding; that is, until it is successfully decoded. In this case, an *agreement matrix* as depicted in Table 7.6 should be computed. This matrix contains the agreement percentages when comparing the output for successively larger clusters to the encoded data of the smaller clusters in coordination with their CETs. Since six colors are used—red, green, blue, cyan, magenta, and yellow—a minimum agreement of 0.167 is expected due to randomness alone. The data for Table 7.6 show 100% agreement between the 4×4 and 8×8 clusters. The 4×4 clusters contain three times the data density as the 8×8 clusters, accounting for the fact they are one-fourth the size but require 25% of their area for CETs. Thus, the 4×4 clusters will be used for authentication, since they provide higher security than the 8×8 clusters.

The decoded 4×4 and 8×8 clusters are 99.5% and 99.7% in agreement with the decoded 2×2 clusters in Table 7.6. If a relatively small error rate, such as 0.5% or 0.3%, can be tolerated or even compensated for using error-correcting code, then the 2×2 clusters can be used for authentication, since their density is three times that of the 4×4 clusters and nine times that of the 8×8 clusters. However, no amount of error-correcting code will allow the 1×1 clusters to be used for authentication. Their agreement with the 2×2, 4×4, and 8×8 clusters is only 34.5%, 28.4%, and 24.7%, respectively. This indicates that most of the 1×1 clusters will be decoded incorrectly.

Table 7.6, therefore, provides the means to maximize the amount of data that can be encoded into a color barcode implementing CETs and the scaled decoding associated with them. Figure 7.6 illustrates how the Try first-order meta-algorithmic pattern is deployed for this security printing application.

In addition to this example, there are a number of other security printing applications that can use the Sequential Try or Try pattern. One reason these patterns are a good match for security printing is the dependency of security printing on the use of variable data printing technology. For example, the Try pattern can be used to simultaneously decode multiple variable printed features when they are encoded independently. If, instead, the variable printed features are hybridized, or chained, then the output of one variable feature may be used to determine how or where to extract the information encoded in another security mark. The Sequential Try pattern is well suited to this hybridized approach.

7.4 Secondary Domains

The secondary domains chosen for this chapter are intended to provide an example implementation for the last remaining first-order meta-algorithmic pattern (Voting and Weighted Voting).

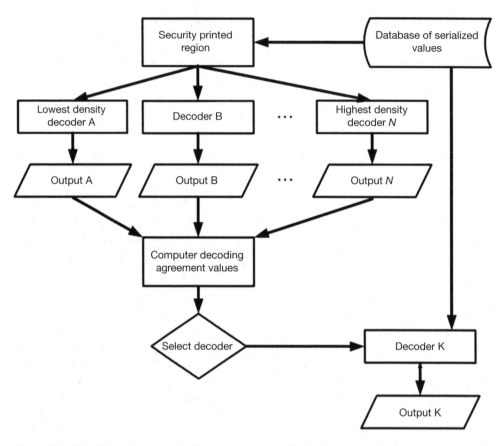

Figure 7.6   Try first-order meta-algorithmic pattern as applied to the staggered, chroma-enhancing tile (CET) technology

This is provided for the NLP task, part-of-speech (POS) analysis. The other two secondary domains, medical signal processing and medical imaging, illustrate further the Constrained Substitute and Predictive Select patterns.

7.4.1   Medical Signal Processing

The medical signal processing task of interest in this section is the analysis of the electro-cardiogram (ECG). The ECG is a recording of the electrical potential on the surface of a human that can be used to monitor the electrical activity of the heart, and from this determine important timing events during the cardiac cycle. There are several different electrical lead configurations. The standard three-lead configuration uses electrodes on the right arm (RA), left arm (LA), and left leg (LL). Lead I is the difference in electrical potential in the LA with respect to the RA; that is, LA-RA. Lead II is LL-RA. Lead III is LL-LA. Leads I, II, and III are at $0°$, $60°$, and $120°$, respectively, by convention. Thus, $0°$ points to the right and $90°$ points up. A typical recording for Lead II is shown in Figure 7.7. In addition to these leads, there are

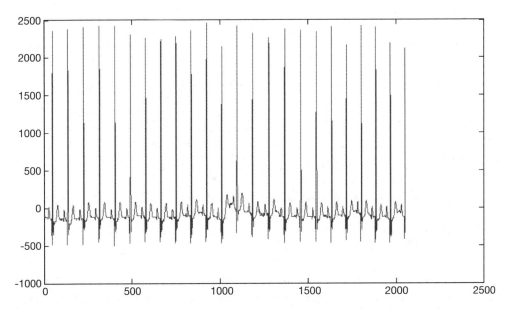

Figure 7.7 A Lead II electrocardiographic recording. The abscissa represents time in centiseconds, and the ordinate represents voltage in microvolts. The recording includes 23.5 cardiac cycles, with the 24 large peaks corresponding to the R wave. The pulse rate is 69 bpm. Between successive R waves are the prominent T wave, then the less prominent P wave. Some baseline deviation, due to motion artifact, is visible between 10 and 12 s of the recording

also three augmented leads, the aV_L, aV_R, and aV_F leads. These leads are at $-30°$, $-150°$, and $90°$, respectively. These six bipolar leads define electrical activity along the frontal plane of the chest. Six more standard leads are unipolar, and help to record surface electrical activity in a plane perpendicular to the frontal plane. These six leads are simply designated V1–V6.

In this section, we are concerned with the three bipolar leads, I, II, and III, for ECG diagnosis. ECG diagnosis is used to determine, among other information, (1) the timing of the P wave, corresponding to atrial contraction, (2) the timing of the QRS complex, corresponding to ventricular contraction, and (3) the timing of the T wave, corresponding to ventricular repolarization. The timing and magnitude of these three electrical events, designated P, R, and T, can be used to determine the following important cardiac cycle events:

1. Heart rate from the frequency of the cardiac cycle. The cardiac cycle is the timing between two consecutive P, R, or T events. If, for example, the P events occur at 11.3 s and then 12.2 s, heart rate HR = 60/(12.2 − 11.3) beats per minute, or 66.7 bpm.
2. PR interval, the time between successive P and R events, which indicates how long it takes electrical activation of the atria and ventricles to occur. The PR interval is also used to ensure that normal cardiac excitation occurs. Anomalies in the PR interval may indicate a wide variety of cardiac disorders, including atrial fibrillation (two successive P events without an interposed R event), ventricular fibrillation (two successive R events without an interposed P event), ventricular escape (longer R-R timing than P-P timing, and unpredictable P-R timing), and so on.

Table 7.7 Set of electrocardiographic features to monitor and the corresponding elements in the ECG necessary to determine these features

Feature	Minimum ECG Elements Necessary to Determine these Features
Heart rate	R wave peak timing
PQ interval	R wave peak timing and P wave peak timing
QT interval	R wave peak timing and T wave peak timing
TP interval	P wave peak timing and T wave peak timing
Vectorcardiogram (VCG)	R, P, and T wave peak magnitudes

3. The relative magnitude of the P, R, and T events on the three leads is used to compute the angle of the mean cardiac vector, which can be used to diagnose a variety of conductive disorders of the heart, such as ventricular hypertrophy and injury currents; for example, caused by cardiac ischemia.

We are interested in using the Constrained Substitution pattern for ECG signal analysis. In order to provide a useful ECG-based system, we require the set of cardiac cycle features listed in Table 7.7. These are the timing and magnitudes of the peaks of the P, QRS, and T waves (generally the P, R, and T peaks). From these, a wide host of disorders—including first-degree atrioventricular (AV) block, Wenckebach periodicity, premature ventricular contraction, atrial flutter, bradycardia, tachycardia, and sinoatrial node delay—can be identified.

Table 7.8 summarizes 15 approaches to analyzing electrocardiographic content. They are arranged in hierarchical order, based on the relative compression of data predicted for each approach. For this particular system the amount of storage required is the cost. Thus, the lowest cost option among those in Table 7.8 is the approach with the lowest number of bytes per cardiac cycle that achieves the analysis goal. The total amount of data, D, is determined from

$$D = \text{Bps} \times f_S \times T,$$

where Bps is the number of bits/sample, f_S is the sampling frequency in hertz, and T is the total amount of signal time in seconds for analysis.

Approach [1] simply counts the number of cardiac cycles, N_{CC}, in time T. Assuming an integer value requires 4 bytes for storage, this approach requires only $4/N_{CC}$ bytes/cardiac cycle. This approach can provide a substitution for a higher-cost algorithm if only the heart rate is to be monitored. This may be valuable for an exercise machine, for example.

Approach [2] counts the number of normal cardiac cycles, N_{NCC}, and the number of abnormal cardiac cycles, $N_{ACC} = N_{CC} - N_{NCC}$. This requires only twice the data of Approach [1], but can identify abnormal cardiac sequences. This may be valuable for alerting hospital staff. It may also be used to trigger a higher-cost ECG analysis approach, if appropriate.

In Approach [3], the cardiac cycle intervals are collected in a histogram. For example, a simple histogram may collect the number of cardiac cycles falling in ranges 0.6–0.65 s, 0.65–0.7 s, and so on. If the histogram has a single narrow peak, it is likely that many arrhythmias such as AV block, atrial flutter, and atrial fibrillation are not occurring. This approach may therefore serve as a low-cost substitute, requiring only about 100 bytes of data, for other AV block and atrial arrhythmia detectors.

Table 7.8 A hierarchical list of ECG representations we can store, in ranked order of expected compression (i.e., maximum compression is achieved for (1)), which generally corresponds to simplicity of the final signal

Number in the Hierarchy	Description	Bytes per Cardiac Cycle
1	Number of cardiac cycles, N_{CC}, over a given time period T	$4/N_{CC}$
2	Number of normal and abnormal cardiac cycles over a given period of time	$8/N_{CC}$
3	Histogram of the cardiac cycle intervals, where the histogram represents N_R different time ranges	$4N_R/N_{CC}$
4	The timing events for P, R, and T waves for each interval	24
5	The timing events of (4), plus Q, S, and U waves	48
6	Tokenized versions of the cardiac cycle, using Bezier curves. Assuming 9 linear sections and 2 arcs for simplest interval, linear QRS, and arc P and T waves, this is 4 floating point values per section—$4 \times 11 \times 8$ bytes total	352
7	(6) + tokenizing of the residual	$\sim352 + 0.001D$
8	(6) + lossy compression of the residual	$\sim352 + 0.01D$
9	(6) + lossless compression of the residual	$\sim352 + 0.1D$
10	Average cycle + tokenizing of the residual	$\sim0.1D + 0.001D$
11	Average cycle + lossy compression of the residual	$\sim0.1D + 0.01D$
12	Average cycle + lossless compression of the residual	$\sim0.1D + 0.1D$
13	Previous cycle + tokenizing of the residual	$\sim0.2D + 0.001D$
14	Previous cycle + lossy compression of the residual	$\sim0.2D + 0.01D$
15	Previous cycle + lossless compression of the residual	$\sim0.2D + 0.1D$

N_{CC} is the number of cardiac cycles in the ECG data record of interest. Integer values are assumed to be stored with 4 bytes of data. Timing values are assumed to be stored with 8 bytes of data. D is the total amount of data in a cardiac cycle at full sampling frequency and the complete set of bits for the voltage measurements. Please see the text for details on computing the bytes/cardiac cycle column.

Approach [4] stores the timing events for all P, R, and T wave peaks. Assuming a floating point value requires 8 bytes, this reduces the data storage requirement to 24 bytes/cardiac cycle. With the addition of these data, the PR, RT, and TP intervals can be calculated, along with their histograms. This allows the discovery of additional cardiac arrhythmias, including SA node block and delayed ventricular repolarization consistent with ischemia or damage due to myocardial infarction. For an additional 24 bytes/cardiac cycle of data storage, Approach [5] keeps the timing data for the Q, S, and U waves. The Q and S wave timing information can help identify left or right axis deviation, bundle branch blockage, and ectopic pacing. The U wave, if, for example, inverted, may identify myocardial ischemia or even left ventricular volume overload. As with Approach [4], this provides a low-cost substitute for saving a less digested ECG signal.

Approach [6] introduces ECG tokenization, in which the ECG signal is represented by a set of Bezier curves. Bezier curves are typically used to describe typefaces—any published typeface consists of the Bezier curves describing each of the 256 characters in its set. Most Bezier curves are lines or arcs on a circle, although in general they are described by any quadratic expression. A Bezier curve is a quadratic equation used to describe the slope of a line or the curve between two points; for example: $y = mx + b$. Scaling is achieved simply by bringing a multiplier into the equation: $y = M \times (mx + b)$. P waves are typically semicircular (an arc of 180°) in shape, and this is a very straightforward Bezier (center point of the circle, radius, and start/end angles) curve. More complex P waves can be described by, among other combinations, (a) two linear pieces and one arc, (b) two linear pieces and three arcs, or (c) three arcs. The T waves and U waves (if present) can be described similar to the P waves. The PQ, ST, TU (if present), and TP/UP intervals can be described by a single line segment each. This is also a very simple Bezier curve, described by two points. The remaining portion of the cardiac cycle is the QRS complex, which in normal conditions can be described with six line segments, two for each of the three waves. Altogether, the P and T waves, QRS complex, and PR, RT, and TP segments require 352 bytes/cardiac cycle. More is required if the U wave is visible. If the ECG is sampled at 1000 Hz and stored in floating point, and the pulse rate is 72 bpm, this still represents a 94.7% reduction in data storage. However, additional diagnostic information can be gleaned from this information, such as baseline (evident in Figure 7.7) and lengthened P and/or T waves. Approach [6] thus provides additional information over Approach [5], while still providing a significantly lower-cost storage solution than storing the entire ECG waveform.

The next three approaches, [7], [8], and [9], introduce the use of a residual signal. After the Bezier curve representation of the ECG is formed for Approach [6], the Bezier curve representation is subtracted from the actual signal. The difference signal, or residual, is then itself tokenized, usually using straight line segments, for Approach [7]. Approach [8] provides lossy compression of the residual from [6], and Approach [9] provides lossless compression of the residual. Each of these requires 352 bytes/cardiac cycle for the Bezier curve, and an additional percentage of the original data for tokenization (0.1%), lossy compression (1%), or lossless compression (10%). These three approaches provide successive better approximations to the actual ECG signal, and so may provide additional diagnostic information, especially for biometrics or more subtle arrhythmias such as minor ischemias or conductive path blockages.

The next three approaches, [10], [11], and [12], use the average cardiac cycle in place of a Bezier curve representation. Since cardiac cycles vary in length, the average cycle will be time compressed or stretched to fit each successive cycle. The same three residual representations—tokenization [10], lossy compression [11], or lossless compression [12]—are used along with the average cardiac cycle starting point. Approaches [13], [14], and [15] repeat these three residual representations for the previous cycle, which will tend to require more data since no averaging is used. As with Approaches [7]–[9], Approaches [10]–[15] represent increasingly higher-cost alternatives to storing the original waveform. Approaches [9], [12], and [15], in fact, provide the means to reconstruct exactly the original signal, and thus are equivalent.

The Constrained Substitute pattern is implemented in the form shown in Figure 7.8 to allow the substitution of a lower-numbered approach in Table 7.8 for a higher-numbered approach. This version of the Constrained Substitute pattern provides three exact solutions—Approaches [9], [12], and [15]—and compression percentages ranging from 70% to more than 99%. Since it generally offers higher compression, Approach [9] is usually the preferred method for replacing the exact original ECG, although in practice the relative bytes/cardiac cycle are dependent on Bps, f_S, and T.

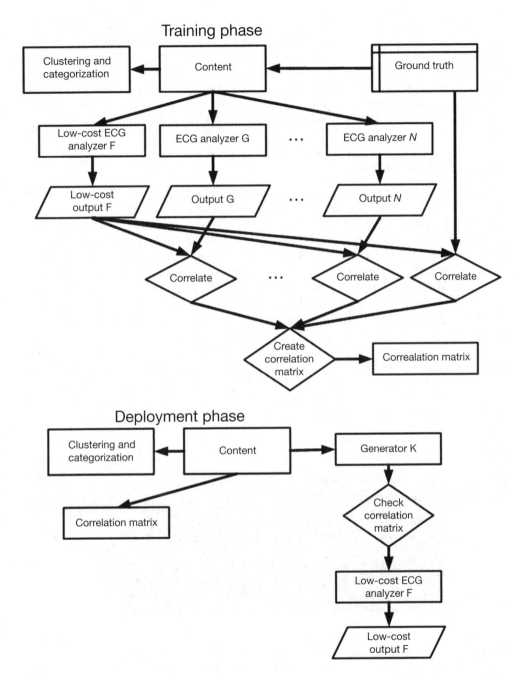

Figure 7.8 Constrained Substitute first-order meta-algorithmic pattern as applied to the ECG analysis as described in this section

As discussed above, more and more diagnoses are possible as the number increases from [1] to [9], allowing the lower-cost, lower-numbered storage approaches to effectively substitute for salient aspects of the full-data storage approaches, [9], [12], and [15]. Because a number of ECG monitoring applications require only a modest amount of information, these low-cost approaches are broadly applicable and offer a huge storage cost savings. While the example focuses on data compression as the system cost, it should be noted that the size of the data after compression could also directly impact bandwidth costs if the ECG monitoring is remote.

7.4.2   Medical Imaging

Compared to medical signal processing, medical imaging adds one or two dimensions to the mixture. In this section, we consider the use of the Predictive Select pattern, introduced in Section 6.2.4, for classifying biomedical images. The images selected in this case are of craniofacial implant materials, manufactured from TiO_2, along with the surrounding craniofacial bone, which, if biocompatible, grows into the implant (bone ingrowth) and grows directly against the surface of the implant (bone apposition). Three separate image analyzers were used to classify each of the biomedical images. Four classes of images were identified: (1) excellent ingrowth and excellent apposition; (2) poor ingrowth and excellent apposition; (3) excellent ingrowth and poor apposition; and (4) poor ingrowth and poor apposition. Class [1] corresponds to an implant with proper pore structure and biocompatibility. Class [2] indicates a biocompatible implant with nonoptimal porous structure. Class [3] identifies a nonbiocompatible implant with good pore structure. Finally, class [4] indicates a nonbiocompatible implant with nonoptimal pore structure. Figure 7.9 illustrates an example belonging to class [1] and to class [4].

In order to determine both an optimal pore architecture—which includes both pore size and percent porosity of the implant material—and an optimal biocompatibility for the

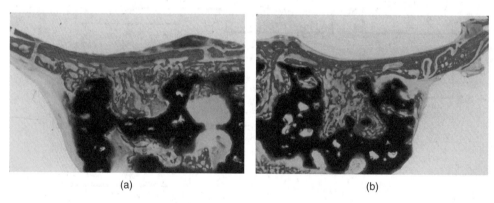

(a)   (b)

Figure 7.9 Sample images analyzed using two-dimensional (2D) medical imaging as described in this section. The black areas are the implant material, TiO_2, into which the bone (red areas) should grow and against which the bone areas should abut if biocompatibility is achieved. The lighter areas are voids, or porosities within the bone tissue. (a) An example of poor bone ingrowth and implant apposition, belonging to class [4], generally implying poor biocompatibility. (b) An example of good bone ingrowth and implant apposition, belonging to class [1], generally indicative of biocompatibility. See text for details

implant, a wide variety of implant porosities, and surface coatings were investigated. For each (porosity, coating) combination, a minimum of 100 two-dimensional (2D) cross-sections of implant/craniofacial material were captured. Training was performed on half of the images, and the remaining 50% were used for the later classifier evaluation. The meta-algorithmic approach chosen (Figure 7.10) requires the calculation of a predictive set of features, each of which can be used to determine classifier precision on the training data. These features are meant to be calculated relatively quickly from the image, and in so doing select one and only one image analyzer, which in turn provides the image classification.

Six features are computed for each new image during run-time: (1) the percentage of orange pixels within the convex hull of the implant region (%OrangeWithin), indicative of areas where porosity rather than bone material is within the implant; (2) the percentage of red pixels within the convex hull of the implant region (%RedWithin), indicative of areas where bone material has grown into the implant; (3) the percentage of orange pixels along the implant's external and internal boundaries, also known as the percentage of orange pixels apposed (%OrangeApposed), indicative of areas where porosity rather than bone material abuts the implant; (4) the percentage of red pixels apposed to the implant's external and internal boundaries (%RedApposed), indicative of areas where bone material abuts the implant; (5) the mean nonblack run lengths within the convex hull of the implant region (MeanNonBlackRunWithin), correlated with the mean porosity channel width in the implant; and (6) the mean pore size (MeanPoreRegionSizeWithin, estimated by created connected component regions from the nonblack pixels) within the convex hull of the implant region. For each image, each of these features will have a sample belonging to one of three or four subranges generally corresponding with different relative ingrowth, apposition, or pore size. The precisions of each of the three image analyzers for each subrange of each feature are collected in Table 7.9.

In Table 7.9, where possible, the subranges for the six features were designed such that roughly the same percentages of images belonged to each subrange. However, the distributions of, for example, the bone apposition values (%OrangeApposed and %RedApposed), which had three obvious subpopulations of different percentages, precluded this for some measurements. Nevertheless, using these percentages, it is left as an exercise for the reader to show that Analyzer A provides overall precision of 0.81, 0.81, 0.76, 0.71, 0.53, and 0.39 for the features %OrangeWithin, %RedWithin, %OrangeOpposed, %RedOpposed, MeanNonBlackRunWithin, and MeanPoreRegionSizeWithin, respectively, by simply computing the sum of the percentage values in parenthesis multiplied by the subrange precision, for all of the feature subranges. For Analyzer B, the reader can readily compute the same feature precision values as 0.88, 0.84, 0.75, 0.56, 0.66, and 0.35, respectively; for Analyzer C, these values are 0.84, 0.70, 0.84, 0.83, 0.46, and 0.39, respectively. We will assume independence among the features and equal weighting of each feature for the overall predictive value, even though we might be better served to use an approach summarized in Table 6.6 as a better means of estimating the relative weights of the six features based on the precisions calculated above. However, I have left out optimized weighting in this stage to keep the example below focused on the elements salient to the Predictive Selection pattern. Given this simplification, the estimated precision of Analyzers A, B, and C are 0.67, 0.67, and 0.68, respectively, when deployed by themselves. The overall precision of the system (selecting the best of the three analyzers) is therefore estimated to be 0.68.

However, when Predictive Selection is used, we wish to select the analyzer with the highest precision for each subrange of each feature. These are the underlined precisions in

Implant biocompatibility: statistical learning phase

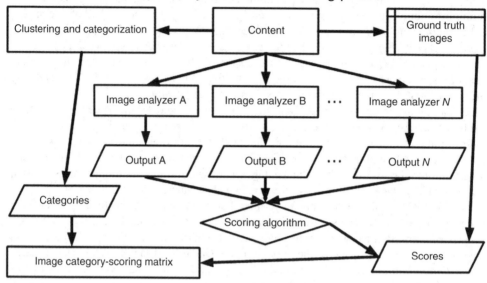

Implant biocompatibility: run-time phase

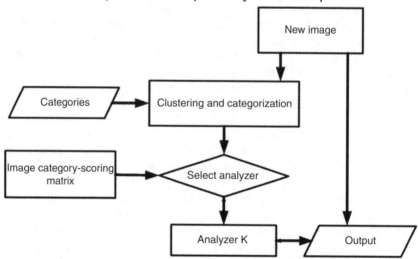

Figure 7.10 Predictive Select first-order meta-algorithmic pattern as applied to the image analysis problem described in this section. Each of the image analyzer engines produces its specific content, which is then associated with each of the image classes during training. During run-time, the new image is analyzed for a predictive set of features, and the analyzer that offers the highest precision for the given set is selected to make the final classification decision

Table 7.9 Precision for each of the three image analyzers—A, B, and C—for 20 different image feature subranges belonging to the six image features analyzed during the predictive portion of the pattern

Image Feature Subrange	Analyzer Precision		
	Analyzer A	Analyzer B	Analyzer C
%OrangeWithin I (0.27)	0.78	0.83	*0.85*
%OrangeWithin II (0.40)	0.81	*0.88*	0.85
%OrangeWithin III (0.33)	0.83	*0.91*	0.83
%RedWithin I (0.31)	0.77	*0.84*	0.64
%RedWithin II (0.39)	0.83	*0.86*	0.78
%RedWithin III (0.30)	*0.82*	0.81	0.66
%OrangeApposed I (0.52)	0.67	0.75	*0.81*
%OrangeApposed II (0.20)	*0.84*	0.76	0.83
%OrangeApposed III (0.28)	0.88	0.73	*0.91*
%RedApposed I (0.61)	0.65	0.55	*0.85*
%RedApposed II (0.19)	0.77	0.61	*0.79*
%RedApposed III (0.20)	*0.84*	0.52	0.83
MeanNonBlackRunWithin I (0.24)	0.44	*0.58*	0.42
MeanNonBlackRunWithin II (0.27)	0.61	*0.66*	0.46
MeanNonBlackRunWithin III (0.23)	0.57	*0.70*	0.51
MeanNonBlackRunWithin IV (0.26)	0.49	*0.68*	0.47
MeanPoreRegionSizeWithin I (0.22)	0.33	*0.41*	0.40
MeanPoreRegionSizeWithin II (0.31)	*0.42*	0.29	0.38
MeanPoreRegionSizeWithin III (0.27)	*0.40*	0.38	0.37
MeanPoreRegionSizeWithin IV (0.20)	0.38	0.33	*0.42*

The highest precision for each row, representing one feature and one subrange for the feature, is in italics. The percent of images falling in each subrange for the given features are indicated in parentheses; for example, (0.34) implies 34% of images map to that subrange for the given feature. The image analyzer to be used for making the final decision (see Figure 7.10) is determined by selecting the analyzer with the largest sum of precision values for all image feature subranges associated with the image to be classified.

Table 7.9. When these are selected, the estimated precision for the features %OrangeWithin, %RedWithin, %OrangeOpposed, %RedOpposed, MeanNonBlackRunWithin, and Mean-PoreRegionSizeWithin become 0.88, 0.84, 0.84, 0.84, 0.66, and 0.41, respectively. This represents an absolute maximum for precision, since in practice the recommended analyzer will vary for different features, and so we will not achieve this precision. Using the same simplifications as above for weighting the six features while combining, this results in an overall system precision estimate of 0.75. When deployed against the test set, this system achieved 70% accuracy, as opposed to the 66% accuracy actually observed for the best of the three individual analyzers (Analyzer C), a reduction in error from 34% to 30%, or a relative reduction in error rate of 12%. The 70% accuracy is well below the predicted meta-algorithmic accuracy for reasons cited in this paragraph, but still represents a valuable improvement.

Let us now consider one example of how to deploy the system. Suppose a new image has %OrangeWithin in subrange III, %RedWithin in subrange I, %OrangeApposed in subrange III, %RedApposed in subrange II, MeanNonBlackRunWithin in subrange II, and MeanPoreRegionSizeWithin in subrange III. Summing the appropriate analyzer precisions from Table 7.9, Analyzers A, B, and C sum to 4.26, 4.13, and 4.00, respectively. Dividing by the number of features, the expected precisions of Analyzers A, B, and C are 0.71, 0.69, and 0.67, respectively. We choose, therefore, Analyzer A with an expected improvement in precision of 0.02 (the difference of 0.71 and the mean of the three analyzers, or 0.69).

It is left as an exercise for the reader to show that, when using weighting based on the inverse of the error rate (see Table 6.6 and the associated discussion), the estimated precision of Analyzers A, B, and C are 0.73, 0.76, and 0.76, respectively, and the overall system precision is estimated to be 0.75. This is a powerful statement for the Weighted Voting meta-algorithmic approach, separate from the Predictive Selection pattern described above (which, coincidentally, also improved the overall system precision estimate to 0.75).

When this weighting approach is used in combination with the Predictive Select pattern (choosing the output from the analyzer with the highest precision for each subrange), the overall system precision estimate is 0.82 (left as an exercise for the reader, with the hint that the weights for the features in order presented in Table 7.9 are 0.23, 0.17, 0.17, 0.17, 0.06, and 0.02, respectively). This is a huge improvement over the 0.67 precision estimated without meta-algorithmics. When deployed against the test set, this system achieved an actual 76% accuracy, a reduction in error from 34% (the error rate for the best of the three individual analyzers) to 24%, or a relative reduction in error rate of 29%. Again, the 76% accuracy is below the predicted accuracy of 82% due to the simplifications we had to make to predict the final precision without knowledge of the correlation between features for different images. The reduction in error rate, nevertheless, is impressive.

This example shows the power of being able to apply knowledge from multiple meta-algorithmic patterns simultaneously, a topic we will return to in the next two chapters. Additionally, the severity of the image classification errors made is significantly reduced. In this section, the focus was on the classification accuracy, where meta-algorithmic approaches were able to reduce the error rate by 29% during deployment. However, it is important to point out that the nature of the classification errors also was changed by the meta-algorithmics. The most severe type of misclassification errors possible for this medical imaging system are when either (a) class [1] is mistaken for class [4], or (b) class [4] is mistaken for class [1]. This is true because these misclassifications actually represent two errors, one about optimal porosity of the implant and the other about biocompatibility of the implant. Without the use of meta-algorithmics, these types of misclassifications occurred on 3–4% of the images for the three analyzers when used alone. With the use of meta-algorithmics, these types of errors disappeared completely (on a test set of 100 images).

7.4.3 Natural Language Processing

NLP is the broad field concerned with internalizing human language (spoken and written) for use in other computer applications, services, and/or workflows. POS tagging is an important part of NLP. In it, each word in a set of text is tagged, or logically labeled, as a verb, noun, adverb, article, and so on. These tags are very important, because they identify important context (usually nouns), sentiment (usually verbs and adverbs), and cardinality.

Because the definition of words is relatively unambiguous in comparison to, for example, segmenting an object from an image, a simple Voting pattern can be used to decide the output of multiple POS tagging engines, commonly referred to as "taggers." Tagger combination approaches were introduced many years back—see, for example, Brill and Wu (1998), Tjong Kim Sang (2000), and Halteren, Zavrel, and Daelemans (2001)—but a detailed study of the advantages of voting was not provided in these references. In our work on the Voting and Weighted Voting patterns (introduced in Section 6.2.3), we decided to perform tagging on the well-known and publicly available Penn Treebank (Marcus, Santorini, and Marcinkiewicz, 1993) data set.

To perform our investigation, we chose three taggers. The first tagger is the transformation-based learning system commonly referred to as the Brill tagger (Brill, 1992), at the time (2001) we performed this research, perhaps the best known tagger. It first tags each word with its highest probability tag and then modifies the tagging according to learned transformation rules. The second tagger is the Infogistics NLP Processor (www.infogistics.com; accessed January 25, 2013). According to its Web site at the time, the system implements the traditional trigram model. The third tagger is the QuickTag from Cogilex R&D Inc. (www.cogilex.com; accessed January 25, 2013), whose Web site at the time explicitly claimed "tagging is not done on a statistical basis but on linguistic data (dictionary, derivational and inflectional suffixes, prefixes, and derivation rules)." These three taggers are hereafter referred to as the Brill, Infogistics, and Cogilex taggers.

We performed calibration on the three taggers using the University of Pennsylvania tag set (Marcus, Santorini, and Marcinkiewicz, 1993), and found the Brill tagger provided the highest accuracy. We therefore employed the simplest form of Weighted Voting (Figure 7.11) wherein

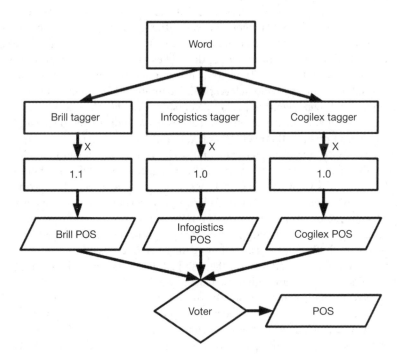

Figure 7.11 Figure showing Weighted Voting approach to part-of-speech (POS) tagging. The Brill tagger is weighted 1.1, and the other two taggers are weighted 1.0. If there is no agreement, then, the Brill tagger output is the chosen output

Table 7.10 Part-of-speech (POS) error rates for the individual POS taggers and the weighted voting of the tagging with three different corpora from the Amalgam project (http://www.comp.leeds.ac.uk/amalgam/amalgam/amalghome.htm; accessed January 25, 2013) as described in Lin (2002)

Corpus	Error Rate (percentage)				Error Reduction (%)
	Brill	Infogistics	Cogilex	Weighted Voting	
A	13.30	13.50	16.09	10.09	24.1
B	8.66	8.86	9.74	6.79	21.6
C	5.03	5.38	13.75	4.29	14.7

the Brill tagger was given the tie-breaking weight of 1.1 and each of the other two taggers was given the weight of 1.0. In case of complete disagreement, then, the Brill tagger's output was chosen. If, and only if, the other two taggers agreed but differed from the Brill tagger, was the Brill tagger's output not chosen.

Table 7.10 presents the results of the tagger combination when applied to novel corpora not associated with the training set (Lin, 2002). Corpus A was quite challenging, resulting in error rates of 13–16% for the individual taggers. The Brill tagger was again the most accurate, but only by a slight percentage. The Weighted Voting meta-algorithmics, however, significantly reduced the error rate, to 10%. The relative effect on error was 24.1% compared to the best individual classifier. Corpus B, less challenging than Corpus A, nevertheless followed its trends: the Brill tagger provided the lowest error rate, and Weighed Voting reduced the error rate by 21.6% compared to the Brill tagger. The least difficult corpus, Corpus C, had very mixed results, with tagger error rate varying by as much as 275% in comparing the Brill and Cogilex taggers (with the Brill tagger again providing the highest accuracy). Nevertheless, Weighted Voting reduced the error rate by 14.7% in comparison to the Brill tagger. These data seem to indicate that weighted voting provides a larger relative improvement on more challenging data sets; however, such a broad conclusion is not defensible with only three test sets.

A separate test of challenging data sets, in fact, seems to indicate the opposite. We added errors to the text (comprising all three Corpora A, B and C) by changing 0.5%, 1%, 2%, 3%, 4%, 6%, 8%, or 12% of the characters randomly. The results are shown in Table 7.11. For the error-free text, the Brill, Infogistics, and Cogilex taggers have POS error rates of 6.60%, 6.65%, and 10.66%, respectively. The Brill tagger outperformed the other two taggers for every character error rate. The Weighted Voting meta-algorithmic tagger outperformed the Brill tagger for character error rates of 0–3%. The Brill tagger, however, outperformed the Weighted Voting approach for character error rates of 6% or more. At this level of character error rate, the tagger POS error rates are all more than 1 in 6 words.

Clearly, this is a different trend than for the data in Table 7.10. There, increasingly difficult to tag text benefits from meta-algorithmics. This may be due to the fact that the POS tagging errors are relatively small; that is, relatively similar in POS. With the data of Table 7.11, however, character error rates of 6% or more correspond with the expectation that more than one-fourth of the words will have one of more character error. At this high of an error rate, the advantage of the meta-algorithmic approach disappears. This may be due to the POS tagging errors being more random in nature (in the mean, more distant from the actual tag), obviating the central tendency advantages of the Weighted Voting pattern.

Table 7.11 Part-of-speech (POS) error rates for the individual POS taggers and the weighted voting of the tagging with different percentages of simulated OCR errors occurring in the text (Lin, 2002). The original (0% error rate) text is the combined Corpora A, B, and C from Table 7.10

Simulated OCR Character Error Rates (%)	Brill Tagger Error Rates (%)	Infogistics Tagger Error Rates (%)	Cogilex Tagger Error Rates (%)	Weighted Voting Error Rates (%)
0.0	6.60	6.65	10.66	5.05
0.5	7.89	8.07	11.98	6.43
1.0	9.52	10.32	13.77	8.46
2.0	10.41	12.01	14.80	9.72
3.0	12.95	14.75	17.11	12.72
4.0	14.00	17.22	19.07	14.02
6.0	16.80	21.45	21.95	17.45
8.0	21.49	26.26	26.17	21.67
12.0	26.16	33.53	31.25	27.65

7.5 Summary

This chapter provided examples of how to employ first-order meta-algorithmic patterns. A broad set of domains were addressed, and for each example the improved results obtained were tabulated. The Tessellation and Recombination pattern was shown valuable for domains in which repartitioning of output may reduce error. The Predictive Select pattern was shown valuable for using domain expertise incorporated into rules to improve generator selection. Constrained Substitution was shown to be valuable for reducing a wide range of operational costs—both overall system and data storage in its two examples. Simultaneous analysis of multiple sets of content, such as in security printing applications, is well served by the Sequential Try/Try pattern. Weighted Voting was shown to be valuable both for combining features and for combining generators. Together, these five patterns provide a powerful set of tools to increase overall system accuracy, cost, and/or robustness.

References

Brill, E. (1992) *A Simple Rule-Based Part-of-Speech Tagger*. Proceeding of the Third ACL Conference on Applied NLP, Trento, Italy, pp. 152–155.

Brill, E. and Wu, J. (1998) *Classifier Combination for Improved Lexical Disambiguation*. COLING '98 Proceeding 17th International Conference on Computational Linguistics, pp. 191–195.

Halteren, H.V., Zavrel, J., and Daelemans, W. (2001) Improving data driven word class tagging by system combination. *Computational Linguistics*, **27** (2), 199–230.

Lin, X. (2002) Impact of imperfect OCR on part-of-speech tagging. HP Labs Technical Report HPL-2002-7R1, 5 pp.

Marcus, M.P., Santorini, B., and Marcinkiewicz, M.A. (1993) Building a large annotated corpus of English: the Penn Treebank. *Computational Linguistics*, **19** (2), 313–330.

Revankar, S.V. and Fan, Z. (1997) Image segmentation system. US Patent 5,767,978, filed January 21, 1997.

Shi, J. and Malik, J. (2000) Normalized cuts and image segmentation. *IEEE Transactions on Pattern Analysis Machine Intelligence*, **22** (8), 888–905.

Tjong Kim Sang, E.F. (2000) *Noun Phrase Recognition by System Combination*. Proceeding of ANLP-NAACL, pp. 50–55.

Wahl, F.M., Wong, K.Y., and Casey, R.G. (1982) Block segmentation and text extraction in mixed/image documents. *Computer Vision Graphics and Image Processing*, **2**, 375–390.

8

Second-Order Meta-algorithmics and Their Applications

What is important is to spread confusion, not eliminate it.

—Salvador Dalí

8.1   Introduction

Salvador Dalí apparently understood the central importance of the confusion matrix in the science of meta-algorithmics. In this chapter, we will see the confusion matrix in action, and discover how its confusion can indeed be spread to determination of precision, decision tree classification optimization, targeted classification, and ground truthing optimization.

This chapter provides more in-depth applications of the second-order meta-algorithmic patterns, introduced in Section 6.3. All nine of the second-order patterns are demonstrated in this chapter, applied once more to the usual four primary domains: (1) document understanding, (2) image understanding, (3) biometrics, and (4) security printing. In addition, second-order patterns are applied to the domains of image segmentation and speech recognition.

Second-order meta-algorithmics are in general more complicated than the first-order meta-algorithmics. A greater number of deployment options are available, enabling more tunable and robust systems to be architected. Most of the patterns employ two or more decisioning approaches, enabling more design flexibility through trade-off of the parameterization of these multiple components. These patterns use a new set of analysis tools: confusion matrix, output space transformation, expert decisioners, and thresholded confidence.

The confusion matrix is a very powerful tool for analysis, and can be used not just as a component in a variety of second-order meta-algorithmic patterns, but also as a filtering mechanism for assigning different partitions of an input space to different meta-algorithmic patterns. This approach will be introduced in the next section.

Output space transformation is an important tool for normalizing the output of multiple generators. In particular, complicated analysis engines—which are often treated as black boxes—often provide idiosyncratic confidence values for the set of possible outputs. Some

Meta-algorithmics: Patterns for Robust, Low-Cost, High-Quality Systems, First Edition. Steven J. Simske.
© 2013 John Wiley & Sons, Ltd. Published 2013 by John Wiley & Sons, Ltd.

engines may provide only a ranked set with no confidence values, others may provide actual probabilities, and still others may provide relative confidence values. Output space transformation may be used to create similar relative behavior among the engines, but more importantly it is used to create better behavior of the engines in combination for a particular meta-algorithmic pattern. Thus, output space transformation will typically provide a different approach for each meta-algorithmic pattern in which it is deployed.

Expert decisioners are used to internalize domain expertise in a pattern. The domain expertise may be a set of allowed ranges or other constraints, co-occurrences of specific data or patterns of data, or ratios of data. One important aspect of expert decisioners is that they can be introduced into a system even when there is no training data available. For example, the occurrence of the word "Feuerstelle" in an invoice may be used to make the decision that the document is in German, which then simplifies the downstream task of determining the company that created the invoice.

Finally, a thresholding approach compares the confidence in a first result to a relevant minimum or maximum value. If the confidence is sufficient, no further analysis is required and the output is this first result. Otherwise, additional analysis is performed and an alternate output is created.

8.2 Second-Order Meta-algorithmics and Targeting the "Fringes"

Why might a system architect decide to use a second-order meta-algorithmic pattern in place of a first-order pattern? One reason is the data to be analyzed represent two or more relatively distinct populations—once the input is assigned to the correct population, downstream analysis is greatly simplified. This means that there must be a good way to identify the existence of these populations. Typically, the smaller populations are obscured by the larger population(s). One of the reasons to deploy meta-algorithmics is to be able to identify these smaller populations. Another reason is to be able to target each smaller population with an additional generator. Each of these tasks can be assisted by the confusion matrix.

The general form of the confusion matrix is shown here. The "origin class" is also referred to in the literature as the "actual" or "true" class. The "assigned class" is also referred to as the "predicted" class. I use the term "origin" class here since it more closely captures the plasticity of origin of the data—especially in this section where we will see origin classes combined, thereby changing the "actuality" or "truth" of the original rows. I use the term "assigned" class since it is more general than "predicted" and as origin classes are recombined the original predictions are somewhat altered in context:

$$\text{Origin class} \begin{bmatrix} C_{11} & C_{12} & C_{13} & \cdots & C_{1N} \\ C_{21} & C_{22} & C_{23} & \cdots & C_{2N} \\ C_{31} & C_{32} & C_{33} & \cdots & C_{3N} \\ \vdots & \vdots & \vdots & \ddots & \vdots \\ C_{N1} & C_{N2} & C_{N3} & \cdots & C_{NN} \end{bmatrix}$$

Assigned class

For $X = 1, \ldots, N$, the sum $C_{X1} + C_{X2} + C_{X3} + \cdots + C_{XN} = 1.0$, since these values represent how the origin class X is assigned to all of the classes in the input space. In order to use the confusion matrix for evaluation of potential secondary populations, it may need to be reorganized. One way is to use maximum confusion:

$$\text{Maximum confusion} = \text{argmax}\,(\{i,\, j \neq i\} \in 1 \ldots N[C_{ij} + C_{ji}]).$$

If it then makes sense to organize classes $\{i,j\}$ as a single aggregate class (which can be later targeted by a specific binary classifier that can distinguish Class i from Class j), we have reduced the order of the confusion matrix and identified a smaller population within the overall population. Thereafter, this class can be treated as an individual class and later disaggregated with a targeted binary or series of binary classifiers to recover the original classes. The process continues so long as the aggregation of the two classes (and/or aggregate classes) with maximum confusion results in a measurable and significant improvement in the overall predicted system performance.

A simple example is used to introduce these concepts. Suppose we have the following confusion matrix for a 4-class problem:

$$
\begin{array}{c}
\text{Assigned class} \\
\begin{array}{cc}
\text{Origin} \\
\text{class}
\end{array}
\begin{bmatrix}
0.7 & 0.2 & 0.1 & 0.0 \\
0.15 & 0.8 & 0.05 & 0.0 \\
0.05 & 0.05 & 0.8 & 0.1 \\
0.0 & 0.05 & 0.05 & 0.9
\end{bmatrix}.
\end{array}
$$

Then, as shown in Table 8.1, the maximum confusion is between class 1 and class 2, for which it is 0.35. The overall accuracy of the table is 0.8, the percentage of classifications on the diagonal of the confusion matrix.

If we then combine the original classes 1 and 2 into an aggregated class, we obtain the following simplified confusion matrix:

$$
\begin{array}{c}
\text{Assigned class} \\
\begin{array}{cc}
\text{Origin} \\
\text{class}
\end{array}
\begin{bmatrix}
0.925 & 0.075 & 0.0 \\
0.1 & 0.8 & 0.1 \\
0.05 & 0.05 & 0.9
\end{bmatrix}.
\end{array}
$$

Table 8.1 Confusion (sum of paired off-diagonal elements $\{i, j\}$, with $j > i$, which is the error of mistaking class i with class j, or vice versa). The maximum confusion is for $\{i = 1, j = 2\}$, for which the confusion is 0.35

i	j (where $j > i$)	Confusion $= (C_{ij} + C_{ji})$
1	2	$0.15 + 0.2 = 0.35$
1	3	$0.1 + 0.05 = 0.15$
1	4	$0.0 + 0.0 = 0.0$
2	3	$0.05 + 0.05 = 0.1$
2	4	$0.05 + 0.0 = 0.05$
3	4	$0.05 + 0.1 = 0.15$

The larger confusion matrix is transformed into this one remembering that each row sums to 1.0. The overall accuracy of the aggregated confusion matrix is now 0.875. The overall system accuracy can then be improved by focusing on creating a high-accuracy binary classifier for original classes 1 and 2.

This approach is recursive. Next, the aggregate of original classes 1 and 2 and the original class 3 can be aggregated, based again on this combination having the maximum confusion, 0.175, of the three confusions in the simplified confusion matrix above—the other two being 0.5 and 0.15. This aggregation is used to create a new 2×2 confusion matrix as follows:

$$
\begin{array}{cc}
 & \text{Assigned} \quad \text{class} \\
\begin{array}{c} \text{Origin} \\ \text{class} \end{array} & \begin{bmatrix} 0.95 & 0.05 \\ 0.1 & 0.9 \end{bmatrix} .
\end{array}
$$

This further simplified confusion matrix has an accuracy of 0.925 for distinguishing class 4 from the aggregation of classes 1, 2, and 3. In order to make this step worthwhile, a highly accurate classifier for distinguishing class 3 from the aggregate of classes 1 and 2 must be created.

The equation for maximum confusion, while simple, is quite powerful. The example above shows how it can be used to aggregate classes into a single cluster of classes, suitable for a binary-decision-tree-based classification approach. However, it can also be used to create multiple clusters sequentially. To illustrate this, consider the following 6-class confusion matrix:

$$
\begin{array}{cc}
 & \text{Assigned} \quad \text{class} \\
\begin{array}{c} \\ \\ \text{Origin} \\ \text{class} \\ \\ \\ \end{array}
\begin{array}{c} 1 \\ 2 \\ 3 \\ 4 \\ 5 \\ 6 \end{array}
&
\begin{bmatrix}
0.71 & 0.07 & 0.04 & 0.01 & 0.15 & 0.02 \\
0.01 & 0.82 & 0.01 & 0.13 & 0.01 & 0.02 \\
0.04 & 0.03 & 0.75 & 0.05 & 0.02 & 0.11 \\
0.02 & 0.07 & 0.03 & 0.85 & 0.01 & 0.02 \\
0.04 & 0.01 & 0.02 & 0.02 & 0.90 & 0.01 \\
0.03 & 0.04 & 0.24 & 0.05 & 0.03 & 0.61
\end{bmatrix} .
\end{array}
$$

In this confusion matrix, the mean of the diagonal elements is 0.773, which is the overall accuracy of the 6-class classification. The sum of each individual row is 1.0, corresponding to 100% of the samples of each origin class. Since this is a normalized confusion matrix, the accuracy is also equal to the recall. The sum of the columns, however, shows that some classes act as *attractors* and others as repellers. Attractors have sums greater than 1.0, meaning that they are more likely to be false positives, but also in general more likely to have higher recall, than repellers if their accuracies are equal. Attractors could also be interpreted in terms of classifier bias.

Repellers, meanwhile, have sums less than 1.0, meaning that they are more precise than attractors if their accuracies are equal. In practice, the accuracies are not equal, but these are the expected trends. To that end, Table 8.2 shows that, for the 6-class confusion matrix above, the two repellers have higher relative precision and much lower relative recall than the four attractors.

In Table 8.3, the confusion values for all pairs of classifiers are shown. The maximum is for $i = 3, j = 6$. Thus, the obvious first choice is to combine classes 3 and 6 into an aggregate

Table 8.2 Classification statistics extracted from the 6-class confusion matrix. Since the confusion matrix is normalized (all classes equally probable, and each row sums to 1.0), the recall is the same as the accuracy. Precision is the ratio of the on-diagonal element to the sum of the column to which it belongs. The recall of the repellers (66%) is much lower than that of the attractors (83%), while the precision of the repellers (81%) is slightly higher than the precision of the attractors (77%)

Class	Column Sum	Accuracy	Precision	Recall	Type of Class
1	0.85	0.71	0.84	0.71	Repeller
2	1.04	0.82	0.80	0.82	Attractor
3	1.09	0.75	0.69	0.75	Attractor
4	1.11	0.85	0.77	0.85	Attractor
5	1.12	0.90	0.80	0.90	Attractor
6	0.79	0.61	0.77	0.61	Repeller

class. After combining these two classes, the maximum confusion is for the original classes 2 and 4, and then the original classes 1 and 5.

Each of these seems to be an obvious combination, based on the relative confusion within these aggregates compared to the confusion between classes of different aggregates (see, e.g., the F-score discussion in Section 1.9), and so we redefine the set of six classes to be a set of three aggregate classes as follows:

$$\text{Aggregate class A} = \text{Original class 1} + \text{Original class 5};$$
$$\text{Aggregate class B} = \text{Original class 2} + \text{Original class 4};$$
$$\text{Aggregate class C} = \text{Original class 3} + \text{Original class 6}.$$

Table 8.3 Confusion (sum of paired off-diagonal elements $\{i, j\}$, with $j > i$, which is the error of mistaking class i with class j). The maximum confusion is for $\{i = 1, j = 2\}$, for which the confusion is 0.35

i	j (where $j > i$)	Confusion $= (C_{ij} + C_{ji})$
1	2	0.08
1	3	0.08
1	4	0.03
1	5	0.19
1	6	0.05
2	3	0.04
2	4	0.20
2	5	0.02
2	6	0.06
3	4	0.08
3	5	0.04
3	6	0.35
4	5	0.03
4	6	0.07
5	6	0.04

The confusion matrix for the new aggregate-class-based, or simplified, confusion matrix is constructed from the elements of the original 6-class confusion by combining the appropriate elements. For example, the new aggregate class A, comprised original classes 1 and 5, thus collects original confusion matrix elements (1,1), (1,5), (5,1), and (5,5) into the appropriate element. The intersection of aggregate classes A and B, which becomes the new (1,2) element of the simplified matrix, collects original elements (1,2), (1,4), (5,2), and (5,4), while the intersection of aggregate classes B and A collects original elements (2,1), (2,5), (4,1), and (4.5). Proceeding along these lines, we collect nine new elements for the simplified confusion matrix as follows:

$$
\begin{array}{c}
\text{Aggregate} \\
\text{class}
\end{array}
\begin{array}{c}
\\
\text{A} \\
\\
\text{B} \\
\\
\text{C}
\end{array}
\left[
\begin{array}{ccc}
\begin{pmatrix} 0.71 & 0.15 \\ 0.04 & 0.90 \end{pmatrix} & \begin{pmatrix} 0.07 & 0.01 \\ 0.01 & 0.02 \end{pmatrix} & \begin{pmatrix} 0.04 & 0.02 \\ 0.02 & 0.01 \end{pmatrix} \\
\begin{pmatrix} 0.01 & 0.01 \\ 0.02 & 0.01 \end{pmatrix} & \begin{pmatrix} 0.82 & 0.13 \\ 0.07 & 0.85 \end{pmatrix} & \begin{pmatrix} 0.01 & 0.02 \\ 0.03 & 0.02 \end{pmatrix} \\
\begin{pmatrix} 0.04 & 0.02 \\ 0.03 & 0.03 \end{pmatrix} & \begin{pmatrix} 0.03 & 0.05 \\ 0.04 & 0.05 \end{pmatrix} & \begin{pmatrix} 0.75 & 0.11 \\ 0.24 & 0.61 \end{pmatrix}
\end{array}
\right]
$$

$$\text{Assigned class}$$

When these elements are added, the simplified confusion matrix now becomes

$$
\begin{array}{c}
\\
\text{Aggregate} \\
\text{class}
\end{array}
\begin{array}{c}
\\
\text{A} \\
\text{B} \\
\text{C}
\end{array}
\left[
\begin{array}{ccc}
0.90 & 0.055 & 0.045 \\
0.025 & 0.935 & 0.04 \\
0.06 & 0.085 & 0.855
\end{array}
\right]
$$

$$\text{Assigned class}$$

The accuracy of this simplified confusion matrix is once again the mean of the elements in the diagonal, which is 0.897. This means that the error rate for distinguishing aggregate classes A, B, and C is now only 10.3%, as opposed to 22.7% for the original 6-class case. However, this is not the complete story, as we now have to follow this 3-class problem with three separate binary classifications: class 1 versus class 5, class 2 versus class 4, and class 3 versus class 6. But, there are some significant advantages to this aggregation. Firstly, it is usually much easier to create an effective binary classifier than a higher-order classifier, since statistical interaction does not occur with only two variables. Secondly, the primary sources of error have been isolated through aggregation, improving the probability of reducing the overall error rate.

The subsections of the confusion matrix corresponding to the aggregations are used to determine the binary classification accuracy that must be achieved to improve the overall system accuracy. If this accuracy cannot be obtained, then the original classification system, with more classes and less aggregate classes, should be used to make this secondary binary classification. This original, more simply aggregated, system would provide the same or better accuracy with almost certain improvement in processing time.

Table 8.4 Classification statistics extracted from the 3-class simplified confusion matrix. As before, the attractor has higher recall, and the repellers have higher precision

Aggregate Class	Column Sum	Accuracy	Precision	Recall	Type of Class
A	0.985	0.90	0.91	0.90	Repeller
B	1.075	0.935	0.87	0.935	Attractor
C	0.940	0.855	0.91	0.855	Repeller

For the example shown, the aggregate class accuracies are determined directly from their confusion matrices. For the example, these reduced-order confusion matrices are all 2×2:

$$A \begin{bmatrix} 0.71 & 0.15 \\ 0.04 & 0.90 \end{bmatrix} \quad B \begin{bmatrix} 0.82 & 0.13 \\ 0.07 & 0.85 \end{bmatrix} \quad C \begin{bmatrix} 0.75 & 0.11 \\ 0.24 & 0.61 \end{bmatrix}.$$

For aggregate A, the accuracy is $(0.71 + 0.90)/(0.71 + 0.15 + 0.04 + 0.90) = 0.894$. For aggregates B and C, the accuracies are 0.893 and 0.795, respectively. Thus, if a binary classifier for distinguishing the original two classes in A, B, or C with better than 89.4%, 89.3%, or 79.5% accuracy, respectively, can be crafted, the overall system accuracy will be improved. Note that if no binary classifier can be found for aggregate A, B, or C, then the overall system accuracy is determined from

$$\text{Accuracy} = \frac{(0.90)(0.894) + (0.935)(0.893) + (0.855)(0.795)}{3} = 0.773.$$

This is of course the same accuracy as for the original 6-class problem. Table 8.4 shows classification results for the simplified confusion matrix. Here again, the two repellers have higher relative precision and lower relative recall than the attractor.

In comparing the precision values in Tables 8.2 and 8.4, we see that as the aggregate classes were formed, the mean precision (μ_p) rises from 0.776 to 0.898 and the standard deviation (σ_p) of the precision values decreases from 0.050 to 0.024. The ratio of σ_p/μ_p decreases from 0.064 to 0.027. With perfect accuracy classification, the ratio σ_p/μ_p moves to 0.0, since all of the groups will have perfect precision. However, as the number of classes or aggregate classes becomes smaller, this ratio may become very small coincidentally. For example, let us consider further aggregation of the problem above. There are three cases: (1) aggregating B and C; (2) aggregating A and C; and (3) aggregating A and B. The further simplified confusion matrices are

$$\begin{array}{c} A \\ B+C \end{array} \begin{bmatrix} 0.90 & 0.10 \\ 0.0425 & 0.9575 \end{bmatrix} \quad \begin{array}{c} A+C \\ B \end{array} \begin{bmatrix} 0.935 & 0.065 \\ 0.07 & 0.93 \end{bmatrix} \quad \begin{array}{c} A+B \\ C \end{array} \begin{bmatrix} 0.9575 & 0.0425 \\ 0.145 & 0.855 \end{bmatrix}.$$

For case (1), the precisions are 0.955 and 0.905, for which σ_p/μ_p is 0.038. For case (2), the precisions are 0.930 and 0.935, for which σ_p/μ_p is a much lower 0.004. For case (3), the precisions are 0.868 and 0.953, for which σ_p/μ_p is 0.066. Assuming that we reject aggregations where σ_p/μ_p increases from the previous confusion matrix, this means that only case (2) would be an allowable step forward.

However, another heuristic for further aggregation of classes using the confusion matrix data has proven useful. This metric is the ratio of the required accuracy for the downstream binary classification to the accuracy of the classification of the aggregation itself. This ratio, designated $R_{\text{aggregate}}$, is simply computed once the two classes to aggregate—that is, classes i and j—are identified from the maximum confusion equation and the appropriate matrix element of the aggregate class:

$$R_{\text{aggregate}} = \frac{C_{ii} + C_{jj}}{C_{ii} + C_{ij} + C_{ji} + C_{jj}} \bigg/ C_{i+j,i+j},$$

where C_{ii}, C_{ij}, C_{ji}, and C_{jj} are the matrix elements from the original ith and jth classes, and $C_{i+j,i+j}$ is the matrix element from the reduced-order aggregate class. In general, if $R_{\text{aggregate}}$ < 1.0, the aggregation is promising and so it should be carried out. For the aggregation of classes 1–6 into aggregates A, B, C, these values are

$R_{\text{aggregate}}(A) = ((0.71 + 0.90)/(0.71 + 0.90 + 0.04 + 0.15))/0.90 = 0.8944/0.90 = 0.994;$
$R_{\text{aggregate}}(B) = 0.893/0.935 = 0.955;$
$R_{\text{aggregate}}(C) = 0.7953/0.855 = 0.930.$

Since each of these values of $R_{\text{aggregate}}$ < 1.0, and the ratio σ_p/μ_p becomes smaller, the heuristics suggest that these are aggregations worth performing. However, for the new cases, the same values are

$R_{\text{aggregate}}(\text{case } 1) = (1.79/1.915)/0.9575 = 0.977;$
$R_{\text{aggregate}}(\text{case } 2) = (1.755/1.860)/0.93 = 1.015;$
$R_{\text{aggregate}}(\text{case } 3) = (1.865/1.945)/0.9575 = 1.002.$

The heuristic for the metric $R_{\text{aggregate}}$ fails for case (2)—and again for case (3)—meaning that none of the three possible aggregations of A, B, and C should occur.

This section, therefore, has introduced some of the manipulations of the confusion matrix that can be used to partition an input space. The combined use of maximum confusion, the ratio σ_p/μ_p, and the metric $R_{\text{aggregate}}$ provides an effective approach to deciding whether or not to partition, and if so which partition(s) to make. After the partitioning has occurred, the aggregates formed can be analyzed together or individually using whatever algorithm or meta-algorithm is appropriate. Importantly, this allows the "fringes" of an input set to be identified, and so thereafter be separately analyzed, as appropriate.

8.3   Primary Domains

The primary domains of this book are now used to demonstrate the second-order meta-algorithmics.

8.3.1 Document Understanding

The first domain used to exemplify the second-order meta-algorithmics is document understanding (Simske, Wright, and Sturgill, 2006). In this previously reported set of research, first- and second-order meta-algorithmic patterns were used to overcome some of the limitations of three individual document classification engines using meta-algorithmic design patterns. These patterns, collectively, were shown to explore the error space of three different document analysis engines, and to provide improved and in some cases "emergent" results in comparison to the use of voting schemes or to the output of any of the individual engines. A variety of first- and second-order meta-algorithmic patterns were used to reduce the document classification error rates by up to 13% and to reduce system error rates by up to 38%.

A broad range of the meta-algorithmic design patterns introduced in Chapter 6 can improve the document classification. However, in our earlier research, we were also able to show some that did not. Because a large and publicly available data set was used for these studies, it makes a good starting point for considering how to select the right meta-algorithmic pattern for a problem space.

In Simske, Wright, and Sturgill (2006), the primary measure of improvement was a decrease in the classification error rate, that is, improving the precision, recall, and accuracy of document classification. This results in significant reduction of overall system costs, since mistakes in document classification as part of a document management system incur significant expense. If the document is not indexed correctly, it cannot be retrieved from the electronic file system or a loan is denied because incorrect salary information was extracted from financial documents. Common current approaches to this indexing/extraction process are primarily manual and highly error-prone: PC-based indexing stations are staffed with a cadre of users who view the scanned image and type in the corrected index or workflow data.

In addition to reducing the overall indexing system error rate, meta-algorithmics were used to decrease the mean number of indexing attempts required to obtain a correct classification and thus avoid human intervention. In order to achieve the best possible classification, we needed to increase the likelihood of obtaining the best initial classification, and to increase the relative rank of the actual (correct) classification when this initial classification is wrong.

A document understanding system was developed comprising three different classifier technologies (one neural-net-based, one Bayesian-based, and one natural-language-processing-based) to identify which meta-algorithmic design patterns provide the desired improvements. We chose a commercially available neural-net document classifier (Mohomine text classifier, subsequently purchased by Kofax (www.kofax.com; accessed January 28, 2013) and integrated into Indicius (http://www.kofax.com/support/products/indicius/6.0/index.php; accessed February 1, 2013)), an open source Bayesian classifier (Divmod Reverend; http://www.ohloh.net/p/divmod-reverend; accessed February 1, 2013), and our own classifier that uses basic TF*IDF (term frequency multiplied by inverse document frequency; Salton, Fox, and Wu, 1983) calculations to classify documents. These are designated the Mohomine, Bayesian, and TF*IDF engines. The set of documents used for training and testing was the "20 Newsgroups" data set collected on the University of California, Irvine Knowledge Discovery in Databases Archive (http://kdd.ics.uci.edu/databases/20newsgroups/20newsgroups.html; accessed January 28, 2013), which provides automatic author-specified classification. The classifiers were trained and tested with 2.5%, 5%, ..., 17.5%, and 20% of the corpus used as training data, and the remaining 97.5%, 95%, ..., 82.5%, 80% as test data. There were 1000

Table 8.5 Document classification statistics for each of the classification engines. Eight different training percentages, varying from 2.5% to 20% of each class, were investigated. The Mohomine engine was the most accurate for every training set, followed by the TF*IDF engine for every training set except for the 2.5% set

Number of Training Documents Per Class (and %)	Mohomine Engine Accuracy	Bayesian Engine Accuracy	TF*IDF Engine Accuracy
25 (2.5%)	0.611	0.494	0.470
50 (5.0%)	0.681	0.547	0.564
75 (7.5%)	0.721	0.582	0.618
100 (10.0%)	0.734	0.603	0.649
125 (12.5%)	0.751	0.615	0.674
150 (15.0%)	0.764	0.627	0.698
175 (17.5%)	0.774	0.642	0.716
200 (20.0%)	0.782	0.637	0.723

documents per newsgroup, so that training took place on 500–4000 documents, and testing on the residual 16 000–19 500 documents. Classifier output confidence values, or probabilities, were normalized to a range of 0.0–1.0.

Document classification accuracy was separately determined for each of the eight different training percentages for each of the three classification engines (Table 8.5). The Mohomine engine is the most accurate engine, and the TF*IDF engine has the greatest improvement in accuracy with increasing amounts of training.

If the three engines were mutually independent, then two or more of the engines would provide the correct classification with the following probability:

$$MBT + MB \times (1.0 - T) + MT \times (1.0 - B) + BT \times (1.0 - M),$$

where M is the Mohomine accuracy, B is the Bayesian accuracy, and T is the TF*IDF accuracy. For the 20% training set, this value is

$$0.360 + 0.138 + 0.205 + 0.100 = 0.803.$$

Since this 80.3% value is above that of any individual engine—the highest being 78.2% as shown in Table 8.5—the combination of these three engines appeared to provide a good opportunity for the application of meta-algorithmics. However, Voting and Weighted Voting patterns did not improve the accuracy of the system above that of the best engine (Mohomine). This may have been due to the fact that the three engines were not, in fact, independent. In fact, the correlation coefficients (r^2 values) for the three engines were calculated, providing the following:

Bayesian/Mohomine : $r^2 = 0.4522$, 45% correlated, $p < 0.05$;
Bayesian/HP1 : $r^2 = 0.6434$, 64% correlated, $p < 0.005$;
Mohomine/HP1 : $r^2 = 0.2514$, 25% correlated, $p > 0.10$, not statistically significant.

Table 8.6 Document classification statistics for each of the three classification engines and the six different second-order meta-algorithmic patterns investigated. The test set comprised 15 997 newsletters, with all header information removed. The values shown are for the 20% training set. Meta-algorithmic pattern accuracies outperforming the best engine's accuracy are highlighted in boldface

Engine/Meta-algorithmic Pattern Deployed	# Correct	# Incorrect	Accuracy (%)
Mohomine engine	12 436	3 561	77.7
Bayesian engine	9 983	6 014	62.4
TF*IDF engine	11 408	4 589	71.3
1. Confusion Matrix and Weighted Confusion	12 724	3 273	**79.5**
Matrix (0.56, 1.28, 0.96) (Section 6.3.1)	12 777	3 220	**79.9**
2. Confusion Matrix with Output Space	12 839	3 158	**80.3**
Transformation (Section 6.3.2)			
3. Majority Voting or Weighted Confusion	12 299	3 698	76.9
(Section 6.3.6)			
4. Majority Voting or Best Engine (Section 6.3.7)	12 265	3 732	76.7
5. Best Engine with Absolute Confidence (0.75)	12 725	3 272	**79.5**
or Weighted Confusion Matrix (threshold			
confidence) (Section 6.3.9)			
6. Best Engine with Differential Confidence	12 767	3 230	**79.8**
(0.25) or Weighted Confusion Matrix			

Thus, more sophisticated meta-algorithmic approaches were required to better explore the input space for the engines. The following set of second-order meta-algorithmic patterns were deployed:

1. Confusion Matrix and Weighted Confusion Matrix (Section 6.3.1).
2. Confusion Matrix with Output Space Transformation (Section 6.3.2).
3. Majority Voting or Weighted Confusion (Section 6.3.6).
4. Majority Voting or Best Engine (Section 6.3.7).
5. Best Engine with Absolute Confidence or Weighted Confusion Matrix (threshold confidence) (Section 6.3.9).
   In addition, a variation on the pattern introduced in Section 6.3.9 was deployed:
6. Best Engine with Differential Confidence or Weighted Confusion Matrix.

These are now discussed in more depth. For purposes of comparison, we use the 20% training results, summarized in Table 8.6.

1. The Confusion Matrix pattern was generated using the output probabilities matrix (OPM) as shown in Section 6.3.1 (Figure 8.1). This approach was found to provide better overall accuracy than the best engine under virtually all training/testing combinations. In Table 8.6, the Confusion Matrix pattern outperformed the best classifier by 1.8%, for a reduction in error 8.1%. The Weighted Confusion Matrix pattern (Figure 8.2) performed even better. The weights used in Table 8.6—0.56 for the Bayesian engine, 1.28 for the Mohomine engine, and 0.96 for the TF*IDF engine—are derived using the results in Lin *et al.* (2003).

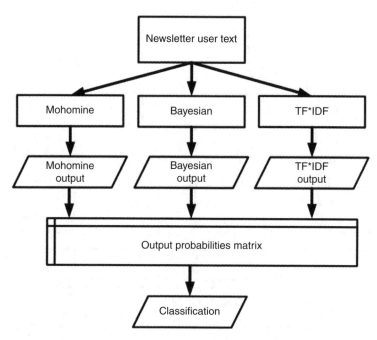

Figure 8.1   Confusion Matrix second-order meta-algorithmic pattern as applied to the document classification problem described in the text

However, nearly identical results were obtained at each training level when the weights used were the engine accuracies: at 20%, these are 0.62, 0.78, and 0.71, respectively. The Weighted Confusion Matrix accuracy was 79.9%, fully 2.2% higher than the best engine. This reduced the overall error rate—which is directly correlated with system cost as described above—by 9.6%.

2. Since the Confusion Matrix patterns were so promising, the Confusion Matrix with Output Space Transformation pattern (Figure 8.3) was applied next. The transformation investigated, as described in Section 6.3.2, mapped an output probability p, to a new probability, p^α, prior to applying the Weighted Confusion Matrix meta-algorithmic pattern to the transformed Mohomine engine in combination with the untransformed Bayesian and TF*IDF engines. For the 20% training set, transforming p to $p^{0.208}$, where p is the Mohomine engine probability output, removes 43 additional errors. When the TF*IDF engine is then independently transformed, the square root operator ($\alpha = 0.5$) removed the most (14) additional errors. Next, we evaluated simultaneously α_{Mohomine} and $\alpha_{\text{TF*IDF}}$. The iterative procedure to determine these coefficients (and the fact that optimizing α_{Mohomine} affected the error count much more so than optimizing $\alpha_{\text{TF*IDF}}$) indicates that the transformation power (i.e., α) should be set for the Mohomine engine first, then for the TF*IDF engine. In so doing, the following transformation powers were identified: $\alpha_{\text{Mohomine}} = 0.208$, $\alpha_{\text{TF*IDF}} = 0.185$. We found that the optimum α for the Bayesian engine when $\alpha_{\text{Mohomine}} = 0.208$ and $\alpha_{\text{TF*IDF}} = 0.185$ was 1.0. The procedure for determining these powers was exactly in keeping with Section 6.3.2. Deploying these two transformations (with the output of the Bayesian engine

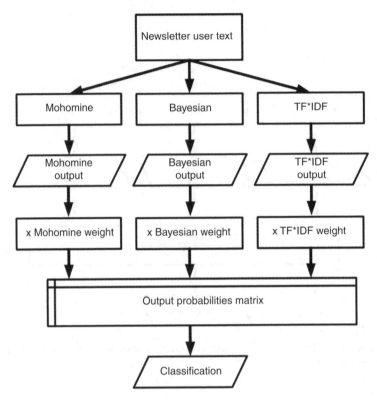

Figure 8.2   Weighted Confusion Matrix second-order meta-algorithmic pattern as applied to the document classification problem described in the text

left untransformed) provided an optimal 12 839 correct answers, 62 more than the Weighted Confusion Matrix pattern. The error rate was reduced by 11.3%.

3. No improvement in accuracy was observed when the Majority Voting or Weighted Confusion Matrix pattern (Figure 8.4) was employed. The accuracy rate was 0.8% lower than the accuracy of the best engine, and 3.0% lower than the Weighted Confusion Matrix pattern when used by itself. Thus, the Majority Voting portion of the pipelined pattern (originally presented in Section 6.3.6) was responsible for this 3.0% drop in accuracy. This implies that two or more engines were often correlated with each other on the wrong classification.

4. A similar result was observed for the Majority Voting or Best Engine pattern (Figure 8.5), originally described in Section 6.3.7. The accuracy was 1.0% less than the accuracy of the best engine. This result establishes the fact that the Majority Voting pattern failed to improve overall system accuracy because of the behavior of the Bayesian and TF*IDF engines in combination: when they disagreed with the Mohomine engine, they were more often wrong than correct.

5. The fifth of the second-order meta-algorithmic patterns implemented was the Best Engine with Absolute Confidence or Weighted Confusion Matrix pattern (Figure 8.6), introduced in Section 6.3.9. Absolute confidence was varied from 0.50 to 1.00 over the training set, and the best results were observed for absolute confidence equal to 0.75. During later testing,

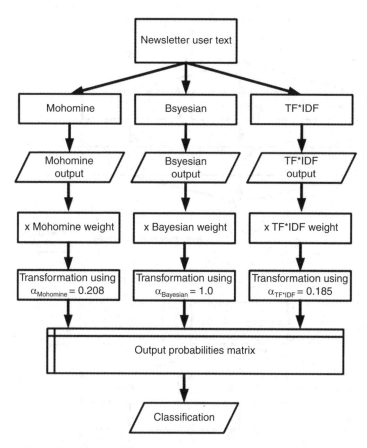

Figure 8.3 Weighted Confusion Matrix with Output Space Transformation second-order meta-algorithmic pattern as applied to the document classification problem described in the text

when the Mohomine engine reported a confidence value of 0.75 or higher, its output was accepted. Otherwise, the Weighted Confusion Matrix pattern was employed. This threshold confidence approach improved the system accuracy by 1.8%, the same as for the Confusion Matrix approach, but 0.4% less than the Weighted Confusion Matrix pattern by itself. These results indicated that the absolute confidence pattern was not an effective means of selecting the output.

6. The final second-order meta-algorithmic pattern deployed was the Best Engine with Differential Confidence or Weighted Confusion Matrix pattern (Figure 8.7). Here, the thresholded confidence was a relative threshold: when the Mohomine engine reported a confidence value 0.25 or more higher than both the Bayesian and TF*IDF engines, its output was accepted; otherwise, the Weighted Confusion Matrix pattern was selected. A small improvement (0.3% greater accuracy) was observed over the Confusion Matrix pattern. The results were essentially the same (0.1% less accuracy) as the Weighted Confusion Matrix. The differential confidence, by itself, was not an effective means of selecting the output.

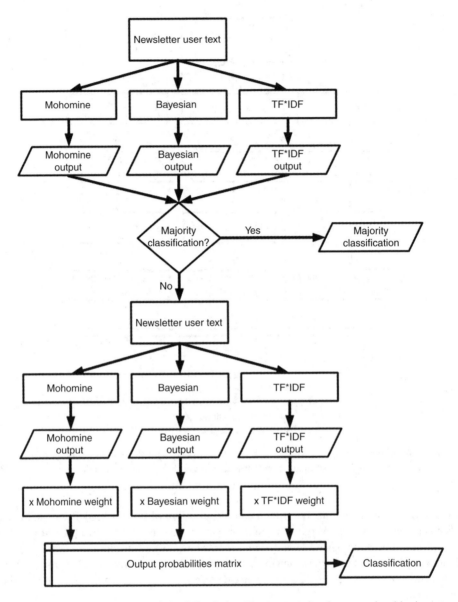

Figure 8.4 Majority Voting or Weighted Confusion Matrix second-order meta-algorithmic pattern as applied to the document classification problem described in the text

After considering the different patterns used for this problem, it is clear that the Confusion Matrix and Weighted Confusion Matrix were effective in reducing document classification error. However, the overall most effective second-order meta-algorithmic pattern deployed was the Weighted Confusion Matrix with Output Space Transformation. These results illustrate the value of considering multiple meta-algorithmic patterns at the same time. First off, some patterns will be useful, while others may be less so or even deleterious. Secondly, since

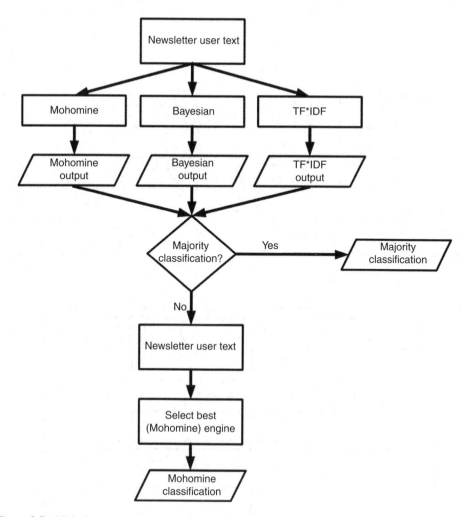

Figure 8.5 Majority Voting or Best Engine second-order meta-algorithmic pattern as applied to the document classification problem described in the text

second-order patterns are generally combinations of two simpler (usually first-order) patterns, the relative impact of different components can be compared as above. Third, the relative effect of the best second-order patterns can be used to guide both the decision to investigate third-order meta-algorithmics, and perhaps even which third-order meta-algorithmic patterns to implement.

Classification accuracy is, of course, only one reason to use meta-algorithmics. In terms of the overall system costs, we also wish to increase the relative classification rank of the true classification even when the initial classification is incorrect. This is termed *improving the central tendency* of the classification, and can be very important for a number of real-world systems in which a human editor is used to make final corrections. It is also important in systems in which different, more expensive downstream applications are set in motion

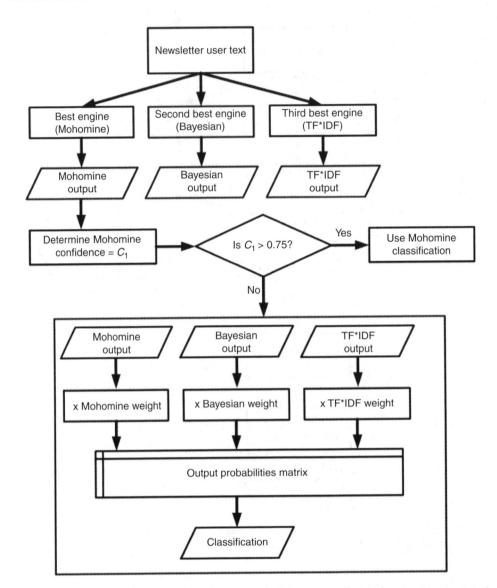

Figure 8.6 Best Engine with Absolute Confidence (of 0.75) or Weighted Confusion Matrix second-order meta-algorithmic pattern as applied to the document classification problem described in the text

by the classification decision—these downstream applications will eventually flag upstream classification errors, but only after further processing. Face recognition following face detection is one example of this type of downstream processing.

Given these considerations, the optimal settings for the meta-algorithmic patterns—for example, the power, α, used in output space transformation or the weights used in the Weighted Confusion Matrix pattern—may be slightly different since this is a different goal than classifier accuracy in and of itself. The rank of the correct document classification is a prediction of the

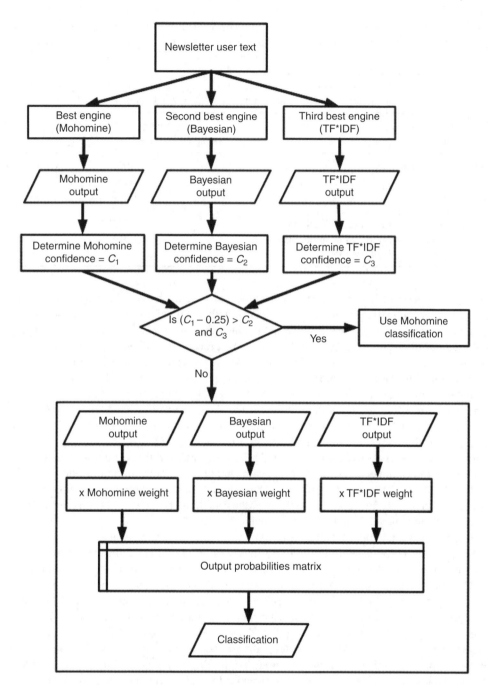

Figure 8.7   Best Engine with Differential Confidence (of 0.25) or Weighted Confusion Matrix second-order meta-algorithmic pattern as applied to the document classification problem described in the text

Table 8.7 Mean number of classifications to evaluate before finding correct classification for different document classification engines or meta-algorithmic patterns. Note that the TF*IDF engine is actually slightly better than the Mohomine engine, and both of these are significantly better than the Bayesian engine. The top meta-algorithmic patterns are significantly better than any single engine. The Weighted Confusion Matrix pattern provides the best overall results of any meta-algorithmic pattern, and a 18.9% reduction in mean rank of correct classification compared to the best individual engine, TF*IDF

Approach	Mean Number of Classifications Attempted until the Correct One Is Obtained
100% "perfect" classifier	1.000
Bayesian	2.877
Mohomine	1.904
TF*IDF	1.891
Weighted Confusion Matrix (0.56, 1.28, 0.96)	1.533
Weighted Confusion Matrix with Output Space Transformation (1.0, 0.208, 0.185)	1.554

amount of processing time a downstream document processing task will take: for example, the identification and extraction of indices associated with a particular form (name, address, loan amount, patient ID, etc.). If we assume that the incorrect classifications can be automatically detected during the processing, then the amount of processing is directly proportional to the mean rank of the correct classification. This value is presented in Table 8.7 for each of the three engines and for the two most effective meta-algorithmic patterns investigated.

Table 8.7 illustrates the central tendency, or mean rank of the correct classification, results. The best results for any system are those for the Weighted Confusion Matrix pattern with optimal engine weighting: 0.56 for Bayesian, 1.28 for Mohomine, and 0.96 for TF*IDF. For this pattern, the correct classifier had a mean ranking of 1.533, only 53.3% higher than a perfect classifier. The second-order meta-algorithmic pattern with the highest classification accuracy—the Weighted Confusion Matrix with Output Space Transformation where $\alpha_{Bayesian} = 1.0$, $\alpha_{Mohomine} = 0.208$, and $\alpha_{TF*IDF} = 0.185$—did not, however, provide a better mean rank of classification than the Weighted Confusion Matrix pattern. Its value, 1.554, was nevertheless also significantly better than that of any individual engine.

Another way of looking at the mean ranking of the correct classification of data is to determine what percentage of incorrect classifications remains after the first M (where $M \leq N$, the total number of classes) ranked classes have been considered. This is shown in Tables 8.8 and 8.9 for the two top meta-algorithmic patterns. The best results, in agreement with Table 8.7, are obtained for the Weighted Confusion Matrix pattern as summarized in Table 8.8. The results describe what percentage of classification errors remain after the first M classes, when $M = 1$–8. The total number of classes, N, is 20. Thus, when $M = 8$, a full 40% of the possible classes have been evaluated. Nevertheless, almost 10% of the correct classifications are not in the set of eight highest confidence outputs of the Bayesian engine. This value is smaller for the Mohomine engine (3.8%) and still smaller for the TF*IDF engine (2.6%). However, applying the weighted confusion matrix approach reduces this error to less than 1.2%, a huge improvement over even the TF*IDF pattern.

In Table 8.9, the meta-algorithmic pattern considered is the Weighted Confusion Matrix with Output Space Transformation. For this pattern, the optimum Bayesian, Mohomine,

Table 8.8 Remaining classification errors after the first M classes have been considered for the Weighted Confusion Matrix pattern and each of the three individual document classification engines. The last column provides the percent improvement for the meta-algorithmic pattern over the best single engine (which is Mohomine for $M = 1$–5 and TF*IDF for $M = 6$–8). The meta-algorithmic pattern is always significantly better than any single engine, and its relative improvement increases with M

M	Weighted Confusion Matrix Pattern (%)	Bayesian (%)	Mohomine (%)	TF*IDF (%)	Improvement, Meta-algorithm (%)
1	20.13	37.59	22.69	28.69	11.3
2	9.61	27.12	12.00	16.16	19.9
3	6.13	21.70	8.21	10.98	25.3
4	4.28	18.07	6.60	7.63	35.2
5	3.08	15.70	5.44	5.70	43.4
6	2.28	13.61	4.58	4.56	50.0
7	1.73	11.73	4.19	3.59	51.8
8	1.19	9.85	3.81	2.63	54.8

Table 8.9 Remaining classification errors after the first M classes have been considered for the Weighted Confusion Matrix with Output Space Transformation pattern and each of the three individual document classification engines. The last column provides the percent improvement for the meta-algorithmic pattern over the best single engine (which is Mohomine for $M = 1$–5 and TF*IDF for $M = 6$–8). This meta-algorithmic pattern is always significantly better than any single engine, but lesser so than the Weighted Confusion Matrix pattern of Table 8.8. The relative improvement of the meta-algorithmic pattern compared to the best single engine is maximum for $M = 6$ (46.9%), also unlike for the Weighted Confusion Matrix pattern (which continues to rise with M, reaching 54.8% by $M = 8$ in Table 8.8)

M	Weighted Confusion Matrix with Output Space Transformation pattern (%)	Bayesian (%)	Mohomine (%)	TF*IDF (%)	Improvement, Meta-algorithm (%)
1	19.74	37.59	22.69	28.69	13.0
2	9.36	27.12	12.00	16.16	21.9
3	5.78	21.70	8.21	10.98	29.6
4	4.10	18.07	6.60	7.63	37.9
5	3.13	15.70	5.44	5.70	42.5
6	2.42	13.61	4.58	4.56	46.9
7	1.97	11.73	4.19	3.59	45.1
8	1.56	9.85	3.81	2.63	40.7

and TF*IDF alpha values are 1.0, 0.208, and 0.185, respectively. For $M = 1$–4, this meta-algorithmic pattern outperforms any of the three individual engines, and also the Weighted Confusion Matrix. However, for $M = 5$–8 (as in the mean), it is outperformed by the Weighted Confusion Matrix pattern. The Weighted Confusion Matrix with Output Space Transformation pattern, in fact, provides its maximum percent improvement over the single best engine for $M = 6$.

Clearly, the two meta-algorithmic patterns tested that provided the best overall classification results, as shown in Tables 8.8 and 8.9, significantly increase the likelihood of providing the correct classification (thus enabling the automatic extraction of correct index values) for a given value of M. The improvement percentages provided in the last column of these tables, however, is if anything an underestimate of the system value. As Table 8.7 shows, a "perfect" classifier requires 1.0 mean attempts at classification before finding the correct classification. The Bayesian engine takes an extra 1.877 classification attempts, in the mean, before providing the correct class and thereby leading to the correct extraction of document information. The Mohomine engine, however, requires only an extra 0.904 classification attempts before providing the correct classification. This is less than half as many as the Bayesian engine. The TF*IDF engine, with a mean value of 0.891 extra classifications, is still better. However, in comparison, the Weighted Confusion Matrix pattern, requiring a mean of just 0.533 extra classification attempts, thus requires 40.2% less than the TF*IDF engine, 41.0% less than the Mohomine engine, and 71.6% less than the Bayesian engine. In many real-world systems, this metric—mean number of extra classifications expected before determining the correct class—is more important than simple accuracy (the percent of highest ranked classifications that are correct).

Another interesting aspect of the results in this study was the fact that many emergent correct results occurred due to the application of the meta-algorithmic patterns. Since the number of classes is large ($N = 20$) relative to the number of classification engines ($N_E = 3$), it is not surprising that, for many difficult-to-classify samples, none of the three engines provides the correct class as its highest confidence classification. However, there is no guarantee that the combination of these three engines will provide the correct classification under this circumstance—in other words, lead to an "emergent" correct result. Table 8.10, however, shows that, on average, approximately one-ninth of the errors removed by the meta-algorithmic

Table 8.10 Number and percentage of errors removed by emergence for the three best meta-algorithmic patterns (20% training condition). Note that 18 of the 62 additional errors removed by output space transformation when coupled with the weighted confusion matrix approach (29%) are due to emergence

Meta-algorithmic Pattern	Number of Errors Less than Mohomine Engine by Itself	Number of Errors Removed by "Emergence"	Percentage of Errors Removed by "Emergence"
Confusion Matrix	288	39	13.5%
Weighted Confusion Matrix	341	29	8.5%
Weighted Confusion Matrix with Output Space Transformation	403	47	11.7%

patterns are due in fact to emergence. A disproportionately large fraction—approximately two-seventh—of the additional errors removed by adding output space transformation to the Weighted Confusion Matrix pattern are due to emergence.

The results of this section show that, for given classification problems, not all meta-algorithmic algorithms provide improvement over the best individual engine. This does not preclude the use of meta-algorithmics; rather, it argues for the investigation of several meta-algorithmic patterns at the same time. Many commercial document classification systems, however, currently make no overt use of meta-algorithmics. Past research has, however, shown the effectiveness of voting schemes (Ruta and Gabrys, 2000; Sebastiani, 2002; Lin *et al.*, 2003) for combining the output of several classifiers. Classifier fusion techniques, including Bayesian/fuzzy integral classifier combinations (Ruta and Gabrys, 2000), are also reported in the literature. Decision profiles and "product of experts" approaches share much with the Confusion Matrix approach described herein (Ruta and Gabrys, 2000). Thus, where previous work has deployed meta-algorithmic approximations, improved classification accuracy has been reported.

8.3.2   Image Understanding

We next turn to image understanding and in particular object extraction using the Tessellation and Recombination with Expert Decisioner pattern, introduced in Section 6.3.3. In the template for this pattern (Figure 6.10), the expertise is injected during the tessellation step and again just after the recombination step. In the example provided in this section, I modify the pattern through the addition of an information pathway to the template pattern, allowing an *expert decisioner* to operate upstream from the tessellation step. As Figure 8.8 illustrates, for the object extraction problem I am addressing, expert decisioning is incorporated into the meta-algorithmic pattern in two places: (1) the post-recombination step as originally introduced in Figure 6.10, and (2) upstream from the actual tessellation, as a means of communication between the two object segmentation algorithms.

The application of the meta-algorithmic pattern described in Figure 8.8 will be explained through example: Figure 8.9 illustrates the application of the pattern for a specific object extraction. In this example, the object to be segmented from the original image (Figure 8.9a) is a red toy ladder. Figure 8.9b provides the output of the "red finder," which is one of the two segmentation engines incorporated into the system. This segmentation engine is used to perform specific color segmentation. The image is a 24-bit color image with three 8-bit channels: red, green, and blue. By analyzing the peaks in the red channel for the training set of images ($N = 10$ images in the training set), the expert decisioner found a peak with red channel values from 200 to 255. Thus, a red threshold of 200 was chosen first. The green and blue histograms of these red pixels were then computed, and large green and blue peaks in the range of 0–120 were identified for each of these two color channels. Pixels were then classified as belonging to red objects if their (red, green, blue) values, or (r,g,b) values belonged, simultaneously, to the following three ranges:

$$200 \le r \le 255 \,;$$
$$0 \le g \le 120 \,;$$
$$0 \le b \le 120 \,.$$

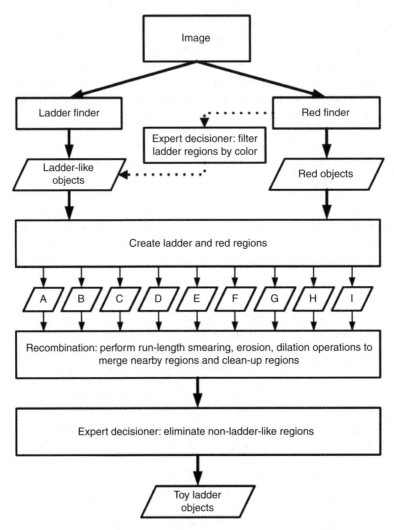

Figure 8.8   Tessellation and Recombination with Expert Decisioner second-order meta-algorithmic pattern as applied to the object segmentation problem as described in the text

These red pixels are shown in Figure 8.9b as light gray. Both ladders, another red toy object, and several smaller pieces are so identified. Importantly, these three ranges (the expert input) are passed to the "ladder finder" segmentation algorithm and used to prune the original sets of regions found by the ladder finder, which included not only the red ladders but also the train tracks and the rooftop. This is because the ladder finder searched for structured sets of perpendicular edges that combined are used to identify rectangular regions. These object defining criteria are consistent with the logs of the roof and building and of course the railroad tracks; however, they are inconsistent with the obstructed ladder lying on the ground in the right center of the image. Regardless, the expert decisioner ranges readily removed these track

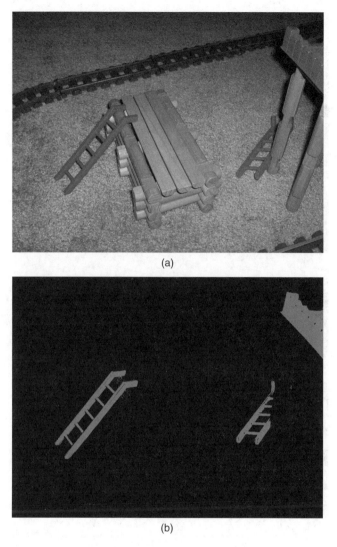

Figure 8.9 Tessellation and Recombination with Expert Decisioner second-order meta-algorithmic pattern as applied to the object extraction problem described in the text. (a) The original image, including mixed objects against a complex textured background. (b) The red finder output pixels are shown in light gray. The red finder is also used to create an expert decisioner that incorporates three thresholds, one for each of the red, green, and blue color channels, to winnow all but the red objects from the other segmentation algorithm, the ladder finder. (c) The extracted objects so matching the "ladder" object definition after expert decisioner pruning are shown. (d) The results of the recombination step are shown, resulting in two ladder-like objects

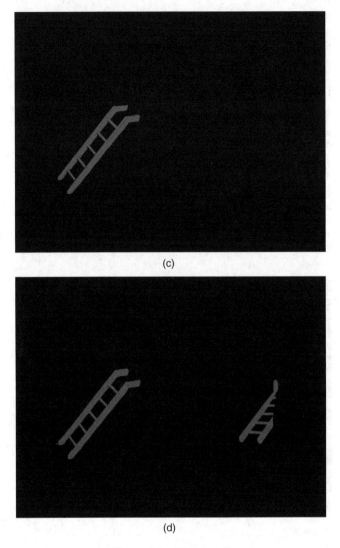

Figure 8.9   (*Continued*)

and rooftop false positives, as shown in Figure 8.9c, and only a single, unobstructed ladder region is identified. It is clear that the set of red finder regions is a superset of the ladder-like regions. When the regions are combined, then, there are two types of regions:

1. Ladder regions
2. Red regions not identified as ladder regions.

The regions in set (1) are then used as templates for comparison for the remaining regions in set (2) during the recombination step of the algorithm. Only one intact ladder object is identified as a possible ladder candidate by this algorithm, but a lot of information on size and structure of this region can be gained. Specifically, the ladder length and width, rung length and width, and number of rungs can be identified. The partially obstructed ladder, even without any depth or perspective, is a close match (within 30%, 15%, 15%, 10%, and 0%, respectively) for this set of five structural features. It is therefore declared a match and added to the set of ladder regions during the recombination. This serves as the second expert decisioner in this example, and results in the correct identification of two ladder regions as shown in Figure 8.9d.

For conciseness, I have left out some further considerations for this image understanding application. As one example, further expert decisioning may be required if there are no unobstructed ladder regions. Regardless, this section highlights the interplay between expert domain knowledge and the first-order Tessellation and Recombination pattern in the second-order meta-algorithmic pattern, Tessellation and Recombination with Expert Decisioner.

8.3.3   Biometrics

The field of biometrics is used to exemplify the deployment of the Single Engine with Required Precision second-order meta-algorithmic pattern, introduced in Section 6.3.5 and Figure 6.12. In this example, we desire to periodically establish the identity of a person using a computing system that includes at least one of the following: keyboard, microphone/audio input, and camera/video input. The system will also have one or more traditional "static" biometric systems—such as a fingerprint reader, voice prompt, or camera prompt—that can be used when the automatic biometric authentication fails.

The proposed system is shown in Figure 8.10. The required precision—that is, probability that the identity of the person is who the system identifies—is a function of a number of factors, including the following:

1. The effective sampling rate of the authentication. For example, if the system is authenticating the person every 5 s, the precision can be much lower than if authentication is performed every minute.
2. The desired statistical level of security. The higher the statistical security required, the more the precision required, and the more likely the higher accuracy (and usually higher processing cost) biometrics must be performed.
3. The difficulty of replicating or spoofing the measured behavior. If only timing between consecutive keyboard entries is used, for example, the odds of successfully imitating the behavior of another person are much higher than for voice imitation, which in turn is higher than facial appearance imitation.
4. The odds of events corresponding to the behavior occurring during the sampling window. Smaller sampling windows, for example, are much less likely to have audio information.

In Figure 8.10, the generators are the engines for computing the following three dynamic biometrics: (1) keyboard kinetics, (2) voice recognition, and (3) facial recognition. For this meta-algorithmic pattern, the generators are considered in order of a certain criterion. The order of (1), (2), and (3) above is consistent with scaling by cost and mean processing time. The order would be inverted if, for example, overall accuracy or training maturity/robustness were the criterion.

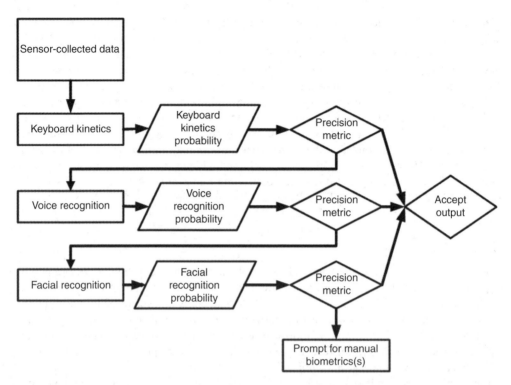

Figure 8.10 Single Engine with Required Precision second-order meta-algorithmic pattern as applied to the biometric system described in the text

In Figure 8.10, our computing system monitors the timing and forces associated with the typing. Biometrics can be based on the timing between consecutive letters in common words such as "the," "and," and "some"; the time between pressing and releasing the haptic pad/keyboard; and the peak force measured during typing. If the user has performed enough typing during the sampling interval, it is possible that these features alone will identify the user with sufficient statistical certainty. However, if the authentication confidence is below that required by the application, then the voice recognition generator will be used as a second biometric. Here, specific phonemes can be matched to the templates generated during training and/or previous sampling for the user. If the match is of sufficient statistical certainty, either alone or in combination with the keyboard kinetics, then no further processing is required. If, however, further authentication is needed to achieve the desired level of statistical certainty, then the facial recognition generator is invoked. This requires additional processing for video recording, face detection, and face recognition. Face recognition relies on the identification and measurement of several dozen landmarks, or nodal points, on the face, including nose width, eye socket depth, cheekbone shape descriptors, jaw line length, and intraocular distance. The numerical code, or faceprint, of the person can be compared to a template generated during training or prior sampling, and a modified Hamming distance can be used to provide a precision value for biometric authentication.

If there is still insufficient statistical certainty for the security requirements of the task even after all three biometrics in the system of Figure 8.10 are computed, the user may be prompted for manual biometric input. The manual biometrics include using a fingerprint reader, saying a particular word or phrase, or posing for the device video camera. Presumably, the manually entered biometric will provide sufficient statistical confidence: if not, a nonbiometric means of authentication will need to be performed.

8.3.4 Security Printing

For the security printing example, the Weighted Confusion Matrix pattern, originally described in Section 6.3.1, is modified and used to determine an optimal *authentication strategy* for a printed item with multiple variable data regions. We wish, in this case, to be able to determine which of four print service providers (PSPs) produced a label, which is a separate means to validate a supply chain. In addition, low-confidence classifications can be indicative of a quality problem with a PSP or even counterfeiting within the supply chain.

In the example illustrated in this section, a set of security labels are printed with three different variable data features, each of which is a potential means of classifying the label. These three features are a 2D Data Matrix barcode, a 3D color tile barcode, and a guilloche pattern—referred to hereafter as "barcode," "3D color tile," and "guilloche," respectively. Specific image feature sets of each of these three printed regions are used for the classification. Section 5.2.2.2 describes the set of 10 features used for the experiment in this section: entropy, mean intensity, image percent edges, mean edge magnitude, pixel variance, mean image saturation, and so on. A simple weighted binary classifier, as described in Simske, Li, and Aronoff (2005), was used to perform the classifications.

The simpler examples for deploying the Weighted Confusion Matrix given in Sections 6.3.1 and 8.3.1 use only a single data source for the classification. However, in this security printing example, we have the three separate variable data printed (VDP) regions as data sources. This allows us to produce seven different, partially uncorrelated, classifiers based on the feature set from the three individual regions separately, the three combinations of features from exactly two of the regions, and the one combination of features from all three regions. The training phase of Figure 8.11 illustrates this process, which generates these separate, partially uncorrelated classifiers. These classifiers are described not by their OPMs, as introduced in Section 6.3.1, but by their true confusion matrices.

In our experiment, 100 labels from each PSP were used to generate the confusion matrices. These are described next. The first confusion matrix for class assignment is for when only the barcodes are analyzed:

$$\text{Barcodes:} \quad \begin{array}{cc} & \begin{array}{cccc} \text{Assigned} & \text{class} \\ \text{A} & \text{B} & \text{C} & \text{D} \end{array} \\ \begin{array}{c} \\ \text{Origin} \\ \text{class} \end{array} \begin{array}{c} \text{A} \\ \text{B} \\ \text{C} \\ \text{D} \end{array} & \left[\begin{array}{cccc} 0.84 & 0.05 & 0.07 & 0.04 \\ 0.13 & 0.76 & 0.07 & 0.04 \\ 0.11 & 0.09 & 0.68 & 0.12 \\ 0.15 & 0.08 & 0.06 & 0.71 \end{array} \right] \end{array} .$$

For this confusion matrix, there are two sets of important data: the precisions of classes {A, B, C, D}, which are {0.683, 0.776, 0.773, 0.780}, and the weighting of the classifier itself. The

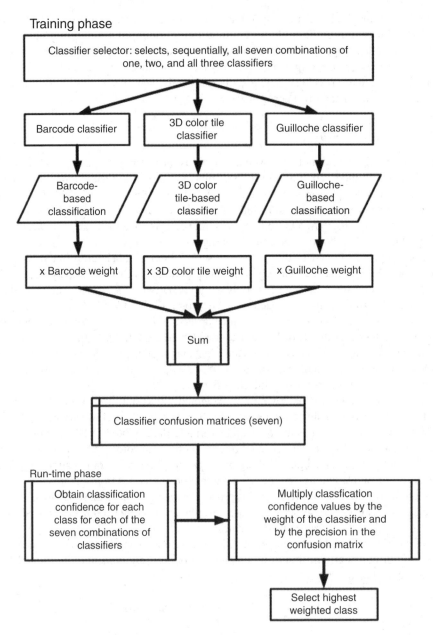

Figure 8.11   Weighted Confusion Matrix second-order meta-algorithmic pattern in context of the larger security printing authentication system described in the text. The complexity of this pattern is significantly increased over the template Weighed Confusion Matrix pattern as introduced in Figure 6.8. See text for details

accuracy of the classifier is the mean of the diagonal elements in the confusion matrix; that is, 0.748. We will use that value to determine the overall weighting after all of the confusion matrices are introduced. The second confusion matrix for class assignment is when only the 3D color tiles are used:

$$
\text{3D color tiles:} \quad
\begin{array}{c}
\\
\text{Origin} \\
\text{class}
\end{array}
\begin{array}{c}
\\
\text{B} \\
\text{C} \\
\text{D}
\end{array}
\quad
\begin{array}{c}
\text{Assigned} \quad \text{class} \\
\begin{array}{cccc}
\text{A} & \text{B} & \text{C} & \text{D}
\end{array} \\
\text{A} \begin{bmatrix} 0.89 & 0.03 & 0.05 & 0.03 \\ 0.02 & 0.92 & 0.03 & 0.03 \\ 0.02 & 0.05 & 0.91 & 0.02 \\ 0.02 & 0.05 & 0.06 & 0.87 \end{bmatrix}
\end{array}
$$

For this confusion matrix, the precisions of classes {A, B, C, D} are {0.937, 0.876, 0.867, 0.916}, and the accuracy of the classifier is 0.898. The third confusion matrix for class assignment is shown next. This confusion matrix is computed for the classifier that only incorporates the features for the guilloche regions:

$$
\text{Guilloche:} \quad
\begin{array}{c}
\\
\text{Origin} \\
\text{class}
\end{array}
\begin{array}{c}
\\
\text{B} \\
\text{C} \\
\text{D}
\end{array}
\quad
\begin{array}{c}
\text{Assigned} \quad \text{class} \\
\begin{array}{cccc}
\text{A} & \text{B} & \text{C} & \text{D}
\end{array} \\
\text{A} \begin{bmatrix} 0.93 & 0.01 & 0.02 & 0.04 \\ 0.05 & 0.90 & 0.01 & 0.04 \\ 0.07 & 0.04 & 0.86 & 0.03 \\ 0.03 & 0.02 & 0.03 & 0.92 \end{bmatrix}
\end{array}
$$

For this confusion matrix, the precisions of classes {A, B, C, D} are {0.861, 0.928, 0.935, 0.893}, while the classifier accuracy is 0.903. The relative accuracy of these individual—barcode, 3D color tile, and guilloche—classifiers is accounted for when generating the four combinational classifiers and their respective confusion matrices, as will be shown next. For these three classifiers, two types of relative weights were considered, as introduced in Section 6.2.3. The weights for the barcode, 3D color tile, and guilloche classifiers as determined using the optimized method introduced in Lin et al. (2003) and, separately, weighting proportional to the inverse of the error rate, are collected in Table 8.11. The latter are chosen since they do not lead to any negative values.

The last four confusion matrices are computed using the weights for barcode (W_{Barcode}), 3D color tile ($W_{\text{ColorTile}}$), and guilloche ($W_{\text{Guilloche}}$) provided in the last row of Table 8.11. The fourth confusion matrix is obtained when we analyze both the barcode and 3D color tile regions. Pooling their features and performing the classification on the training set, we get the following confusion matrix:

$$
\text{Barcodes} + \text{3D color tiles:} \quad
\begin{array}{c}
\\
\text{Origin} \\
\text{class}
\end{array}
\begin{array}{c}
\\
\text{B} \\
\text{C} \\
\text{D}
\end{array}
\quad
\begin{array}{c}
\text{Assigned} \quad \text{class} \\
\begin{array}{cccc}
\text{A} & \text{B} & \text{C} & \text{D}
\end{array} \\
\text{A} \begin{bmatrix} 0.93 & 0.02 & 0.04 & 0.01 \\ 0.02 & 0.94 & 0.03 & 0.01 \\ 0.04 & 0.03 & 0.90 & 0.03 \\ 0.03 & 0.04 & 0.05 & 0.88 \end{bmatrix}
\end{array}
$$

Table 8.11 Weights (normalized to sum to 1.0) obtained using the optimized weighting scheme (Lin *et al.*, 2003) and the proportionality to inverse of the error rate scheme, both introduced in Section 6.2.3. The inverse-error rate proportionality method was adopted since the barcode weight was otherwise less than 0.0 for the seventh classifier

	(Confusion Matrix Number) Classifiers Combined			
Weighting Scheme	(4) Barcode + 3D Color Tile	(5) Barcode + Guilloche	(6) 3D Color Tile + Guilloche	(7) Barcode + 3D Color Tile + Guilloche
Optimized weighting (Lin *et al.*, 2003)	$W_{Barcode} = 0.210$ $W_{ColorTile} = 0.790$	$W_{Barcode} = 0.204$ $W_{Guilloche} = 0.796$	$W_{ColorTile} = 0.491$ $W_{Guilloche} = 0.509$	$W_{Barcode} = -0.005$ $W_{ColorTile} = 0.490$ $W_{Guilloche} = 0.515$
Weighting α inverse of error rate	$W_{Barcode} = 0.288$ $W_{ColorTile} = 0.712$	$W_{Barcode} = 0.278$ $W_{Guilloche} = 0.722$	$W_{ColorTile} = 0.487$ $W_{Guilloche} = 0.513$	$W_{Barcode} = 0.165$ $W_{ColorTile} = 0.407$ $W_{Guilloche} = 0.428$

This confusion matrix has an accuracy of 0.913 and precisions of {0.912, 0.913, 0.882, 0.946}. When we analyze both the barcodes and the guilloche regions, pooling their features, we get the fifth confusion matrix:

$$
\text{Barcodes + guilloche:} \quad
\begin{array}{c}
\\
\\
\text{Origin} \\
\text{class}
\end{array}
\begin{array}{c}
\text{Assigned \quad class} \\
\begin{array}{cccc}
\text{A} & \text{B} & \text{C} & \text{D}
\end{array} \\
\begin{array}{c}
\text{A} \\ \text{B} \\ \text{C} \\ \text{D}
\end{array}
\begin{bmatrix}
0.98 & 0.01 & 0.00 & 0.01 \\
0.04 & 0.91 & 0.02 & 0.03 \\
0.07 & 0.06 & 0.79 & 0.08 \\
0.04 & 0.08 & 0.02 & 0.86
\end{bmatrix}
\end{array} .
$$

For the barcode + guilloche confusion matrix, the accuracy is 0.885 and the precisions are {0.867, 0.858, 0.952, 0.878}. Next, we analyze the pooled features from both the 3D color tile and guilloche regions, obtaining another confusion matrix:

$$
\text{3D color tiles + guilloche:} \quad
\begin{array}{c}
\\
\\
\text{Origin} \\
\text{class}
\end{array}
\begin{array}{c}
\text{Assigned \quad class} \\
\begin{array}{cccc}
\text{A} & \text{B} & \text{C} & \text{D}
\end{array} \\
\begin{array}{c}
\text{A} \\ \text{B} \\ \text{C} \\ \text{D}
\end{array}
\begin{bmatrix}
0.94 & 0.02 & 0.02 & 0.02 \\
0.01 & 0.91 & 0.03 & 0.05 \\
0.02 & 0.03 & 0.93 & 0.02 \\
0.02 & 0.01 & 0.01 & 0.96
\end{bmatrix}
\end{array} .
$$

The confusion matrix has an accuracy of 0.935 and precisions of {0.949, 0.938, 0.939, 0.914} for classes {A, B, C, D}. The final confusion matrix was generated when the features of all three of the regions—barcode, 3D color tile, and guilloche —were considered together:

$$
\text{Barcodes + 3D color tiles + guilloche:} \quad
\begin{array}{c}
\text{Origin} \\
\text{class}
\end{array}
\begin{array}{c}
A \\ B \\ C \\ D
\end{array}
\begin{array}{c}
\overset{\displaystyle \text{Assigned \quad class}}{\overset{\displaystyle \begin{array}{cccc} A & B & C & D \end{array}}{\begin{bmatrix} 0.93 & 0.02 & 0.03 & 0.02 \\ 0.03 & 0.90 & 0.03 & 0.04 \\ 0.02 & 0.01 & 0.94 & 0.03 \\ 0.01 & 0.03 & 0.01 & 0.95 \end{bmatrix}}}
\end{array}.
$$

The accuracy is 0.930 and the precisions are $\{0.939, 0.938, 0.931, 0.913\}$.

The computation of these seven confusion matrices completes the training phase of the Weighted-Confusion-Matrix-based second-order meta-algorithmic pattern shown in Figure 8.11. Note that any of these seven confusion matrices corresponds to a single classifier. Of the seven, the 3D color tile + guilloche combination (the sixth confusion matrix) resulted in the highest overall accuracy on the training set, and so if we are only concerned with deploying the best classifier we would simply use this combination and deploy the system.

The testing accuracies are, as expected, somewhat lower than the training accuracies, since the training data was also used for model validation. However, the relative order of the classifiers is unchanged, and the accuracies are a maximum of 0.023 lower in the test set. These data are collected in Table 8.12.

In Table 8.12, overall classification accuracy of 92.3% was observed for the test set when the 3D color tile and guilloche features were used. This is 4.0% better than the best individual classifier, that based on the guilloche features alone, and a corresponding 34.2% reduction in error rate.

However, a plurality of classifiers creates the possibility to use a different classifier (from the seven) or different means of combining classifiers for each sample. In this case, we can use one or more of the confusion matrices in place of the single confusion matrix. This is effectively a form of boosting, or overloading, in which the same classifier can be used multiple times depending on which matrices are chosen. A simple means of incorporating all seven classifiers at once is outlined in the *run-time phase* portion of Figure 8.11. Here, each classifier reports its confidence in each class, collecting this in an OPM. Next, these values are multiplied by the

Table 8.12 Accuracy of each of the seven classifiers on the training and test sets ($N = 400$ each, 100 from each of the four classes)

Classifier (or Classifier Combination)	Accuracy, Training	Accuracy, Test Set
1. Barcode	0.748	0.733
2. 3D color tile	0.898	0.875
3. Guilloche	0.903	0.883
4. Barcode + 3D color tile	0.913	0.890
5. Barcode + guilloche	0.885	0.868
6. 3D color tile + guilloche	0.935	0.923
7. Barcode + 3D color tile + guilloche	0.930	0.915

weight of the classifier, which is determined for the final set of seven classifiers in the same manner as shown in Table 8.11. Again, using the inverse-error rate proportionality method to avoid negative weights, the weights for the seven classifiers named in Table 8.12 are 0.054, 0.133, 0.139, 0.155, 0.118, 0.208, and 0.193, respectively. Finally, these values are multiplied by the appropriate precision value for the particular class and classifier.

An example illustrates this approach. Suppose that the barcode + 3D color tile classifier reports confidences in classes {A, B, C, D} of {0.21, 0.43, 0.19, 0.17}. Then, these values are multiplied by the weight of this classifier, which is 0.155, to produce the values {0.033, 0.067, 0.029, 0.026}. Finally, these values are multiplied by the appropriate precision values of {0.912, 0.913, 0.882, 0.946} to yield the final set of classification confidences of {0.030, 0.061, 0.026, 0.025}. Similarly, the other six classifiers will produce a confidence for each of the four classes. These confidences are not normalized since they reflect the weighting of the individual classifiers. Summing over the set of classifiers produces a new meta-classifier that assigns the class based on the argmax of the final classification confidences. This is the "select highest weighted class" block in Figure 8.11.

When this meta-pattern was deployed, its accuracy on the test set was 93.3%, a modest improvement over the best Weighted Confusion Matrix pattern at 92.3%. This reduced the error rate by 13.0%. No improvement over this accuracy was observed when leaving out one or more of the classifiers and performing the same procedure, although leaving out classifiers (1) and (5), individually or together, resulted in the same 93.3% accuracy.

There are additional approaches that can be taken after the seven weighted confusion matrices are computed. These approaches can use one or more of the following variations on the pattern:

1. The weighting of the individual classifiers. For example, four other methods for weighting not considered here are summarized in Table 6.6.
2. The manner in which to treat the classification confidence values. For example, only the top one, two, or three values can be kept.
3. Thresholding the precision values. For example, if precision is below a threshold value, it is assigned a value of 0.0 and so does not contribute to the final weighting.
4. Leaving out one or more of the combinational classifiers, as described above.

For the problem set in this section, applying classifier (3) and ignoring the classification confidence values when the precision was below 90% modestly increased (one more result was correct out of 400) the test accuracy to 93.5% (and also increased the training accuracy). This was the highest overall test accuracy achieved.

In summary, this section highlights the power of the Weighted Confusion Matrix approach. It also shows that hybridization of classifiers offers new opportunities to deploy meta-algorithmic expertise garnered from more than one pattern to create a more accurate overall system. In this case, Weighted Voting and Predictive Selection pattern principles were used to improve the overall accuracy of the classification system.

8.4 Secondary Domains

The two remaining second-order meta-algorithmic patterns are illustrated for image segmentation and speech recognition applications. As opposed to the relatively complicated meta-algorithmic application described in Section 8.3.4, these represent relatively straightforward, unaltered use of the pattern templates defined in Section 6.3.

8.4.1   Image Segmentation

An image segmentation application is used to illustrate the implementation of the Predictive Selection with Secondary Engines pattern introduced in Section 6.3.4. The schematic for the application is given in Figure 8.12, which is a straightforward implementation of Figure 6.11, with no modification in the architecture of the components (note that the training phase is omitted).

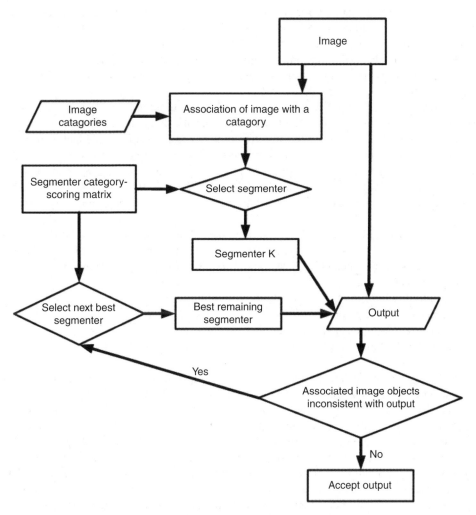

Figure 8.12   Predictive Selection with Secondary Engines second-order meta-algorithmic pattern as applied to the image segmentation task described in the text

For the Predictive Selection with Secondary Engines pattern, the image is immediately associated with a given image category. In this case, medical images comprising the superset of those shown in Figure 7.9 were used. This set included three different types of implants—Ti/TiO_2, NiTi aka nitinol, and $Ca_3(PO_4)_2$ that is also designated tricalcium phosphate or TCP. The latter, being a light colored ceramic, is readily distinguishable from the first two. The first two, however, both being metals largely based on titanium, are much more difficult to distinguish. A simple decision tree algorithm, involving only two binary decisions, is used for the predictive assignment of the image to the category:

1. Determine if a $(Ca_3(PO_4)_2)$ or a $\{(Ti/TiO_2),(nitinol)\}$ implant.
2. If a $\{(Ti/TiO_2),(nitinol)\}$ implant, determine if (Ti/TiO_2) or (nitinol) implant.

The first decision is easily performed using a threshold operation on the histogram of the image, and had an accuracy of 100% for the images tested (approximately 150/implant type). If the implant type was determined to be $Ca_3(PO_4)_2$, then segmenter A was chosen to identify $Ca_3(PO_4)_2$ regions distinct from bone and non-bone (e.g., intracortical void) regions of the image. Performing image analysis with segmenter A requires roughly 80 times the processing of the predictive step of the process.

If, however, the implant was found to have a Ti/TiO_2 or nitinol implant, then the images are analyzed separately to distinguish the two types of metals. Differences in intensity—nitinol was much lighter than the Ti/TiO_2—and differences in porosity—the nitinol had more pores within the metallic areas—were used together, again affording 100% accuracy for the images tested. This step required roughly 20 times the processing of the predictive step of the process. This step, which identified metallic implant regions, is segmenter B. Conveniently, it provided all of the metallic regions as output, obviating the need to identify them in the downstream step, described next.

If the implant material was Ti/TiO_2 or nitinol, then all that remained was to find the bone and nonbone regions, a process that required roughly 30 times the processing of the predictive step of the process (segmenter C). The overall relative processing requirements using this approach are as follows:

1. $Ca_3(PO_4)_2$: $1\times$ + segmenter A = $1\times + 80\times = 81\times$
2. Ti/TiO_2: $1\times$ + segmenter B + segmenter C = $1\times + 20\times + 30\times = 51\times$
3. Nitinol: $1\times$ + segmenter B + segmenter C = $1\times + 20\times + 30\times = 51\times$.

If the image classes each represent one-third of the input, then the mean processing time is $61\times$ the processing of the predictive step of the process.

Originally, the image segmentation software was written without the benefit of meta-algorithmics. The original software thus had to identify and segment all five potential region types at once. This added complexity resulted in a mean processing time of approximately $135\times$ the processing of the predictive step of the meta-algorithmic process.

Unlike many of the other examples in this chapter, this pattern is focused on reducing the processing time. Accuracy is not a consideration, since 100% accuracy during segmentation was obtained both with and without applying the meta-algorithmic pattern (Figure 8.12). Robustness, also, was not an explicit consideration, although it is clear that the decision tree approach used for the predictive selection portion of the pattern does make the system more robust for the segmentation of additional implant types, should they be added in the future. The processing time, however, as the main design consideration, was in fact reduced by 54.8%.

8.4.2 Speech Recognition

The final example uses the Best Engine with Differential Confidence or Second Best Engine pattern originally introduced in Section 6.3.8 for an automatic speech recognition (ASR) task. The schematic of the system is provided in Figure 8.13.

For this problem, three commercially available ASR engines were fed voice data associated with call center tasks: predominantly numbers, menu options, names, and addresses. The first and second most accurate engines were two well-known ASR engines in optimized configuration, while the third most accurate engine was the first of these two engines used in a high-speed configuration. Accuracy was determined simply from the ratio of ASR errors

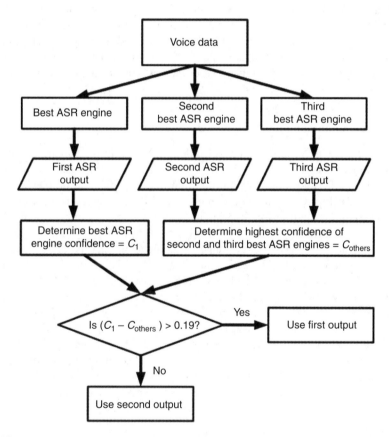

Figure 8.13 Best Engine with Differential Confidence or Second Best Engine second-order meta-algorithmic pattern, run-time phase only. The training phase—omitted from the diagram—was used to determine the threshold value of 0.19. The Best Engine is the ASR engine with the overall highest accuracy in the call system task

to ASR voice tasks attempted. The three ASR engines were ranked in order of accuracy on the training set. As for most of the illustrative examples in this chapter, the data was assigned equally to training and test sets. The three main purposes of the training set were to (1) rank the ASR engines in order of accuracy; (2) validate—and normalize to the range [0,1]—the confidence settings of the engines; and (3) determine the optimum setting of the threshold to use as shown in Figure 8.13. On the training set, the threshold value was varied in increments of 0.01 from 0.00 to 0.50. The values from 0.17 to 0.21 provided the same peak accuracy, and so the threshold value chosen was the mean of this range, 0.19, assuming this would make it the most insensitive to variability in input data.

For the test set, each ASR engine provided its output and its confidence value. If the best engine reported a confidence value, C_1, satisfying the following requirement, its output was accepted:

$$(C_1 - 0.19) \geq C_2 \, \| C_3$$

Here C_2 and C_3 are the confidence values reported by the second and third best ASR engines. Note that if either C_2 or C_3 is within 0.19 of C_1, the output of the second best engine is chosen. The third best engine, of course, does not directly provide the system output, but it can serve to select the second best engine's output in lieu of the best engine's output. Applying this pattern resulted in higher accuracy (85.6%) than that of the best engine (83.4%), reducing the error rate by 13.3%.

8.5   Summary

This chapter began with an in-depth consideration of how confusion matrices can be used powerfully to aid in classification, especially for converting many-class classification problems into more manageable and more targeted decision tree classification problems. Confusion matrices can be used to determine an optimum system architecture, since the output of the attractor/repeller approach discussed in depth in the chapter are a set of aggregate classes that can be treated with independent meta-algorithmic approaches.

Next, this chapter cataloged a large range of applications incorporating both relatively straightforward (Confusion Matrix and Weighted Confusion Matrix, Predictive Selection with Secondary Engines, Single Engine with Required Precision, Majority Voting or Weighted Confusion, Majority Voting or Best Engine, Best Engine with Differential Confidence or Second Best Engine, Best Engine with Absolute Confidence or Weighted Confusion Matrix) and more complicated (Confusion Matrix with Output Space Transformation, Tessellation and Recombination with Expert Decisioner) second-order meta-algorithmic patterns. For a security printing application, it was shown how elements of several meta-algorithmic patterns could be brought to bear on a single intelligent system problem. This type of complexity naturally leads to the third-order meta-algorithmic patterns, which is the topic of the next chapter.

References

Lin, X., Yacoub, S., Burns, J., and Simske, S. (2003) Performance analysis of pattern classifier combination by plurality voting. *Pattern Recognition Letters*, **24**, 1959–1969.

Ruta, D. and Gabrys, B. (2000) An overview of classifier fusion methods. *Computing and Information Systems*, **7** (1), 1–10.

Salton, G., Fox, E.A., and Wu, H. (1983) Extended Boolean information retrieval. *Communications of the ACM*, **26** (11), 1022–1036.

Sebastiani, F. (2002) Machine learning in automated text categorization. *ACM Computing Surveys*, **34** (1), 1–47.

Simske, S.J., Li, D., and Aronoff, J.S. (2005) *A Statistical Method for Binary Classification of Images*. ACM Symposium on Document Engineering, pp. 127–129.

Simske, S.J., Wright, D.W., and Sturgill, M. (2006) *Meta-algorithmic Systems for Document Classification*. ACM DocEng 2006, pp. 98–106.

9

Third-Order Meta-algorithmics and Their Applications

There is no such thing as failure, only feedback.

—Michael J. Gelb

The sensitivity of men to small matters, and their indifference to great ones, indicates a strange inversion.

—Blaise Pascal

An algorithm must be seen to be believed.

—Donald Knuth

9.1 Introduction

Third-order meta-algorithmic patterns are distinguished from first- and second-order meta-algorithmic patterns by their complexity in both architecture and *time*. Unlike the meta-algorithmic patterns described in Chapters 7 and 8, these patterns can adapt to changing input sets—culminating with the Generalized Hybridization pattern, in which the *set* of meta-algorithmic patterns used may change over time. To call this a *meta*-meta-algorithmic may be self-indulgent, but not necessarily hyperbolic.

Since there are 16 first- and second-order meta-algorithmic patterns provided in the past two chapters, along with guidelines for generalized hybridization, one might ask why third-order meta-algorithmic patterns are needed at all. First off, first- and second-order meta-algorithmic patterns tend to be most valuable for optimizing system accuracy and/or cost, whereas the adaptability of third-order patterns allows them to optimize system robustness. Secondly, meta-algorithmic patterns do not always improve the system accuracy, cost, or robustness. This was exemplified in Section 8.3.1, wherein several meta-algorithmic patterns based on the confusion matrix were shown to significantly improve system accuracy, while others—based on voting or differential confidence, for example—did not improve system accuracy. It can be said that, for a given system, *a meta-algorithmic pattern is not guaranteed to work, but meta-algorithmics can be made to work*. The third-order patterns are not just adaptable—they are reconfigurable. Thus, they can be configured to make the meta-algorithmics work.

Meta-algorithmics: Patterns for Robust, Low-Cost, High-Quality Systems, First Edition. Steven J. Simske.
© 2013 John Wiley & Sons, Ltd. Published 2013 by John Wiley & Sons, Ltd.

Two of the most important tools for enabling this adaptability are feedback and sensitivity analysis. In keeping with Gelb's quote, when the meta-algorithmic pattern fails, its error signal can be fed back to the meta-algorithmic pattern, ultimately providing an optimized meta-algorithmic configuration and improved system behavior. Feedback, therefore, is a powerful method to prevent failure of a meta-algorithmic approach. Sensitivity analysis, meanwhile, is used to obviate the type of inappropriate attention to irrelevant details that Blaise Pascal bemoaned. With sensitivity analysis coupled to adaptive meta-algorithmics, an appropriate focus on the most important factors in system optimization will be provided.

As Knuth astutely noted, however, "an algorithm must be seen to be believed." I would extend that to say that "a meta-algorithm must be seen to be believed." Let us therefore make these third-order meta-algorithmic patterns visible.

9.2 Third-Order Meta-algorithmic Patterns

Third-order meta-algorithmic patterns indeed provide three levels of adaptation: (1) the meta-algorithmic components, often first order, comprising the pattern; (2) the sequence of these components; and (3) the manner in which output, especially errors, are fed back to the input, altering the architecture of the pattern.

9.2.1 Examples Covered

As in Chapters 7 and 8, the examples herein will include ones for the primary four domains: (1) document understanding, (2) image understanding, (3) biometrics, and (4) security printing. The additional domains for applying the patterns, as shown in Section 6.4, are surveillance, optical character recognition (OCR), and security analytics. As in previous chapters, the decision on which pattern to use for which domain was somewhat arbitrary: I tried to match them based on what I thought would provide the widest range of exposure to third-order pattern deployment considerations.

The Confusion Matrix for Feedback pattern is the first deployed, since it builds on the pattern originally described in Section 6.4.3 and the confusion matrix techniques elaborated in Section 8.2. Next, the Regional Optimization pattern, which is an intelligent extension of the Predictive Selection pattern, is applied to image understanding. The Expert Feedback pattern of Section 6.4.4 is then applied to biometrics. Next, the Generalized Hybridization pattern is used to further extend the authentication and aggregation capabilities of security printing. The power of Sensitivity Analysis for optimizing a meta-algorithmic-based system is then shown for a surveillance system. The last two examples focus on feedback. Ecosystem feedback is provided by the Proof by Task Completion pattern, and its value in an OCR system is illustrated. Finally, training-gap-targeted feedback is used in the domain of security analytics. This exemplifies the ability to identify specific subsets of the overall domain to use the meta-algorithmic patterns, and as such deserves a brief overview. This is provided in the next section.

9.2.2 Training-Gap-Targeted Feedback

One of the key aspects of meta-algorithmic system design is adventitious parallelism. This is the art of meta-algorithmics: finding the decisions and intelligent processes that can be broken down into relatively clean separate subprocesses. Usually the system architect designs for task parallelism, spatial or component-based parallelism, and other forms of structural parallelism. However, feedback systems—in feeding back with a delay the output of the system—provide the means for temporal parallelism. Training gap targeting is using the output to better assess how the system should handle input it is less familiar with and in effect close any gaps in its learning.

The Feedback pattern, originally overviewed in Section 6.4.1, provides the means for two algorithms—usually a specific algorithm and its inverse/converse—to collaborate in the production of highly accurate output. As with any feedback system, the primary components of the feedback system are:

1. *Input*: This is the content to be analyzed, which can contain its own noise, separate from the disturbances to the system in point (4) below.
2. *Comparator and associated error*: This is the difference between the input signal and its transformation into output after being analyzed and after being fed back to the comparator.
3. *Control element*: This is the generator (algorithm, system, engine) used to analyze the content and convert it into output.
4. *Disturbances to the system*: This is error introduced at any point along the pathway from input to output.
5. *Output*: This is the transformed input produced by the generator.

In Figure 6.17, the Feedback third-order meta-algorithmic pattern was introduced in a simple, single-pass system in which the content analyzed was restructured into its components, allowing semantics to be associated with the input content. This results in a better compression ratio for the content, along with the identification and filtering of noise.

This pattern template was therefore reliant on the design and development of an inverse transformation of the algorithm (an OCR system was used as an exemplar), which can be used to reconstruct the input from the compressed, semantically- or context-tagged, transformed output. This "inverse" or feedback algorithm can be directly applied to regenerate a second, preferably highly accurate, copy of the original input. The comparator is then used to compare the regenerated output to the original input document to determine the differences. An error signal is generally then fed back to the algorithm to correct the output, if possible. However, the Feedback pattern shown in Figure 6.17 can be simplified as follows: the error signal, instead of being fed back to the input, is simply compared to a threshold. If the error exceeds a given threshold, then the analysis is deemed to provide insufficient quality, and fails. Instead of affecting the algorithm, this failure requires a separate algorithm to be performed. Because of this, the Feedback pattern of Section 6.4.1 is considered only *partially closed-loop*.

More generally, however, feedback is *fully closed-loop*. One important application for closed-loop feedback in meta-algorithmics is when one or more partitions of the training set data, or input space, is/are poorly analyzed by an existing pool of generators. This "gap" in the training data can be targeted—if not identified in the first place—by deploying the Feedback pattern. An example of how this can be implemented is given for security analytics in Section 9.4.3.

More broadly, the training-gap-targeted feedback approach is based on feedback over time, meaning that the training gap itself can change. Changes in the training gap are noted, thus providing feedback on the feedback.

9.3   Primary Domains

The primary domains will be used to show how to customize four of the third-order meta-algorithmic patterns to solve intelligent system problems. We begin, as usual, with document understanding.

9.3.1   Document Understanding

The Confusion Matrix for Feedback pattern was introduced in Section 6.4.3, and important manipulations of the confusion matrices are described in Section 8.2. In this example, we work in the opposite direction—finding the best confusion matrix reduction based on the binary classifiers. The binary classifier results are then incorporated into the reduction of the confusion matrix originally obtained when three different document indexing engines are used to extract keywords from a set of documents. This 3-class system has its confusion matrix defined by

$$
\text{Origin } \begin{array}{c} \\ A \\ B \\ C \end{array}
\begin{array}{ccc}
A & B & C \\
\left[\begin{array}{ccc}
0.65 & 0.19 & 0.16 \\
0.13 & 0.70 & 0.17 \\
0.11 & 0.20 & 0.69
\end{array}\right].
\end{array}
$$

An equal number of documents from each set (200) were used to create the confusion matrix. The overall accuracy, then, is the mean of the elements in the main diagonal, or 0.68. Class B is the most likely class to be assigned to a sample (36.3%), followed by class C (34.0%) and class B (29.7%). As described in Section 6.4.3, this pattern assesses the set of reduced-order (thus binary) confusion matrices associated when each pair of classes is aggregated.

The three reduced-order confusion matrices are given here:

1. The AB confusion matrix is

$$
\text{Origin } \begin{array}{c} A \\ B \end{array}
\begin{array}{cc}
A & B \\
\left[\begin{array}{cc}
0.65 & 0.19 \\
0.13 & 0.70
\end{array}\right].
\end{array}
$$

2. The AC confusion matrix is

$$
\text{Origin } \begin{array}{c} A \\ C \end{array}
\begin{array}{cc}
A & C \\
\left[\begin{array}{cc}
0.65 & 0.16 \\
0.11 & 0.69
\end{array}\right].
\end{array}
$$

3. The BC confusion matrix is

$$
\text{Origin } \begin{array}{c} B \\ C \end{array}
\begin{array}{cc}
B & C \\
\left[\begin{array}{cc}
0.70 & 0.17 \\
0.20 & 0.69
\end{array}\right].
\end{array}
$$

The AB confusion matrix accuracy is $(0.65 + 0.70)/(0.65 + 0.19 + 0.13 + 0.70) = 0.808$. The AC confusion matrix accuracy is 0.832, and the BC confusion matrix accuracy is 0.790. The relative accuracies of these three 2-class subsystems are relatively similar for this "throw-away" prototyped example (and the classification accuracies in general are not very high). This allows the meta-algorithmic system architect to target any of the three possible binary classifications to improve the overall system accuracy. The overall accuracy of the system therefore depends on the accuracy of the classifier for distinguishing either A versus B, A versus C, or B versus C in comparison to the accuracy of determining the accuracy between these pairs of classes and the appropriate third class. In order to prepare for these comparisons, we produce the following set of reduced-dimension confusion matrices:

1. The (A + B) C confusion matrix is

$$\text{Origin} \begin{array}{cc} & \begin{array}{cc} A+B & C \end{array} \\ \begin{array}{c} A+B \\ C \end{array} & \begin{bmatrix} 1.67 & 0.33 \\ 0.31 & 0.69 \end{bmatrix} \end{array}.$$

2. The (A + C) B confusion matrix is

$$\text{Origin} \begin{array}{cc} & \begin{array}{cc} A+C & B \end{array} \\ \begin{array}{c} A+C \\ B \end{array} & \begin{bmatrix} 1.61 & 0.39 \\ 0.30 & 0.70 \end{bmatrix} \end{array}.$$

3. The (B + C) A confusion matrix is

$$\text{Origin} \begin{array}{cc} & \begin{array}{cc} B+C & A \end{array} \\ \begin{array}{c} B+C \\ A \end{array} & \begin{bmatrix} 1.76 & 0.24 \\ 0.35 & 0.65 \end{bmatrix} \end{array}.$$

The accuracies of these three confusion matrices are 0.787, 0.770, and 0.803, respectively. The aggregated classes (A + B), (A + C), and (B + C) have 55.7%, 53.7%, and 58.7%, respectively, of the samples. Thus, the overall system accuracies of the three above confusion matrices are

1. $0.69/3.00 + 0.557 \times a(A + B)$
2. $0.70/3.00 + 0.537 \times a(A + C)$
3. $0.65/3.00 + 0.587 \times a(B + C)$.

Here and in the following, the notation a(*) denotes the accuracy of the aggregate class *. The above equations reduce to

1. $0.230 + 0.557 \times a(A + B)$
2. $0.233 + 0.537 \times a(A + C)$
3. $0.217 + 0.587 \times a(B + C)$.

The next step in optimizing the deployment is to determine the minimum values of the accuracies for a(A + B), a(A + C), and a(B + C) to improve the overall system accuracy. The governing equations for this are

1. $0.230 + 0.557 \times a(A + B) \geq 0.680$
2. $0.233 + 0.537 \times a(A + C) \geq 0.680$
3. $0.217 + 0.587 \times a(B + C) \geq 0.680$.

Here, 0.680 is taken from the overall system accuracy. These reduce to

1. $0.557 \times a(A + B) \geq 0.450$
2. $0.537 \times a(A + C) \geq 0.447$
3. $0.587 \times a(B + C) \geq 0.463$.

And, thus

1. $a(A + B) \geq 0.808$
2. $a(A + C) \geq 0.832$
3. $a(B + C) \geq 0.789$.

This final set of inequalities describes the minimum binary classification accuracies required to warrant the initial aggregations. The results indicate that creating a good binary classifier for the original classes B and C is the likely best path forward. This is borne out by the following, which show the overall system accuracy when binary classifiers with 90% accuracy are created:

1. $0.230 + 0.557 \times 0.90 = 0.731$
2. $0.233 + 0.537 \times 0.90 = 0.716$
3. $0.217 + 0.587 \times 0.90 = 0.745$.

More pragmatically, once the accuracies of all available binary classifiers are determined, the highest predicted system accuracy, obtained from one of these three equations: (1) $0.230 + 0.557 \times a(A + B)$; (2) $0.233 + 0.537 \times a(A + C)$; and (3) $0.217 + 0.587 \times a(B + C)$, is used to dictate which binary classifier, if any, to deploy.

In the example above, the three best binary classifiers found were $a(A + B) = 0.881$, $a(A + C) = 0.904$, and $a(B + C) = 0.877$. For these, the overall system accuracies are 0.721, 0.718, and 0.732, respectively. Thus, the binary classifier for distinguishing between classes B and C is deployed after the original classifier is used to distinguish class A from the combined set of classes B + C. This system has a 5.2% higher accuracy than the initial classifier and a 16.2% reduction in the error rate. The system schematic is shown in Figure 9.1.

9.3.2 Image Understanding

In order to illustrate the application of the Regional Optimization third-order meta-algorithmic pattern introduced in Section 6.4.6, a traffic image segmentation and object identification system was built. Three different segmentation engines (differentiated largely by the

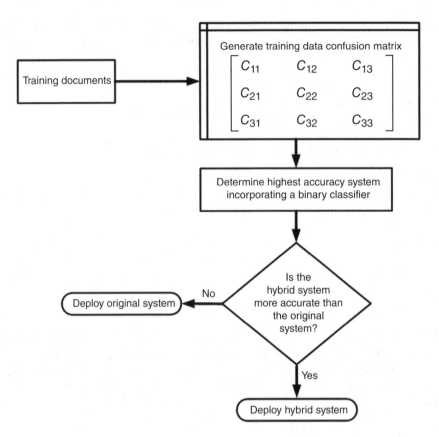

Figure 9.1   Schematic for deploying the Confusion Matrix for Feedback third-order meta-algorithmic pattern for document classification as described in the text

thresholding they employed for object segmentation) were used to extract objects associated with movement during a short (1.0 s burst) video of 5 frames/s. The first and last frames were then used to assess moving objects. The traffic objects, all moving, belonged to five classes that were roughly equally represented in the videos: pets, people, bicycles, cars, and trucks. Half of the 100 burst video samples were assigned to training and half were assigned to testing. The average burst video featured approximately two moving objects.

The Regional Optimization pattern was originally described in Figure 6.22, and its deployment in this example is illustrated in Figure 9.2. As in Chapter 6, the training and run-time phases are straightforward. Training content is collected for every "region" of input space to be optimized during the training phase, and assigned to the appropriate region from among a plurality of regions. The appropriate meta-algorithmic patterns are then configured for each region, or subclass or input. Rather than selecting a single classifier for each subclass when deployed for the test data, however, this pattern selects the best individual meta-algorithmic pattern from the possible set.

The "regions" for which the output-determining meta-algorithmic pattern was selected were specific subranges in the range of movement artifact. Motion artifact was determined

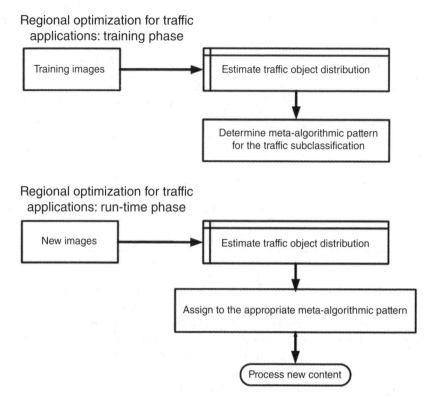

Regional optimization for traffic
applications: training phase

Training images

Estimate traffic object distribution

Determine meta-algorithmic pattern
for the traffic subclassification

Regional optimization for traffic
applications: run-time phase

New images

Estimate traffic object distribution

Assign to the appropriate meta-algorithmic pattern

Process new content

Figure 9.2   Schematic for deploying the Regional Optimization third-order meta-algorithmic pattern
for traffic image segmentation and object identification as described in the text

by aligning the first and the last frame of the burst video, subtracting one from the other,
and then thresholding the resulting "difference" image. For this "throw-away" example, a set
threshold—differences of more than 24 in intensity, which varied from 0 to 255—was used
for all of the images. Roughly 20% of all the frames reported percent suprathreshold pixels in
each of the following five ranges: 0–1%, 1–5%, 5–10%, 10–25%, and 25% or more. For each
range, six different meta-algorithmic approaches were then deployed to determine which of
these provided the highest accuracy object identification on the training set. The algorithms
chosen were the Voting (Section 6.2.3), Weighted Voting (Section 6.2.3), Tessellation and
Recombination (Section 6.2.5), Confusion Matrix (Section 6.3.1), Weighted Confusion Matrix
(Section 6.3.1) and Single Engine with Required Precision (Section 6.3.5) patterns.

The results for each of the six meta-algorithmic patterns on each of the five regions of the
input space are given in Table 9.1. For the 0–1% region, the Single Engine with Required
Precision pattern provided the highest accuracy (54%); for the 1–5% region, the Weighted
Voting pattern provided the highest accuracy (60%); for the 5–10% region, the Tessellation
and Recombination pattern provided the highest accuracy (62%); and for both the 10–25% and
25% or more ranges, the Weighted Confusion Matrix pattern provided the highest accuracies,
at 78% and 68%, respectively. Using these optimum meta-algorithmic patterns for each of the
five regions results in 64.4% overall object identification accuracy.

Table 9.1 Accuracy results of the six different meta-algorithmic patterns for each of the five regions of input space. The highest accuracy meta-algorithmic pattern is generally different for each of the regions. Selecting the optimum meta-algorithmic pattern for each region results in an overall accuracy of 64.4%, higher than the best single meta-algorithmic pattern, Weighted Confusion Matrix, which provides 61.2% accuracy. This is an 8.2% reduction in the relative error rate

Meta-algorithmic Pattern Deployed	Region (Range of Percent Motion in the Image) of Input				
	0–1%	1–5%	5–10%	10–25%	25% or more
Voting	0.34	0.42	0.46	0.48	0.46
Weighted Voting	0.46	0.60	0.58	0.66	0.64
Tessellation and Recombination	0.44	0.46	0.62	0.68	0.62
Confusion Matrix	0.50	0.54	0.56	0.74	0.66
Weighted Confusion Matrix	0.48	0.54	0.58	0.78	0.68
Single Engine with Required Precision	0.54	0.54	0.56	0.58	0.62

Had we simply used a single meta-algorithmic pattern for object identification for each of the five regions of the input space, the accuracies would have been lower—often substantially lower. The single-pattern accuracies are 43.2% for Voting, 58.8% for Weighted Voting, 56.4% for Tessellation and Recombination, 60.0% for the Confusion Matrix approach, 61.2% for the Weighted Confusion Matrix approach, and 56.8% for the Single Engine with Required Precision approach. The improvement of 3.2% in accuracy over the best individual meta-algorithmic equates to an 8.2% reduction in relative error rate.

Interestingly, in this example, an even greater (though not statistically significant) improvement was observed for the test data: 63.8% for the combined approach, 42.2% for Voting, 54.6% for Weighted Voting, 55.8% for Tessellation and Recombination, 59.2% for the Confusion Matrix approach, 60.4% for the Weighted Confusion Matrix approach, and 55.6% for the Single Engine with Required Precision approach. Here, the accuracy improved by 3.4% over that of the best individual meta-algorithmic pattern, and the error rate was reduced by 8.6%.

These results, given for a small set of burst video clips all taken within a city block of one another, are unlikely to be indicative of the best meta-algorithmic pattern to generally employ for such an object recognition task. However, the approach is illustrative of how the Regional Optimization pattern is deployed, and the relative improvement in error, while modest, may still be useful in a number of tracking, surveillance, and object recognition tasks.

9.3.3 Biometrics

The biometrics example illustrating the implementation of the Expert Feedback third-order meta-algorithmic pattern, as introduced in Section 6.4.4, is voice recognition. In this example, the metrics I am using are quite simple to compute—and are certainly not the best set of metrics to use to identify a speaker. I am using them to expedite the set-up of another "throwaway" example meant primarily to illustrate how the Expert Feedback pattern can be used. Each speaker is asked to read aloud a particular sentence 200 times. The noise level is varied by having the person move one step further away from a fan after each 10 readings of the sentences, and then to reverse the pattern by repeating the sentence 10 times at the farthest distance and then take one step closer for each of the remaining sets of 10 readings. Bandgap filtering of the fan noise is performed, resulting in differences in the amount of speaker signal lost. One vocalization at each of the 10 distances from the fan is added to the training and, separately, the test set. Seven individuals were tested and four of them were assigned to the "other people" class, with data for this composite individual created from the adding together of all four voice Fourier transforms (FTs). Thus, the system was designed to recognize three people, presumably who are authorized to access a system or other privileges, and not confuse them with each other or with a composite of some other potential users of the system. It is hoped that unanticipated users of the system will be more like the composited voice information, though this was not explicitly tested for or guaranteed. Thus, at worst, this can be considered a 4-class voice recognition classification problem.

The signal is then processed quite simply as follows. An FT of the voice data is performed, and the spectral content is assigned to 10 bands, each containing 10% of the energy spectral density of the FT. The first 10% and last 10% band are ignored, since they are usually the noisiest and the most variable bands. The middle eight bands are then represented as their sequential percentages of the log-frequency range from the 10% to the 90% point of the FT. These are the first eight features recorded. For example, if a signal is evenly distributed across the log-frequency range, these eight values are $\{0.125, 0.125, \ldots, 0.125\}$. These are types of *input range* measurements that are one of the types of data rules we deploy for the Expert Feedback pattern. Co-occurrences of two bands having a high percentage of the log-frequency range are then tested for. For example, if an individual's eight values for input range are $\{0.2, 0.1, 0.1, 0.2, 0.1, 0.1, 0.1, 0.1\}$, then this person has abnormally high first and fourth band values, or a (1,4)-band co-occurrence. For L bands, there are $L(L - 1)/2$ such co-occurrence comparisons. For eight bands, this results in a total of 28 features.

The third type of expert feedback rule used in this meta-algorithmic pattern is ratios. Band ratios also require the pairing of two bands, providing another 28 features. This brings the total to 64 features. In the example of the previous paragraph, the ratios are $\{2.0, 2.0, 1.0, 2.0, 2.0, 2.0, 2.0\}$ for the first band when divided by the values for the other bands. The schematic for the application of this pattern to this task is shown in Figure 9.3.

A trained classifier (Simske, Li, and Aronoff, 2005) is used to generate a confusion matrix, as shown here. The three individuals are classes A, B, and C, whilst the composite individual is class D. The precisions of each class, computed by the columns, are 0.684, 0.688, 0.720, and 0.714, respectively, for A, B, C, and D:

		Assigned	class		
		A	B	C	D
	A	0.80	0.06	0.08	0.06
Origin	B	0.11	0.66	0.07	0.16
class	C	0.20	0.13	0.59	0.08
	D	0.06	0.11	0.08	0.75

The overall accuracy of the classification system is 0.70. We then target the off-diagonal content in the columns in order of the overall precision. Based on previous experience, I have

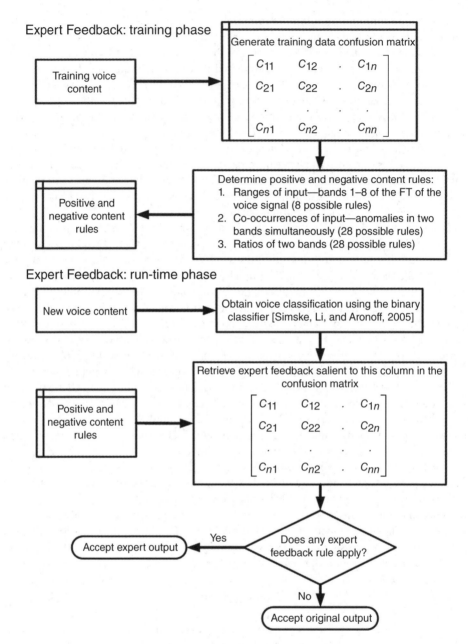

Figure 9.3 Schematic for the Expert Feedback pattern as applied to the voice identification problem described in the text

found it efficient to target columns in the confusion matrix starting with the lowest precision. That column is shown here:

$$
\begin{array}{c}
 & A \\
\begin{array}{c} A \\ B \\ C \\ D \end{array}
\left[\begin{array}{c} 0.80 \\ 0.11 \\ 0.20 \\ 0.06 \end{array}\right]
\end{array} .
$$

Of the three off-diagonal elements in this column, the [A,C] element—corresponding to the misassignment of an element from class C to class A—was considered first since it had the greatest probability of any of the misclassifications. An expert feedback rule about one of the 64 features (the ratio of bands 2 and 7, in this case) was found to correctly reclassify 0.05 of the [A,C] elements to the correct [C,C] element, with lesser *collateral reclassification*. Collateral reclassification is the process by which other elements in the column are also, in this case incorrectly, reassigned to the column associated with the correct reclassification. In this case, 0.01 of each of the elements [A,A], [A,B], and [A,D] is reassigned based on the rules. The "A" and "C" columns are thus altered to the following after this rule is applied:

$$
\begin{array}{c}
 & A \\
\begin{array}{c} A \\ B \\ C \\ D \end{array}
\left[\begin{array}{c} 0.79 \\ 0.10 \\ 0.15 \\ 0.05 \end{array}\right]
\end{array}
\quad \text{and} \quad
\begin{array}{c}
 & C \\
\begin{array}{c} A \\ B \\ C \\ D \end{array}
\left[\begin{array}{c} 0.09 \\ 0.08 \\ 0.64 \\ 0.09 \end{array}\right]
\end{array} .
$$

As can be seen from these columns, some reclassifications—in this case of [A,B] to [C,B] and [A,D] to [C,D]—do not affect the overall system accuracy. They do, however, assign a new set of inputs to column "C" and thus will change the nature of that column. This is important to note since a different set of expert rules may now prove relevant to that column. Regardless, incorporating these new columns, the overall confusion matrix now becomes

$$
\begin{array}{cc}
 & \text{Assigned\quad class} \\
 & \begin{array}{cccc} A & B & C & D \end{array} \\
\begin{array}{c} \\ \text{Origin} \\ \text{class} \\ \\ \end{array}
\begin{array}{c} A \\ B \\ C \\ D \end{array}
\left[\begin{array}{cccc}
0.79 & 0.06 & 0.09 & 0.06 \\
0.10 & 0.66 & 0.08 & 0.16 \\
0.15 & 0.13 & 0.64 & 0.08 \\
0.05 & 0.11 & 0.09 & 0.75
\end{array}\right]
\end{array} .
$$

The precisions of the columns are now 0.725, 0.688, 0.711, and 0.714, respectively. The overall accuracy has improved to 0.710. Column "B," with the lowest current precision, is next probed for expert rules using the 64 metrics collected. We continue along these lines until no further improvement in the overall accuracy of the confusion matrix is obtained after searching all four columns in order of increasing precision. Experience has shown that improvements in overall accuracy must be greater than a certain threshold, to prevent overtraining. For this simple voice classification system, we chose a threshold of 0.004: if applying the rule does not result in an overall accuracy improvement of 0.004 or more, then it is not applied. This heuristic is based on validation experiments that showed values below 0.004 were unlikely to represent an actual improvement as opposed to focusing on a local minimum.

With this process and threshold, a total of seven rules—two each for columns "A," "B," and "C" and one for column "D"—were applied and the final confusion matrix obtained was

$$
\begin{array}{c}
\text{Origin} \\
\text{class}
\end{array}
\begin{array}{c}
 \\
\text{A} \\
\text{B} \\
\text{C} \\
\text{D}
\end{array}
\overset{\displaystyle \text{Assigned} \quad \text{class}}{\overset{\displaystyle \text{A} \quad\; \text{B} \quad\;\; \text{C} \quad\;\; \text{D}}{
\begin{bmatrix}
0.82 & 0.06 & 0.07 & 0.05 \\
0.09 & 0.68 & 0.09 & 0.14 \\
0.15 & 0.10 & 0.66 & 0.09 \\
0.05 & 0.10 & 0.07 & 0.78
\end{bmatrix}}}.
$$

For this final confusion matrix, the column precisions are 0.739, 0.723, 0.742, and 0.736, respectively. The overall accuracy is now 0.735, an improvement of 3.5% and a reduction in the error rate of 11.7%. This result is promising, and indeed held up for the test data, in which the original classification accuracy was 68.8% and the post-Expert Feedback classification rate was 73.0%.

This 13.5% reduction in error rate for test data was only 6.7% if the thresholding requirement (accuracy must increase by 0.004 or more, or do not apply the rule) was not employed. That is, with the threshold set at 0.000 for rules based on training data, several more rules were employed, which in combination reduced the improvement in error rate on test data.

Some other results of the Expert Feedback approach employed are also worth noting. First off, after applying the expert rules, the variability of the column precisions drops significantly. Before applying the expert rules, column precisions (mean ± standard deviation) are 0.702 ± 0.018. After applying the expert rules, column precisions are 0.735 ± 0.008. There is clearly an effect of applying the rules to homogenize the column precisions. This can be viewed, in some ways, as the rules "exploring" the interclass space and thus making decisions more balanced across the classes. This interpretation is also supported by the decrease in variance of the off-diagonal elements (from 0.044 to 0.039) and the decrease in variance of the diagonal elements (from 0.093 to 0.077).

Finally, it should be noted that, while this pattern certainly improved the biometric voice recognition accuracy, it is almost certain to have created a less flexible system. In addition to adding a set of seven rules that must be evaluated, these rules must be evaluated in order. Thus, not only are the variances between training and test data added for all the rules but also for the relative importance, or order, of all of the rules. This means that differences between training and test sets, and changes in input over time, are almost certain to result in obviating the values of the expert rules. As a consequence, this system is nonrobust to change. It is not recommended, therefore, to deploy this pattern to systems in which there is a paucity of training data or in which the system input consistently changes. The good results shown in this section are the exception that proves the rule: the training and test sets were very highly correlated by design.

9.3.4   Security Printing

The Generalized Hybridization pattern, introduced in Section 6.4.7, will be illustrated for the field of security printing. As opposed to the Regional Optimization pattern illustrated Section 9.3.2—which compares a plurality of meta-algorithmic patterns for different partitions of input—the Generalized Hybridization pattern is used to optimize the combination and sequence of first- and second-order meta-algorithmic patterns used for *all partitions* of a given—generally large—problem space.

Few domains are as broad as security printing in terms of machine intelligence applications. Virtually any digitally printed item can be engineered to carry an explicit set of data. In addition, printed regions contain implicit information related to the printing process—from halftoning to ink/substrate interaction—which can be used for inspection, authentication, and/or forensic purposes. For this section, we address generalized hybridization as it applies to a wide plethora of security printing analysis, and show how it can be generalized to incorporate both explicit and implicit data embedded marks.

The generalized approach is shown in Figure 9.4. Before the training phase, the set of meta-algorithmic pipelines to be considered must be determined. Typically, these are second-order meta-algorithmics: if they are all first-order meta-algorithmics, for example, the system will often become, after training, one of the second-order meta-algorithmic systems (Majority Voting or Weighted Confusion Matrix, Majority Voting or Best Engine, etc.). Regardless, Generalized Hybridization obviously is intended to provide a broad pattern that can be used to catch any and all combinations and/or sequences of lower-order meta-algorithmic patterns as possible deployment candidates.

For each meta-algorithmic pipeline to be considered, the correct data and data structures to support them during the run-time must be collected. This includes generator confidence values, error rates, classes and alternative settings, among others, appropriate for each of the meta-algorithmic pipelines.

In order to illustrate the generalized hybridization, both explicit and implicit data carrying regions are considered. However, as opposed to classification accuracy, we are targeting overall system robustness. This means that the configuration of the overall system includes both a default pipeline and one or more pipelines customized to specific data input. As described in Section 8.3.4, hybridized security printing approaches may use more than one type of security printing feature, or deterrent, to provide authentication. If multiple features are used, then each feature can be processed as shown in Figure 9.4 with, potentially, entirely distinct sets of meta-algorithmic pipelines.

In Figure 9.4, the training phase involves collecting image data for the security printing feature. This data will later be used during the run-time phase to correlate with the correct meta-algorithmic pipeline. Ten features are calculated for this simple experiment: (1) mean image hue, (2) mean image intensity, (3) mean image saturation, (4) median red region size, (4) median blue region size, (5) median green region size, (6) percent black pixels after thresholding using (Otsu, 1979), (7) median red pixel run length, (8) median blue pixel run length, and (10) median green pixel run length. The values for each of these features are normalized for the whole training set by subtracting the mean and then dividing by the standard deviation (so each feature was $\mu \pm \sigma = 0.0 \pm 1.0$). For each image, a distance measurement was computed with respect to the set of feature means for each class. The distance of an image sample, S, from the mean feature values of the cluster C is the sum of the

Generalized hybridization for security printing: training phase

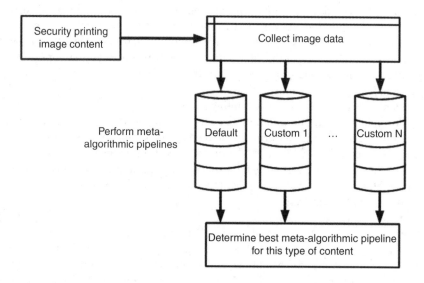

Generalized hybridization for security printing: run-time phase

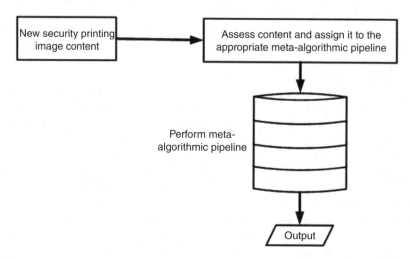

Figure 9.4   Generalized Hybridization third-order meta-algorithmic pattern as deployed for the security printing application described in this section

σ-normalized absolute differences for all N_{features} of the features F used to assign the sample to a cluster:

$$D_S(C) = \sum_{i=1}^{N_{\text{features}}} \frac{F_S(i) - F_C(i)}{\sigma_C(i)}.$$

So far, this approach is similar to the Regional Optimization pattern elaborated in Section 9.3.2. In this case, however, there is incomplete regionalization of the input domain because the run-time phase is not concerned with subranges of the input space. Rather, it is concerned with image data matching between the *new security printing image content* and each of the clusters created, separately, by the ten features above. If there is no good match—that is, if $D_S(C)$ is greater than an appropriate threshold for each cluster C, then the default meta-algorithmic pipeline is chosen during run-time. The summed distance (over the 10 features) corresponding to a good match is dependent on the training data and the type of authentication being performed.

In general, only well-behaved clusters are likely to prove useful for selecting different meta-algorithmic patterns, so relatively small distance thresholds are chosen to create the clusters. The mean distance value is simply $D_S(C)$, as computed above, divided by the number of features *nfeatures*. Past experience with images suggests using a value near 0.1 and varying these values in the range 0.05–0.20 to determine an optimal configuration. However, past research was also performed only for a single meta-algorithmic pattern, not for the more complicated Generalized Hybridization approach. For this example, I chose to select a value of $D_S(C)$/*nfeatures* such that the resulting clusters were of sufficient size (10 or more images) and so that the ratio of the standard deviation within the cluster, σ_{within}, to the value of $D_S(C)$/*nfeatures* over the range was a minimum. Note that these values are different, and the ratio of σ_{within} to $D_S(C)$/*nfeatures* is minimized when the value of $D_S(C)$/*nfeatures* is just below a value that results in the inclusion of a new set of samples belonging to another distribution. For this type of optimal clustering, a value of 0.10 for $D_S(C)$/*nfeatures* was identified.

These 10 simple features were used to create a set of clusters for the training set (Table 9.2). The set of samples belonging to each of these clusters was then evaluated using, separately, four different meta-algorithmic patterns. Samples not belonging to a cluster were evaluated using a default meta-algorithmic pattern, the Weighted Confusion Matrix pattern. A key point for understanding this hybridized pattern is that the clusters formed by the 10 image features above do not correspond with clusters associated with the three actual classes of security printing images, the *3D color tiles* described in Section 8.3.4. They are used, instead, to cluster images that are similar for a small, readily computed set of features, and then perform a more powerful classification on each cluster, independently, using a set of 420 image features (Simske *et al.*, 2009). The rationale for this approach is straightforward: by aggregating images that are

Table 9.2 Percent of the samples belonging to the four clusters defined by the 10 simple features during the training set (and later during testing). Classification accuracy using only these clusters, for later comparison, is 61.7% for the training set and 59.4% for the test set. See text for details

Cluster Number	Samples in Cluster, Training Set (%)	Samples in Cluster, Test Set (%)	Primary Printing Technology Associated with the Cluster
1	19	17	Liquid electrophotography
2	23	20	Dry electrophotography
3	18	19	Inkjet (set 1)
4	17	15	Inkjet (set 2)
5 (The "noncluster")	23	29	Liquid electrophotography, inkjet (set 2)

more alike before the full classification, the hardest challenge—that of finding classification differences within these clusters—is facilitated. In some ways, this is similar to a multi-branch decision tree classification approach, with the 10 features used upfront performing an initial classification and the later, full set of 420 features providing more precise classification. However, complexity is supported through (a) variable number of branches, and (b) variation of the classification approach (different meta-algorithmic patterns) after the initial branching. This means that some of the branches rejoin after the secondary classification, making this unlike a decision tree, and a more robust overall classification system.

The secondary classification in this case is focused on differentiating the printing of 3D color tiles by three different printer technologies: (1) inkjet, (2) dry electrophotography, and (3) liquid electrophotography. In total, 360 images were used: 120 images for one dry electrophotography (dry toner) printer, 120 images for one liquid electrophotographic (liquid toner) printer, and 60 images each from two inkjet printers. Half of the images in each set were used for training, and half for testing ($n_{training} = n_{test} = 180$).

In order to perform the secondary, and final, classification of the printer technology used to create the then-scanned images, three different simple clustering approaches—(1) L1-distance-based classification, (2) L2-distance-based classification, and (3) Gaussian-distribution-based binary classification (Simske, Li, and Aronoff, 2005)—were used in conjunction with the image classification approach described in Simske et al. (2009) as the three generators for several meta-algorithmic patterns, each of which in turn looked to determine the correct classification among the three printing technologies described above. The four nondefault meta-algorithmic patterns used were as follows:

1. Best Engine with Required Precision (Section 6.3.5) or Voting (Section 6.2.3).
2. Best Engine with Required Precision (Section 6.3.5) or Weighted Voting (Section 6.2.3).
3. Best Engine with Required Precision (Section 6.3.5) or Confusion Matrix (Section 6.3.1).
4. Best Engine with Required Precision (Section 6.3.5) or Weighted Confusion Matrix (Section 6.3.1).

The default pattern, deployed when a sample was not within the desired distance of one of the clusters, was the Weighted Confusion Matrix. Since there are several steps in this implementation, the overall process is outlined here:

Step 1: Precluster the images using the set of 10 simple features. For samples not joining a cluster of at least 10 images, assign them to a separate "cluster," which must use a default meta-algorithmic pattern (since required precision for the cluster cannot be achieved).

Step 2: Perform all meta-algorithmic patterns on each cluster. For the nondefault patterns, the required precision is the output confidence of the classifiers for the Best Engine (i.e., "liquid electrophotography" for cluster 1, "dry electrophotography" for cluster 2, etc.). When precision is above a threshold (in this example, 0.75), then use the Best Engine, which is the "primary printing technology associated with the cluster" in Table 9.2. When precision is not above that threshold, then use Voting, Weighted Voting, Confusion Matrix, or Weighted Confusion Matrix for the four nondefault patterns.

Step 3: For images not assigned to one of the clusters (i.e., assigned to the "noncluster" row) in Table 9.2, use the Weighted Confusion Matrix pattern.

Step 4: Compare the mixed approach described in Steps 2 and 3 to the approaches using the same meta-algorithmic pattern throughout.

Table 9.3 Accuracy results (3-class printer technology classification) for the default and four custom meta-algorithmic patterns, along with the results for the use of the optimum meta-algorithmic pattern for each cluster. Results for both the training and test set are shown. The highest classification accuracy rate is observed when the mixed (generalized hybridization) approach is used. Note, however, that the test accuracy (87.2%) was significantly lower than the training accuracy (95.0%)

Meta-algorithmic Approach	Accuracy (Training Set) (%)	Accuracy (Test Set) (%)
(I) Best Engine with Required Precision or Voting	82.2	77.8
(II) Best Engine with Required Precision or Weighted Voting	88.9	82.8
(III) Best Engine with Required Precision or Confusion Matrix	87.2	81.1
(IV) Best Engine with Required Precision or Weighted Confusion Matrix	90.0	83.9
(V) Default (Weighted Confusion Matrix)	85.0	82.2
(VI) Mixed: optimum meta-algorithmic pattern for each cluster	95.0	87.2

The results of the six different classification approaches outlined in Steps 1–4 are given in Table 9.3. The four "Best Engine with Required Precision" patterns are labeled (I) to (IV), the default Weighted Confusion Matrix pattern is labeled (V), and the Generalized Hybridization, or mixed, pattern, is labeled (VI). Pattern (VI) ended up being a combination of patterns (II), (III), (IV), (IV), and (V). No cluster performed best for the meta-algorithmic pattern (I).

As Table 9.3 illustrates, the individual meta-algorithmic patterns provided differing levels of accuracy across the entire set of images. For the training data, the individual meta-algorithmic approaches provided a minimum of 82.2% accuracy and a maximum of 90.0% accuracy. Adding the upfront decision to accept the output of the Best Engine generally improved the accuracy, as the top three meta-algorithmic patterns were (IV), (II), and (III). Specifically, the "Best Engine with Required Precision" reduced the error rate of the Weighted Confusion Matrix pattern by 33.3%.

Substantial reduction in error rate—50.0% in comparison to meta-algorithmic pattern (IV)—was obtained for the training data using the Generalized Hybridization approach. As shown in Table 9.2, 29% of the test samples are not assigned to the four main clusters defined by the training set—a 6% increase over the training data. In Table 9.3, the test set results are presented. Here, the accuracy obtained by the six meta-algorithmic approaches dropped by 4.4%, 6.1%, 6.1%, 6.1%, 2.8%, and 7.8%, respectively, in terms of absolute accuracy. If the accuracy drop for test data compared to training data is due to overfitting or "overtraining," these results suggest that the mixed approach—pattern (VI)—was the most overtrained, followed by the three best individual meta-algorithmic patterns—(II), (III), and (IV). Nevertheless, the benefit of the Generalized Hybridization approach was clear. The test accuracy improved to 87.2% from a maximum of 83.9% for the individual meta-algorithmic approaches. This is still an impressive 20.5% reduction in relative error rate.

The utility of the image preclustering using the set of 10 simple features (Step 1 in the process description above) dictates the overall advantage gained by selecting one pattern versus another. If these features result in a set of clusters that behave similarly, then a single meta-algorithmic pattern is likely to provide accuracy as high as that provided herein by Generalized Hybridization. In order for Generalized Hybridization to provide high value, the preclustering step must act to "mix" the input space so that the different meta-algorithmic patterns can explore the input space more efficiently. In so doing, Generalized Hybridization prevents a solution that is focused on a local, rather than global, optimum, much like genetic and stochastic approaches.

It is also important to note that the simplified feature set used for preclustering will not by itself comprise a highly accurate classifier. Using those 10 features in place of the later 420 features, the meta-algorithmic approaches (I)–(VI) provided a maximum 61.7% accuracy on the training set and a maximum 59.4% accuracy for the test set. This is more than three times the error rate of the 420-feature classification. There is likely a relationship, perhaps even a useful predictive relationship, between the number of features in and the accuracy of the simple clustering and later full classification algorithms. Although this relationship was not explored here, it would likely make useful future work.

In summary, then, the Generalized Hybridization approach was shown to be advantageous for the security printing application posed. The use of a different meta-algorithmic approach for different sets of the input data, moreover, provides no significant effect on performance. Computing the 420 image features is required for each meta-algorithmic pattern (not to mention any of the classifiers), so that the overhead of extra classifiers and meta-algorithmics is less than a 1% increase in processing time.

9.4   Secondary Domains

We now apply third-order meta-algorithmics to three of the "secondary domains." Sensitivity analysis offers broad possibilities, of which only the surface is scratched here. It is applied here to surveillance; in particular, to the matched identification of an object in two different images. Proof by task completion is then applied to OCR, particularly in terms of defining the mapping of a character before printing to the character after printing and scanning. Finally, the Feedback pattern is used for a security analytics application; namely, the anonymization of content.

9.4.1   Surveillance

The Sensitivity Analysis pattern was introduced in Section 6.4.5. Here it is applied to an important component in a surveillance system; namely, to the matched identification of an object in two different images. The strategy for moving the right object higher to the top of the list in classification is not a classification problem per se. Rather, it is a problem for minimizing the mean number of classifications attempted until the correct one is obtained, as was introduced in Table 8.7.

In surveillance problems, there may be a number of objects that need to be tracked across a large set of images. There may, therefore, be a relatively large set of object extraction algorithms, each capable of identifying and tracking, potentially, a different set of object possibilities. This means that each of the extraction algorithms may provide confidence in only a subset of the total set of objects that could be tracked. Even if the tracking algorithm is capable of identifying each of the object type, it may only report output confidence values for only a subset at a time.

Suppose, for example, that 15 objects of interest are to be tracked, but the tracking engine provides confidence values for only its top four choices. This means that if the correct object is not in the top four choices, it is not selected—or given any weighting at all—by this engine.

This is an unsatisfactory situation for several reasons. First, let us consider the trends that are observed when the number of generators in a meta-algorithmic system increases:

1. The generators themselves become more highly correlated in the mean.
2. Clusters of generators that are more closely correlated with each other than with the rest of the generators emerge.
3. Subsets of the generators are often more effectively combined with different meta-algorithmic patterns.

In Section 6.4.5, these observations were used to argue for the removal (or recombination) of some of the generators through one or more of the following mechanisms: correlation, confusion matrix behavior, and selection among multiple meta-algorithmic pattern options. The latter would, in fact, have much in common with the Generalized Hybridization approach just described in Section 9.3.4.

In this section, however, sensitivity analysis will be used in a different way. Here, sensitivity analysis will be used to "repopulate" the confidence values reported in order to give better overall accuracy, as well as tracking robustness, to the surveillance system.

Suppose we have a set of images, each of which is analyzed by one or more surveillance engines. The surveillance engine is tracking Q objects but providing confidence values for only P objects at a time, where $P < Q$. Often, $P \ll Q$. Table 9.4 provides sample output for $P = 4$ and $Q = 15$.

It is not immediately obvious from Table 9.4 which, if any, of the 15 object classes is actually being tracked. The output for the first classifier is particularly unsatisfactory:

Table 9.4 Example of the output probabilities matrix (OPM) when $P \ll Q$. Here $P = 4$ and $Q = 15$. Several images in a stream of video can be used as input to generate these confidence values

Output Probabilities Matrix		Classifier				
		1	2	3	4	Sum
Classifier confidence	1	0.90			0.13	1.03
output for object class	2		0.6			0.6
	3		0.1	0.5		0.6
	4				0.45	0.45
	5	0.06				0.06
	6		0.16			0.16
	7			0.25		0.25
	8	0.03				0.03
	9		0.14			0.14
	10	0.01			0.36	0.37
	11			0.14		0.14
	12			0.11		0.11
	13					0.00
	14				0.06	0.06
	15					0.00

probability(1,1) = 0.9, probability(1,5) = 0.06, probability(1,8) = 0.03, and probability(1,10) = 0.01. Only four objects receive any confidence (probability > 0.0) at all, and based on the overwhelming differential confidence for object class 1, we would expect to see further support for this decision in the next three images. But, we do not. Therefore, it is hard to feel confident about the classification provided by the output probabilities matrix (OPM) of Table 9.4: even though one classification will receive a higher overall weight, there is clearly no "majority" winner.

Having identified a potential flaw in the confidence output, we need to find a way to improve upon it. We are interested in how to change the OPM associated with the given assigned class to one that both (a) ranks all Q objects, not just a smaller set of P objects, and (b) is more likely to rank the correct classification high in the rank from $P + 1$ to Q. In simple terms, when $P < Q$ and the correct classification is not assigned a nonzero confidence value, then by randomly selecting the object classification for these $Q - P$ zero confidence objects, our expected value for ranking the correct classification, $E(\text{Rank})$, is the mean of $P + 1$ and Q; that is,

$$E(\text{Rank}) = \frac{P + Q + 1}{2}.$$

For $P = 4$ and $Q = 15$, we then expect the rank of a "no-confidence" correct classification to improve on the $E(\text{Rank})$ of 10. We next show how this is done.

The 15-element column corresponding to the output probabilities provided by one of the classifiers will be used to illustrate the approach. For example, this is the salient column for Classifier 1 in Table 9.4:

$$\begin{bmatrix} 0.90 \\ 0.00 \\ 0.00 \\ 0.00 \\ 0.06 \\ 0.00 \\ 0.00 \\ 0.03 \\ 0.00 \\ 0.01 \\ 0.00 \\ 0.00 \\ 0.00 \\ 0.00 \\ 0.00 \end{bmatrix}.$$

From this column, we need to find a way to distribute the full weight (1.00) across the Q elements, such that (a) the current confidence ranking where Rank[1,1] > Rank[1,5] > Rank[1,8] > Rank [1,10] > (Rank[1,2]=Rank[1,3]=...=Rank[1,15]=0.00), and (b) we change the no-confidence elements {[1,2], [1,3], [1,4], [1,6], ..., [1,15]} such that the correct classification is relatively highly ranked within this set. This is accomplished in a relatively simple manner using the confusion matrix obtained after the training phase. To begin with, the general form of the confusion matrix for this system is

$$
\text{Origin class}
\begin{bmatrix}
C_{11} & C_{12} & C_{13} & \cdots & C_{1P} \\
C_{21} & C_{22} & C_{23} & \cdots & C_{2P} \\
C_{31} & C_{32} & C_{33} & \cdots & C_{3P} \\
\cdot & \cdot & \cdot & \cdot & \cdot \\
C_{P1} & C_{P2} & C_{P3} & \cdots & C_{PP}
\end{bmatrix}
$$

Assigned class

Zero weightings (i.e., the OPM element $= 0.0$) may prevent certain correct classifications from being made, depending on the meta-algorithmic pattern employed. One way to handle zero weightings, of course, is to randomly rank the classifications that were given zero confidence. However, this results in the random ordering of the correct classification, with an expected ranking of $(P + Q + 1)/2$, as described above. As described in Section 6.3.1, though, there is a direct relationship between the OPM and the confusion matrix. It makes sense, then, to use the appropriate column in the confusion matrix to reweight the sparsely populated column in the OPM. In this example, the OPM column (left below) is sparse but the corresponding column in the confusion matrix (right below) is not:

OPM column
$$
\begin{bmatrix}
0.90 \\
0.00 \\
0.00 \\
0.00 \\
0.06 \\
0.00 \\
0.00 \\
0.03 \\
0.00 \\
0.01 \\
0.00 \\
0.00 \\
0.00 \\
0.00 \\
0.00
\end{bmatrix}
$$

Confusion Matrix column
$$
\begin{bmatrix}
0.532 \\
0.054 \\
0.022 \\
0.068 \\
0.057 \\
0.033 \\
0.018 \\
0.053 \\
0.004 \\
0.031 \\
0.066 \\
0.020 \\
0.017 \\
0.009 \\
0.016
\end{bmatrix}
$$

Note that I have normalized the salient column of the confusion matrix to sum to 1.0: the actual column in this case summed to 1.108. It is normalized to make later scaling of the weighting easier to follow, and in no way affects the approach taken. Regardless, because the OPM column identifies object class 1 as the correct classification, the appropriate column in the confusion matrix is

$$
\begin{bmatrix}
C_{11} \\
C_{12} \\
\cdots \\
C_{1P}
\end{bmatrix}.
$$

Table 9.5 Salient values in the sensitivity analysis approach described in the text. Column A provides the original confidence weightings from the OPM; here, $P = 4$ and $Q = 15$. Column B provides the corresponding confusion matrix column when the assigned class is object class 1. Column C provides the corrective weighting scheme: the manner in which nonzero confidence will be assigned to all object classes and in which the relative ranking of the object classes will not be compromised. Column D provides the corrective weights (summing to 0.33) and the scaled original weights (summing to $1.0 - 0.33 = 0.67$). Column E provides the overall combined weights for the object classes. Column F provides the relative ranking of the object classes after performing columns C, D, and E

Object Class Number	Column A: OPM Column	Column B: Confusion Matrix Column	Column C: Corrective Weighting Scheme	Column D: Scaled Original + Corrective Weighting	Column E: Overall (Combined) Weighting	Column F: Overall Rank
1	0.900	0.532	$11X$	$0.033 + 0.603$	0.636	Rank $= 1$
2	0.000	0.054	$9X$	0.027	0.027	Rank $= 7$
3	0.000	0.022	$7X$	0.021	0.021	Rank $= 9$
4	0.000	0.068	$11X$	0.033	0.033	Rank $= 5$
5	0.060	0.057	$11X$	$0.033 + 0.040$	0.073	Rank $= 2$
6	0.000	0.033	$8X$	0.024	0.024	Rank $= 8$
7	0.000	0.018	$5X$	0.024	0.015	Rank $= 11$
8	0.030	0.053	$11X$	0.015	0.053	Rank $= 3$
9	0.000	0.004	$1X$	0.003	0.003	Rank $= 15$
10	0.010	0.031	$11X$	$0.033 + 0.007$	0.040	Rank $= 4$
11	0.000	0.066	$10X$	0.030	0.030	Rank $= 6$
12	0.000	0.020	$6X$	0.018	0.018	Rank $= 10$
13	0.000	0.017	$4X$	0.012	0.012	Rank $= 12$
14	0.000	0.009	$2X$	0.006	0.006	Rank $= 14$
15	0.000	0.016	$3X$	0.009	0.009	Rank $= 13$

This column is shown to the right of the OPM column above. Any number of strategies can be used to reweight the OPM column from the confusion matrix column. Suppose we wish to preserve the relative ratios of the weighted elements—0.90, 0.06, 0.03, and 0.01 in the example. Then, we wish to relatively rank the unweighted elements based on their relative values in the confusion matrix column. This is accomplished with reference to columns A–F shown in Table 9.5.

In Table 9.5, the column A repeats the OPM, and the column B repeats the corresponding column in the confusion matrix associated with all of the training data. Column C represents the weighting to be added to every element in order to rank the nonweighted elements. This is designated the *corrective weighting scheme*. Since there are 11 nonweighted elements, these elements are given $1X, 2X, 3X, \ldots, 11X$ weightings relative to their rankings in the confusion matrix column. Thus, element [1,4] is given a weighting of $11X$ since the value for [1,4] in the confusion matrix is 0.068, making class 4 the most likely misclassification not receiving a weighting. The same $11X$ weighting is also given to the elements that have a weight in the OPM—namely, OPM elements [1,1], [1,5], [1,8], and [1,10]. This ensures that the original weighted classifications will not be reordered. The sum of column C is $110X$, and X was set to 0.003 so that 1/3 (0.330) of the overall weighting is due to the sensitivity analysis weighting.

The other 0.670 of weighting was used to scale the original weightings of $\{0.90, 0.06, 0.03, 0.01\}$ to $\{0.603, 0.040, 0.020, 0.007\}$. These weights were added to 0.033 for the appropriate elements, as shown in column D, resulting in column E. The ranks of the 15 object classes are then given in column F. Notice that element [1,4] now has a ranking of 5.

With the approach taken here, there is the potential for a modest impact on the accuracy of the meta-algorithmic pattern's primary classification, since effectively the approach is a form of output space transformation. To that end, a more comprehensive sensitivity analysis can be performed. This sensitivity analysis can explore, at least, the following:

1. The relative weightings in the corrective weighting scheme.
2. The relative weighting for the corrective weighting scheme and the scaled original weighting in the OPM column.
3. The approach taken for scaling the original weightings in the OPM.

For (1), Table 9.5 scales these weights linearly with ranking, but another method would be to scale the weightings for the object classes with confidences of 0.000 proportionally to their confusion matrix values. In this case, for the example above, the corrective weighting scheme could be based on the confusion matrix elements of the 11 no-confidence classes. Since these sum to 0.327 and since the weighting of $1X + 2X + 3X + \ldots + 11X = 66X$, we obtain $X = 0.005$, and so the relative weights vary from $0.8X$ to $13.6X$ as shown below in the "corrective weighting" column to the left. Assuming we sum all of the corrective weights to 0.330 again, we obtain the *actual weights* as in the column below to the right:

$$
\text{Corrective weighting} \quad
\begin{bmatrix}
13.6X \\
10.8X \\
4.4X \\
13.6X \\
13.6X \\
6.6X \\
3.6X \\
13.6X \\
0.8X \\
13.6X \\
13.2X \\
4X \\
3.4X \\
1.8X \\
3.2X
\end{bmatrix}
\quad \text{Actual weights} \quad
\begin{bmatrix}
0.037 \\
0.030 \\
0.012 \\
0.037 \\
0.037 \\
0.018 \\
0.010 \\
0.037 \\
0.002 \\
0.037 \\
0.036 \\
0.011 \\
0.009 \\
0.005 \\
0.009
\end{bmatrix} .
$$

Note that we can also vary the relative weighting of the previously weighted object classes to be more than (or even less than!) the relative weighting of the most promising no-confidence object class. This can be done when, for example, we find a given engine overreports a given class at the expense of the others, or when because $Q \ll P$, often the second, third, and so on, most likely classifications are not included in the set of Q possible classifications.

This point leads us to (2); that is, the relative weighting for the corrective weighting scheme and the scaled original weighting in the OPM column. In the example provided here, I have matched the total weight in the corrective weighting scheme (0.330) to the total weight in the (normalized) confusion matrix column (0.327). That, in general, is a good approach, but other approaches may work better depending on the nature of the confidence values, the values of Q and P, and the relative grouping of object classes in the confusion matrix.

The effect of the approaches taken determines the relative distribution of weights in the final OPM compared to the original OPM and the corresponding confusion matrix column. In the example of Table 9.5, the weights of all no-confidence OPM elements increased, as did those of the second, third, and fourth highest weighted classes' OPM elements. This came at the expense of the highest weighted class 1, which nevertheless still has an OPM value above its confusion matrix value. However, its confidence value is now only 0.636, which is significantly less than its reported confidence of 0.900. This will weaken its relative confidence in object class 1 within any larger meta-algorithmic approach, which will generally lead to reduced accuracy, especially in confusion matrix incorporating meta-algorithmic patterns, if the 0.900 value is reliable. However, the approach outlined definitely reduces the expected number of classifications to be evaluated before the correct one is found.

The expected number of classifications to be evaluated before finding the correct one is computed from a column in Table 9.5 by summing the products of the rank and the weight in eachobject class number. For the original sparse OPM, this value is 4.20; for the modified OPM of Table 9.5, column F, it is 3.50. This reduces the mean processing time by 16.7% for the classifier. The confusion matrix itself—Table 9.5, column B—has a value of 3.36, which is only 4.0% less than the modified OPM. Thus, the modified OPM can be seen as moving the correct classification closer to the top with, hopefully, a relatively minor effect on accuracy (correctness of Rank = 1 choice).

Finally, with regard to (3) above, the scaling of the original weightings in the OPM need not be directly proportional, but can accommodate any reasonable transformation, as described, for example, in Section 6.3.2 for confusion matrices. For example, filtering of an unsmooth OPM similar to the one illustrated in this example may also lead to better behavior—especially when the object classes are rearranged to have the classes with the highest interclass confusion next to each other. Methods for performing this task were described in Section 8.2 and elsewhere.

When all four classifiers in Table 9.4 are used together with Confusion Matrix and Weighted Confusion Matrix patterns, the approach outlined here—and diagrammed in Figure 9.5—was found to provide a good trade-off of preserving accuracy while reducing the expected number of classifications to be investigated before the correct one was found. Testing accuracy (the correctness of the highest ranked class) dropped by only 0.7–2.3% while the number of classifications until the correct one was found decreased by 13.3–29.1%. For determining these values, the data sets were split into three equal parts—one-third each for training, model validation, and testing. The validation stage was used to optimize the percent of the weighting to use for the corrective weighting schema and to decide whether to use a ranked or confusion matrix element proportional approach to assigning the corrective weights. In general, this type of Sensitivity Analysis approach will benefit from the additional validation step, since by definition it is exploring the sensitivity of the deployed solution to multiple system factors.

9.4.2 Optical Character Recognition

The Proof by Task Completion meta-algorithmic pattern introduced in Section 6.4.2 is here applied toward a content scanning system that includes a mixture of purchased/licensed and

Sensitivity analysis for object recognition: validation phase

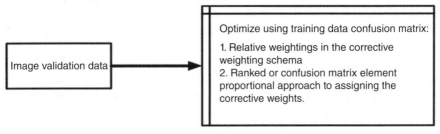

Sensitivity analysis for object recognition: run-time phase

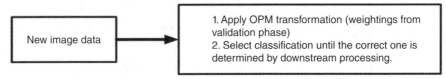

Figure 9.5 Sensitivity Analysis third-order meta-algorithmic pattern as deployed for the surveillance object recognition example described in the text. Note that the correct class can be determined in the deployment phase by downstream processing

open source OCR engines. This pattern is used to dynamically change the weighting of the individual generators, which in this case are the individual OCR engines.

The Proof by Task Completion pattern allows the overall system architecture to be switched between a variety of system deployment patterns as schematically depicted in Figure 9.6. Usually, this pattern is accompanied by a return-on-investment (ROI) model that provides an overall system cost, or cost equivalent, based on the system performance, accuracy, licensing costs, and other requirements.

For a document processing system that includes scanning of paper-based versions of a document workflow, an important ROI consideration is the relative cost of error versus the relative cost of the document analysis technologies, including OCR. In the simplest example, the cost of document analysis error is simply extra processing time. In this case, the expected number of classifications to be evaluated before the correct one is found (also described in the previous section), or E(classifications), is the governing factor in the TotalCost(processing) equation:

$$TotalCost(\text{processing}) = C(\text{proc}) \times E(\text{classifications}).$$

Here, C(proc) is the cost of processing. The other factor in the ROI for this system is the cost of the document analysis engine, which here is confined to the cost of the OCR engine:

$$TotalCost(\text{OCR}) = C(\text{OCR})/ndocuments.$$

Here, the cost of the OCR engine, C(OCR), is prorated over the number of documents, *ndocuments*, that are processed. The total cost of the system is, for this simple model,

$$TotalCost(\text{system}) = C(\text{proc}) \times E(\text{classifications}) + C(\text{OCR})/ndocuments.$$

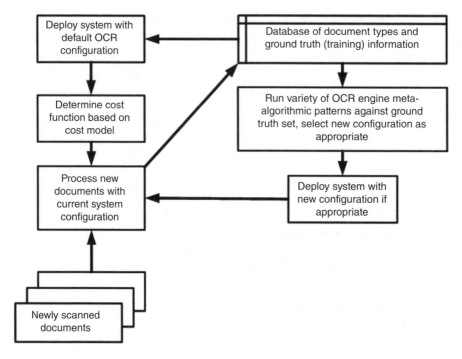

Figure 9.6 Proof by Task Completion third-order meta-algorithmic pattern as deployed for the OCR example described in the text. The architecture of the system is the same as that introduced in Figure 6.18

Three different system configurations are considered for their costs:

1. A commercial OCR engine that costs $4 million/y that leads to an E(classifications) of 1.1. All 100 million documents processed by the client are processed by this OCR engine.
2. A commercial OCR engine that costs $1 million/y but is only allowed to process 25 million documents. These 25 million documents are selected using a Predictive Selection meta-algorithmic pattern, and are in theory the hardest 25 million to process. This leads to an E(classifications) of 1.15. The other 75 million documents are processed using an open source OCR engine with an E(classifications) of 1.4.
3. An open source OCR engine that leads to an E(classifications) of 1.6.

For (1), $TotalCost$(system) $= C$(proc) $\times$ 1.1 $+$ $4 million. For (2), $TotalCost$(system) $= C$(proc) $\times$ (1.15 $\times$ 0.25 $+$ 1.4 $\times$ 0.75) $+$ $1 million. For (3), $TotalCost$(system) $= C$(proc) $\times$ 1.6. Suppose that the cost of processing 1 million documents is $25 000. The cost of processing 100 million documents is thus $2.5 million. Thus, for (1), $TotalCost$(system) $=$ $6.75 million/y; for (2), $TotalCost$(system) $=$ $4.344 million/y; and for (3), $TotalCost$(system) $=$ $4.0 million/y. For this system, then, the most cost-effective route is to use the open source OCR engine.

The Proof by Task Completion pattern requires that these costs be monitored over time. Should the cost of processing rise, for example, to $40 000 per million documents, then the

TotalCost(system) values rise to (1) $8.4 million/y, (2) $6.35 million/y, and (3) $6.4 million/y. With this rise in processing power, the mixed open source/commercial OCR solution becomes the most cost-effective option.

9.4.3   Security Analytics

In this section, a simplified type of security analytic is introduced. The *training-gap-targeted* meta-algorithmic pattern (Section 9.2.2), a fully closed-loop elaboration of the Feedback pattern introduced in Section 6.4.1, is especially useful in one type of security analytic, anonymization. Anonymization is typically defined as the process of obscuring the source and perhaps also the destination of network traffic. In the example here, I define anonymization differently for use in a privacy application. Most internet browsers collect information on a user's browsing behavior (see Figure 9.7 for an example of how to collect both the URL and the content of the URL browsed by a user, using a simple set of Java code), and then can use this information as input for settings when they visit another website. This type of data mining can be useful, but it can also be inappropriate, irritating, or even spooky. For example, I recently downloaded some images to create a slide set, and when I next visited an on-line retailer, the "recommendations" included books, movies, and other items associated with the images I had downloaded from the other sites. Clearly, this implies a privacy risk, and a potential identity-related security risk if, for example, the pattern of behavior associated with the individual is stored and analyzed by the cloud provider—for example, Google, Microsoft, Yahoo!, and so on—and can be used to identify the user by third parties.

```
//    getURL.java
import java.io.*; // we use java.io package
import java.net.*; // we use java.net package

public class getURL {
    public void main (String []args) {
        URL url; // URL is a java class in java.net package
        DataInputStream di_stream; // Used to stream buffered input to
        InputStream i_stream; // Data from the URL enters this Stream
        String s;
        try {
            url = new URL(args[0]);
            i_stream = url.openStream();
            di_stream = new DataInputStream( new BufferedInputStream(i_stream));
            // required in Java for alacrity of stream reading
            while( ( s=di_stream.readLine() ) != null )
            // Handle stream here—save or analyze as needed
        } catch (MalformedURLException MURLe ) {
            // Notify user of malformed URL
            System.exit(1);
        } catch (IOException IOe ) {
            // Notify user of IO exception
            System.exit(2);
        } finally{
            // Close the InputStream and catch any exceptions
        }
    } // End of main() method
} // End of class definition
```

Figure 9.7   Simplified Java code for collecting the information from a URL (universal resource locator)

In order to address this potential risk to privacy, one approach is to anonymize the browsing experience of the user. For simplicity, suppose that all web browsing consists of just six classes of information, as given in this set: [entertainment, news, on-line shopping, sports, travel, weather], abbreviated [E, N, O, S, T, W]. We represent the relative browsing behavior of the user as

$$[p(E), \ p(N), \ p(O), \ p(S), \ p(T), \ p(W)],$$

where $p(E) + p(N) + p(O) + p(S) + p(T) + p(W) = 1.0$ represents the user's *actual on-line interaction* (AOI). Suppose the following "mean," or μ, user is profiled, which represents the average behavior of a large set (e.g., "training set") of users:

$$[p_\mu(E) = 0.1, \ p_\mu(N) = 0.3, \ p_\mu(O) = 0.25, \ p_\mu(S) = 0.1, \ p_\mu(T) = 0.05, \ p_\mu(W) = 0.2].$$

Now suppose we have two different users, A and B. User A browses primarily to shop on-line and read entertainment news, so User A has the profile:

$$[p_A(E) = 0.4, \ p_A(N) = 0.1, \ p_A(O) = 0.4, \ p_A(S) = 0.0, \ p_A(T) = 0.0, \ p_A(W) = 0.1].$$

User B, on the other hand, prefers news, sports, and travel, and so has the profile:

$$[p_B(E) = 0.0, \ p_B(N) = 0.35, \ p_B(O) = 0.05, \ p_B(S) = 0.3, \ p_B(T) = 0.2, \ p_B(W) = 0.1].$$

In order to anonymize each user, we need to add the appropriate amount of *obfuscating on-line interaction* (OOI) for each. To do this, we rewrite the user profiles as ratios of the on-line browsing profiles for the mean user. Thus, for User A, the new ratio-based profile (RBP) becomes:

$$[p_A(E)/p_\mu(E), \ p_A(N)/p_\mu(N), \ p_A(O)/p_\mu(O), \ p_A(S)/p_\mu(S), \ p_A(T)/p_\mu(T), \ p_A(W)/p_\mu(W)].$$

The relative amount of OOI to add, $R = (OOI/AOI)$, is given by

$$R = \frac{OOI}{AOI} = \max\left(\frac{p_A(*)}{p_\mu(*)}\right) - 1.0.$$

For the User A above, the RBP is [4.0, 0.333, 1.6, 0.0, 0.0, 0.5], and so $R = 4.0 - 1.0 = 3.0$. We therefore need to add three times as much OOI as AOI itself. This additional content must contain no information from the highest ratio class in the RBP:

$$RBP[] = [p_A(E)/p_\mu(E), \ \ldots, \ p_A(W)/p_\mu(W)].$$

The OOI values are readily obtained using the following equation:

$$OOI_A(*) = r_{max}p_\mu(*) - p_A(*),$$

which for the samples shown are

$$OOI_A(E) = 4.0^*(0.1) - 0.4 = 0.0;$$
$$OOI_A(N) = 4.0 \times (0.3) - 0.1 = 1.1;$$
$$OOI_A(O) = 4.0 \times (0.25) - 0.4 = 0.6;$$
$$OOI_A(S) = 4.0 \times (0.1) - 0.0 = 0.4;$$
$$OOI_A(T) = 4.0 \times (0.05) - 0.0 = 0.2;$$
$$OOI_A(W) = 4.0 \times (0.2) - 0.1 = 0.7.$$

Thus, in terms of content, $AOI_A[] = [0.4, 0.1, 0.4, 0.0, 0.0, 0.1]$ and $OOI_A[] = [0.0, 1.1, 0.6, 0.4, 0.2, 0.7]$, and the sum $AOI_A[] + OOI_A[] = [0.4, 1.2, 1.0, 0.4, 0.2, 0.8]$, which is exactly equal to $r_{max} \times p_\mu[]$.

To summarize, then, the steps involved in computing the type and amount of OOI to add to the AOI, we have the following:

1. Determine the training set, in this case the mean on-line user profile, $[p_\mu(*)]$, where the sum of all of the $p_\mu(*)$ is 1.0.
2. Determine the user on-line profile, $[p_A(*)]$, where again the sum of all of the $p_A(*)$ is 1.0.
3. Compute the ratio-based, or relative, profile of the user, RBP, given by $[p_A(*)/p_\mu(*)]$.
4. Determine the largest ratio in the RBP profile from: $r_{max} = \max\left(\frac{p_A(*)}{p_\mu(*)}\right)$.
5. Determine the amount of OOI from the equation $OOI_A(*) = r_{max}p_\mu(*) - p_A(*)$.

It is left as an exercise for the reader to show that $AOI_B[] = [0.0, 0.35, 0.05, 0.3, 0.2, 0.1]$ and $OOI_B[] = [0.4, 0.85, 0.95, 0.1, 0.0, 0.7]$.

The schematic of the process is given in Figure 9.8. Browsing content appropriate to the task (period of time, number of users, etc.) of anonymization is collected, then classified. The classification output is compared to that of the mean user, and the difference is the error signal. If the error signal is above a threshold, the browsing behavior is considered distinguishable

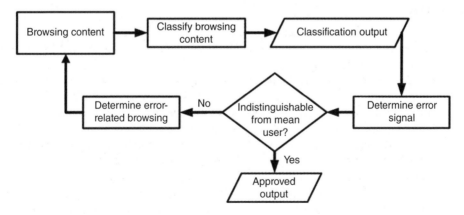

Figure 9.8 Schematic of the closed-loop version of the Feedback third-order meta-algorithmic pattern used for training gap targeting of browsing behavior, as described in the text

from the mean user, and the needed error-related browsing is calculated and thereafter added to the browsing behavior. Otherwise, the output is approved unchanged.

The application of the *training-gap-targeted* meta-algorithmic pattern to the web browsing anonymization task is, in effect, an *inverse analytic*. The OOI is specifically targeted toward the gaps in the user's normal on-line behavior, designated as AOI. When AOI is considered the "training set," we see that the difference in behavior between the mean, and thus anonymized, on-line user, $[p_\mu(E), \ldots, p_\mu(W)]$ and the behavior of the individual user, $[p_A(E), \ldots, p_A(W)]$, must be corrected for to make the user's behavior more anonymous.

The example above is considerably oversimplified from what is required for a robust, deployable, privacy-protecting solution. For example, many more classes of behavior would be considered, and thus the user's on-line behavior—described in AOI above—will likely differ considerably from the mean user behavior for at least one class of browsing. The value of r_{max} will therefore tend to be large, meaning the amount of OOI will significantly overshadow the user's actual on-line behavior. This would obviously require considerable additional usage of potentially valuable Internet bandwidth (assuming browsing behavior the user wishes to anonymize is a considerable amount of internet usage). In addition, regardless of how much OOI is added, the user's behavior will still likely stand out for a few items—for example, a preference for a given sports team or for a specific blog, news site, and so on.

Thus, the method described in this section will benefit from further parallelism; for example, from the use of an additional approach. One such approach is *random emphasis*, in which ectopic browsing behavior on a parallel topic (e.g., browsing behavior for another sports team) is used to obfuscate a specific hard-to-mask browsing behavior. Considering User A above, such a strategy may be used for $p_A(E)$, meaning that an r_{max} of only 1.6 is required to obfuscate the other five classes of browsing behavior. Other parallel strategies are, of course, possible. Regardless of the approach taken, however, this section shows how the *training-gap-targeted* meta-algorithmic pattern can be useful for privacy analytics. In this case, the training set is the "typical user" whose on-line behavior is our user's target—matching some mean user provides a level of anonymity, and lesser so privacy.

9.5   Summary

This chapter, while not terse, only scratched the surface of possibilities for third-order meta-algorithmic patterns. Rather than implement the patterns exactly as templated in Chapter 6, in each of the examples illustrated herein there was an attempt to combine the science of the meta-algorithmic pattern with the art of actually deploying it. The Feedback pattern of Section 6.4.1 was used to create an inverse analytics system based on targeting gaps in the input, or training set. This provides the means for one form of security analytics. The Proof by Task Completion pattern described in Section 6.4.2 was tied to an ROI model for a large document analysis system. The Confusion Matrix for Feedback pattern templated in Section 6.4.3 (and elaborated, in part, in Section 8.2) was used to improve document region typing by creating appropriate aggregate subclassifications. Expert Feedback (Section 6.4.4) was used to define a set of rules to improve biometric classification. Sensitivity Analysis, as introduced in Section 6.4.5, incorporated the results of the confusion matrix to provide a form of output space transformation. Regional Optimization, defined in Section 6.4.6, improved image understanding accuracy by deploying a modified predictive selection to select among multiple meta-algorithmic patterns. Finally, the Generalized Hybridization approach (see Section 6.4.7) was applied to security printing. A small set of metrics was used to precluster the input space, after which a larger set was used to finalize the classification.

These examples illustrate the breadth of third-order meta-algorithmics. They also show how, in many cases, a relatively large set of patterns can be compared for efficacy, with no compromise on run-time performance. This is because the cost of meta-algorithmic pattern processing is usually far less than the processing time required by the individual generators used in the patterns. With this in mind, the next chapter focuses on how meta-algorithmic principles can be used to create more robust systems.

References

Otsu, N. (1979) A threshold selection method from gray level histograms. *Pattern Recognition*, **9** (1), 62–66.

Simske, S.J., Li, D., and Aronoff, J.S. (2005) *A Statistical Method for Binary Classification of Images.* ACM Symposium on Document Engineering, pp. 127–129.

Simske, S.J., Sturgill, M., Everest, P., and Guillory, G. (2009) *A System for Forensic Analysis of Large Image Sets.* WIFS 2009: 1st IEEE International Workshop on Information Forensics and Security, pp. 16–20.

10

Building More Robust Systems

When one has finished building one's house, one suddenly realizes that in the process one has learned something that one really needed to know in the worst way – before one began.
—Friedrich Nietzsche

10.1 Introduction

The four previous chapters have outlined 21 meta-algorithmic patterns, which are meant to provide a system architect with a set of tools to use on any suitable intelligent system problem. These patterns are mapped against the background of four primary and eight secondary domains. If all $21 \times 12 = 252$ combinations of {meta-algorithmic pattern, domain} were described in this book, it would not only exceed the book's targeted length several times over but also miss the point entirely. Building more robust systems is—and likely always will be—a combination of skill and art. The purpose of Chapter 6 was to provide the raw skills for the task; that is, the fundamental meta-algorithmic patterns. Chapters 7–9 were meant to provide examples of the art: how meta-algorithmic patterns may be deployed for specific tasks. Hopefully, these comprise Nietzsche's "something" that needed to be known before we began building the system.

In this chapter, the three forms of parallelism—by task, by component, and especially by meta-algorithm—are used to build more robust systems. Here, as in previous sections, there is no attempt to cover all of the possible ground. Rather, this chapter is meant to show how a "sense" for knowing which parallel approaches are likely to improve system metrics—cost, performance, accuracy, and/or robustness, for example—can be learned through practice. Five different example topics are considered here: (1) summarization, (2) cloud systems, (3) mobile systems, (4) scheduling, and (5) classification.

10.2 Summarization

10.2.1 Ground Truthing for Meta-algorithmics

As the previous chapters have made clear, the application of meta-algorithmics to information extracting tasks often relies on the existence of proper ground truthing (training) data.

Meta-algorithmics: Patterns for Robust, Low-Cost, High-Quality Systems, First Edition. Steven J. Simske.
© 2013 John Wiley & Sons, Ltd. Published 2013 by John Wiley & Sons, Ltd.

Collecting this training data, however, is often expensive. Additionally, in many cases, the ground truth may be collected with a specific type of intelligent system design in mind. This may render the collected ground truth less valuable than other types of ground truth.

Meta-algorithmics are more than just powerful tools for improving overall system accuracy, performance, robustness, and/or cost: they can also be used to influence how data is collected. In this section, the advantages of redesigning how ground truth data is collected are shown in the context of providing a more accurate text summarization system.

In bringing together a set of two or more summarization engines (with no upper limit on the number chosen), we wish to find the means of ranking the individual engines. One means of achieving this, having each summarization engine provide a summary and then having humans evaluate each summary and ranking them relatively, is both time-consuming and provides only a binary datum. This "binary" approach is unsatisfying, as we cannot act on the decision other than to select one summarizer over (an)other summarizer(s). We therefore wish instead to use a meta-algorithmic-enabling approach, wherein the summarizers provide the same type of output (in this case the same number of sentences, $N = 10$). Summarization can be extractive or abstractive (see Section 3.4.5): we are using the more common extractive technique—which simply replicates the original text that is determined to be the most salient for the summary. Thus, the extracted sentences, in order, *are* the summarization. Table 10.1 illustrates the ranked order of sentences for three summarizers for a sample document comprised 35 sentences. Summarizer 1 selected sentence 4 as its most salient sentence, followed by sentence 7, and so on.

The original text (an article) contains 35 sentences, and the three individual summarizers evaluated each select the 10 most salient sentences and they are assigned weights in inverse order to the ranking (the highest ranked sentence receives weight $= 10.0$, the second highest ranked sentence receives weight $= 9.0$, and so forth). Next, human volunteer evaluators are presented with the original text (all 35 sentences) along with the complete set (union) of all sentences ranked by all summarizers. The volunteers then select what they think are the most relevant 10 sentences and order them from 1 to 10. A score of 1 indicates what to them is the most important sentence in the article.

In the example, this set of all significant sentences includes only the 15 sentences {1, 3, 4, 5, 6, 7, 9, 14, 19, 24, 25, 29, 30, 33, 35}. We can see from this that there is a tendency to select

Table 10.1 Sentences selected by three different summarization engines and their relative ranking

Rank	Summarizer 1	Summarizer 2	Summarizer 3	Weighting
1	4	7	1	10.0
2	7	1	14	9.0
3	3	6	4	8.0
4	14	29	3	7.0
5	25	19	7	6.0
6	9	4	19	5.0
7	1	5	25	4.0
8	33	3	30	3.0
9	19	33	9	2.0
10	35	14	24	1.0

Table 10.2   Sentences in ranked order as selected by the human
evaluators, and their corresponding relative human weighting (RHW)

Sentence Number	Relative Human Ranking
1	15.0
7	14.0
3	13.0
14	12.0
4	11.0
6	10.0
9	9.0
19	8.0
33	7.0
35	6.0
5	5.0
25	4.0
29	3.0
24	2.0
30	1.0

sentences from near the beginning and the ending of the article. Table 10.2 provides the ranked ordering of these 15 sentences as judged by the human evaluators. The highest score was given to sentence 1, the next highest score to sentence 7, and so on. The lowest score (1.0) was given to sentence 30, which was deemed the least significant of any of the 15 selected sentences.

Using this approach, we obtain human feedback on the best engine—this is by the way the entirety of what we receive for the "binary" approach described above. A second, quantitative, means of evaluating the summarizers is obtained by this approach—and this cannot be achieved using the binary approach. This is illustrated in Table 10.3, where we find the total weight for each Summarizer j, where $j = 1, \ldots, S$, and $S =$ number of summarizers, by simply performing the following operation:

$$\text{TW}_j = \sum_{i=1}^{N_S} W(i) \times \text{RHW}(S(i, j)),$$

where $\text{TW}_j =$ total weight for Summarizer j, N_S is the number of sentences in each summary; $W(i)$ is the weight associated with rank i (in our example, this is simply the quantity $N_S + 1 - i$, as shown in the "weighting" column of Table 10.1); $S(i,j)$ is the sentence number associated with rank i for Summarizer j (e.g., $S(3,5) = 7$ and $S(8,1) = 33$ in Table 10.1); and $\text{RHW}(S(i,j))$ is the relative human weighting (RHW) of the sentence identified by $S(i,j)$. For example, $\text{RHW}(S(3,5)) = 14.0$ and $\text{RHW}(S(8,1)) = 7.0$ as shown in Table 10.2, in which the left column are the values of $S(i,j)$ and the right column are the values of $\text{RHW}(S(i,j))$. Table 10.3 is therefore populated with the products of the relevant RHW and W rankings. The sum shows that the overall weight of Summarizer 3—that is, 600.0—is slightly greater than that of Summarizer 1—that is, 596.0. The lowest weight is for Summarizer 2, at 564.0. This indicates that, for this document, Summarizer 3 is the best of the three, but Summarizer 1 is

Table 10.3 Weight (weighting of rank multiplied by the relative human ranking) of each individual sentence selected by each summarizer, and total weight of the summarizer (sum of all weights). The overall weight of Summarizer 3, at 600.0, is slightly higher than that of Summarizer 1, at 596.0. The lowest weight is for Summarizer 2, at 564.0

| Rank of the Sentence | (Weighting of Rank) × (Relative Human Ranking) | | |
	Summarizer 1	Summarizer 2	Summarizer 3
1	110.0	140.0	150.0
2	126.0	135.0	108.0
3	104.0	80.0	88.0
4	84.0	21.0	91.0
5	24.0	48.0	84.0
6	45.0	55.0	40.0
7	60.0	20.0	16.0
8	21.0	39.0	3.0
9	16.0	14.0	18.0
10	6.0	12.0	2.0
Total weight	*596.0*	*564.0*	*600.0*

not much different. If the use of one of these summarizers is much less expensive than the use of the other (e.g., it is open source software as opposed to commercial off-the-shelf software), then, that one would be selected.

Clearly, the RHW approach outlined here offers a second level of comparison among *summarization engines*. It also ensures a blind evaluation, since the person providing the sentence order does not know which sentences have been selected by the summarizers. Importantly, the approach gives relative, quantitative comparative data. In the simple example shown above, Summarizers 1 and 3 are shown to be very similar in overall weighting, and relatively speaking more highly weighted than Summarizer 2. For a much larger sample set, such relative differences would be very important—they would indicate that Summarizers 1 and 3 are more or less interchangeable in quality and value. These differences would also indicate that Summarizer 2 should not be used in place of Summarizers 1 and 3.

The third level of value provided by this approach is more subtle. In providing a full ranking to all of the sentences in all of the combined set of summarizations, this method allows us to explore many different combinations of two or more summarizers (i.e., meta-algorithmic patterns). One of the simplest meta-algorithmic patterns is the Voting pattern, which is made possible by the RHW approach. This pattern, when applied to summarization, consists of adding the relative weighting for the ranking of each individual summarizer for each sentence. These values are tabulated in the second column of Table 10.4. To illustrate how this proceeds, consider sentence 7 in the original article. For Summarizer 1, sentence 7 is ranked second (9.0 weighting); for Summarizer 2, it is ranked first (10.0 weighting); and for Summarizer 3, it is ranked fifth (6.0 weighting). The combined weighting, 9.0 + 10.0 + 6.0, is 25.0, and is the highest of any sentence. Similarly, sentence 1 (23.0) and sentence 4 (23.0, with the tie-breaker being the second ranking value) are the next two highest weighted sentences by the combination of summarizers. This {7, 1, 4} is different from the ranking provided by the human evaluators; namely {1, 7, 3} (Table 10.2). If the ranked order of a given summarizer

Table 10.4 Weight (weighting of rank multiplied by the relative human ranking) of each individual sentence selected by the combination of Summarizers 1, 2, and 3, and the total weight of the summarizer (sum of all weights). The combined summarizer substantially outperformed any of the individual summarizers, with a total weight of 627.0, compared to 600.0 or less for each of the individual summarizers

Rank	Sentence (Sum of Ranks)	(Weighting of Rank) × (Relative Human Ranking) of the Nonweighted Combination of Summarizers 1–3
1	7 (25.0)	140.0
2	1 (23.0)	135.0
3	4 (23.0)	88.0
4	3 (18.0)	91.0
5	14 (17.0)	72.0
6	19 (13.0)	40.0
7	25 (10.0)	16.0
8	6 (8.0)	30.0
9	29 (7.0)	6.0
10	9 (7.0)	9.0
Total weight		*627.0*

were the same as the ranked order of the human evaluators, the maximum total weight, TW_{max}, is obtained. This weight is

$$TW_{max} = \sum_{i=1}^{N_S} W(i) \times RHW(i).$$

For the given example, $TW_{max} = 660.0$. As shown in Table 10.4, the (equal) voting combination of Summarizers 1–3 results in a much improved summarizer, for which the total weight is 627.0. This is 45% closer to the ideal score of 660.0 than the best individual summarizer—that is, Summarizer 3 with a score of 600.0.

Importantly, other meta-algorithmic patterns can also be readily applied to the relatively ranked human evaluation data. For example, the Weighted Voting pattern (Section 6.2.3) uses a weighted combination of Summarizers 1–3. If the weights for the individual summarizers in the combination are determined, for example, proportionate to the inverse of the error, e, then the weight of the jth summarizer, W_j, is determined from

$$W_j = \frac{\frac{1}{e_j}}{\sum_{i=1}^{N_{SUMM}} \frac{1}{e_i}},$$

where N_{SUMM} is the number of summarizers and error is defined as

$$e_i = TW_{max} - TW_i.$$

For the specific problem at hand, error $e = 660.0 - TW$.

Table 10.5   Weight (weighting of rank multiplied by the relative human ranking) of each individual sentence selected by the weighted combination of Summarizers 1, 2, and 3, and the total weight of the summarizer (sum of all weights). This combined summarizer also substantially outperformed any of the individual summarizers, with total weight of 628.0, compared to 600.0 or less for each of the individual summarizers

Rank	Sentence (Sum of Ranks)	(Weighting of Rank) × (Relative Human Ranking) of the Weighted Combination of Summarizers 1–3
1	7 (8.074)	140.0
2	4 (8.000)	99.0
3	1 (7.560)	120.0
4	3 (6.390)	91.0
5	14 (6.316)	72.0
6	19 (4.146)	40.0
7	25 (3.756)	16.0
8	9 (2.610)	27.0
9	6 (1.952)	20.0
10	29 (1.708)	3.0
Total weight		*628.0*

For simplicity here, let us assume that the error, e, of each of the three summarizers on the training data is the same as we observed in this example. Then, using the equation above, the weighting of the three summarizers are $\{0.366, 0.244, 0.390\}$ for Summarizers $\{1, 2, 3\}$. The effect of weighting the combination of the summarizers is described as

$$\text{SumOfRanks}_i = \sum_{j=1}^{N_{\text{SUMM}}} W_j \times W(i).$$

These values are shown in the parentheses in the second column of Table 10.5. This Weighted Voting approach results in a total weight of 628.0, which is 46.7% closer to the best possible score of 660.0 than the best of the individual summarizers.

In this example, the meta-algorithmic approaches were shown to improve the agreement between the automated extractive summarization and that provided by human ground truthing. Moreover, the meta-algorithmic approach was shown to be consistent with a different type of ground truthing. Here, the appropriate ground truth is also extractive, meaning that the sentences are ranked for saliency. This is advantageous to the normal ranking of individual summarizations approach because (1) it is scalable to any number of summarizers, (2) it is innately performed with the human expert blind to the output of the individual summarizers, and (3) it supports the advantageous meta-algorithmic patterns illustrated herein.

10.2.2   Meta-algorithmics for Keyword Generation

Weighted Voting provides an excellent means to combine summarizers, as the above example illustrated. In many cases, however, we want the text processing to proceed without the availability of ground truth. Using the Expert Feedback pattern in combination with the output of a summarizer (or meta-algorithmic combination of summarizers as used in Section 10.2.1), we can readily derive an effective set of keywords. These keywords are used for many important text processing applications, including document clustering and categorization, indexing, and search. The basic premise is to use the sentences associated with the extractive summarization differently from the sentences not included as part of the extractive summarization.

Figure 10.1 illustrates this system. The Weighted Voting pattern is used as described in the previous section to generate an extractive summary of the document(s), which consist of the most salient sentences in order of saliency. This output is fed back to the appropriate original document, from which two sets of text are extracted: (a) the set of nonextracted sentences—that is, those not in the summary; and (b) the set of extracted sentences. Next, the words in these two sets are treated; for example, they can be lemmatized/stemmed, weighted based on the weight during the summarization, and filtered if desired. However, for all but

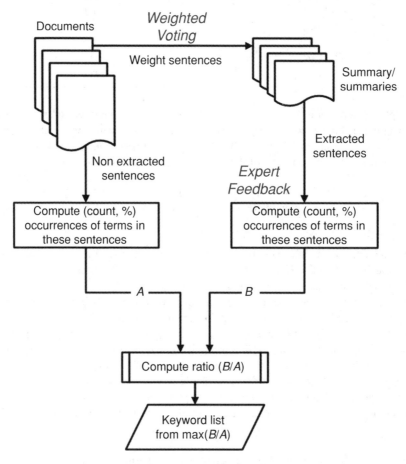

Figure 10.1 System diagram for the keyword generation described in the text. Here, two meta-algorithmic patterns are deployed for different components of the system. The Weighted Voting (Section 6.2.3) pattern is used for the summarization, and the Expert Feedback (Section 6.4.4) pattern is used to select the keywords

Table 10.6 Simple paragraph used to illustrate the Expert-Feedback-driven keyword generation. See text for details

Sentence	Text
1	Kevin has anger control issues
2	He does a lot of great work
3	He sees, however, idiocy in others
4	He sees idiocy rewarded through promotions and bonuses
5	Rewarding idiots makes Kevin angry
6	We are all lucky Kevin believes in gun control

trivial applications, stop words need not be removed as they will generally be automatically filtered by the Expert-Feedback-based keyword selection step, described next.

The absolute word counts and normalized counts (percentages) are computed for both sets of text. The ratio of these two word count statistics for the extracted compared to the nonextracted text is then used to identify the keywords. Table 10.6 provides an example for a very small set of text (a six sentence paragraph). Table 10.7 shows the sentences comprising the extracted summary set.

The words in Table 10.6 are lemmatized, and after this operation there are only eight words that occur more than once. Their statistics are compiled in Table 10.8. The number of occurrences in the extracted and nonextracted sets are labeled N_{ES} and N_{NES}, respectively, and the number of sentences in the entire text is $N_{ES} + N_{NES}$. The ratio (N_{ES}/N_{NES}) for a word should be much greater than expected (1.0) for the word to be a keyword. In the simple example shown, the three keywords are thus "Kevin," "anger," and "control." This simple example, of course, represents far too small of a set of text to provide a reasonable summary (there are only 39 words in the text!), let alone select keywords. But, it does illustrate the power of building an intelligent system from multiple meta-algorithmic patterns.

It is worth noting that the system outlined in Figure 10.1 is distinctly different in architecture from the Generalized Hybridization pattern introduced in Section 6.4.7. There, the hybridization is typically focused on the optimization of a sequence or parallel combination of promising first- and second-order meta-algorithmic patterns for a generally large problem space. In this section, however, the meta-algorithmic patterns are deployed for different parts of the system. For the Generalized Hybridization pattern to be deployed in the current

Table 10.7 The paragraph illustrated in Table 10.6 with the extracted sentences indicated by boldface

Sentence	Text
1	**Kevin has anger control issues**
2	He does a lot of great work
3	He sees, however, idiocy in others
4	He sees idiocy rewarded through promotions and bonuses
5	**Rewarding idiots makes Kevin angry**
6	**We are all lucky Kevin believes in gun control**

Table 10.8   The lemmatized words that occur more than once in the example paragraph, along with the salient occurrences N_{ES} and N_{NES}, along with the ratio (N_{ES}/N_{NES})

Lemma/Word	N_{ES} = # Occurrences in Extracted Set	N_{NES} = # Occurrences in Nonextracted Set	Ratio of N_{ES}/N_{NES}
Kevin	3	0	∞
Anger	2	0	∞
Control	2	0	∞
In	1	1	1.0
Reward	1	1	1.0
Idiot	1	2	0.333
See	0	2	0.000
He	0	3	0.000

example, for instance, we may substitute the Weighted Voting pattern in Section 10.2.1 with the Weighted Voting or Best Engine pattern.

10.3   Cloud Systems

The movement of so many software applications and services to the cloud has created many more opportunities for the use of parallel algorithms, since the intelligent system designer can take advantage of near-infinite computing resources to make the systems more robust. In addition, applications and services are now largely created to be cloud-native. This requires rethinking how parallelism is to be built into these systems from the ground up. This "native-cloud" movement affects all aspects of architecture: hardware, software, networking, security, and information management. In this short section, I mean only to highlight the huge opportunity for meta-algorithmic approaches to aid in the development of cloud-native applications.

Suppose, for purpose of illustration, that we have an application—such as scene analysis—for which the processing requirements are highly variable. The processing time is a function of

1. image complexity;
2. presence or absence of objects of interest;
3. meta-data (location, time of day, image size, user device capabilities).

Image complexity measurements should be selected based on their degree of correlation with processing time. This can, if necessary, be a multi-staged decision, as shown in Figure 10.2. Here, the images are collected on a personal system; for example, a smart phone, a laptop, a workstation, even an intranet. The personal system has more limited resources—cache, processing, storage, and so on—than the cloud, such that the cloud is viewed to have effectively unlimited resources. For each image captured, the image entropy is computed as shown in Figure 10.2. Images with sufficiently low entropy are processed locally, while images with high entropy are sent to the cloud system for the next stage.

Not shown in the figure is the training that has been performed on the images beforehand. This training included the assessment of which image complexity measurements could be

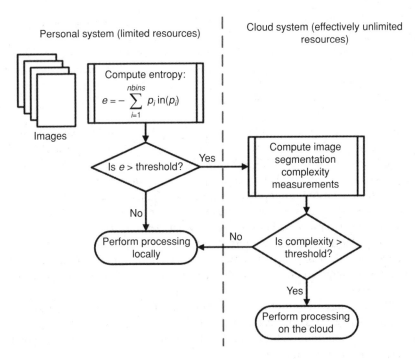

Figure 10.2 System diagram for the cloud system example. Images created local to a personal system are to be processed (scene analysis) locally or packaged for processing on the cloud system, depending the amount of image entropy and, if appropriate, image complexity as defined by a simple image segmentation procedure

performed quickly—this included image histogram entropy as shown, image histogram variance, image contrast, and image chroma variance—and the correlation of these with the actual processing time that is required for the scene analysis. The three measurements {image histogram entropy, image histogram variance, image contrast}, each being computed from the image histogram, have very similar performance requirements. Image chroma variance, however, requires significantly more processing, and does not provide better predictive accuracy. Image histogram entropy was selected as it had the best correlation ($r^2 > 0.90$) with actual processing time.

If the image is not processed locally, it is compressed and/or encrypted and sent to the cloud-based system. There, a second, also relatively simple (requiring modest processing time and resources) image complexity measurement is computed. This measurement (raw number of connected components formed after binarization is performed) was found to be highly correlated ($r^2 > 0.99$) with the processing time required for the scene analysis. This binarization step is performed on the cloud system, since (a) it requires a lot more processing than the computation of entropy, (b) it is unlikely to push the processing back to the personal system, and (c) it is useful as a part of the downstream scene analysis.

Once a scene has been analyzed, the output of the analysis can be used to guide the processing decisions for a length of time afterwards. If, for example, no objects of interest are extracted from a scene, then simple monitoring of image histogram entropy (looking for changes in image entropy) can be used to identify when the system settings need to be reviewed.

The third consideration for processing time, as listed above, is meta-data. There are a number of conditions under which we may choose not to send the data to the cloud—and in so deciding, we have to perform the processing on the personal system with more limited resources. These conditions include the following:

1. *Security*: The data associated with the images may not be allowed to leave the intranet.
2. *Image size/resolution*: The images may be small enough such that they can be readily sent to the cloud system, and thus be logged, processed, and archived there. Alternatively, they may be too large to be effectively transmitted, and so they must be processed locally. In this case, data associated with the images (objects identified, meta-data, etc.) can be transmitted.
3. *Time of image capture*: If the images are captured during times of peak bandwidth usage, and they can be batch processed at a later time, then their transmission can be delayed.
4. *Poor compressibility*: Image measurements such as entropy, variance, and complexity described above can also be used to predict the compressibility of an image. Images that can be highly compressed are more likely to meet the size constraints for transmission to the cloud system.
5. *Bandwidth of connection*: All other factors being equal, the greater the transmission rate, the larger the size of image that can be processed by the cloud system.
6. *Personal system capabilities*: Can the personal system compress the data, encrypt the data, or analyze the data locally most efficiently?

The diagram of Figure 10.2 is relatively straightforward, but it does illustrate that the processing decisions in a cloud environment can be branching, rather than simple binary ones. In this system and other similar cloud connection intelligent systems, a number of the meta-algorithmic approaches can be considered for use in optimizing the overall system performance. Here, let us consider briefly seven of the most salient meta-algorithmic patterns and their potential utility for system architectures related to the one described above.

The Sequential Try pattern (Section 6.2.1) is used to try one algorithm after another until a desired task can be accomplished or specific decision made. The architecture in Figure 10.2 could be replaced with a sequential decision-making approach as follows: instead of making the decision for one image at a time, the decision is made for a sizable set (e.g., a video stream) of images all at one. If the image histogram entropies (try #1) of the images are insufficient to merit sending the image to the cloud, then segmentation may be performed on the image in the set with the median value of entropy. If the segmentation complexity measurement indicates that processing on the cloud is appropriate (try #2), then all of the images are transmitted to the cloud system. If, however, the complexity of the segmentation is low enough, the scene analysis is performed locally. If, at any time, the scene analysis processing time exceeds the limits of the personal system (try #3), then images can be transmitted to the cloud system.

The Constrained Substitute pattern introduced in Section 6.2.2 can be used to find a scene analysis substitute algorithm—a so-called partial generator—that can perform the required elements of the full scene analysis algorithm (the so-called full generator of the pattern) that is present on the cloud system. One example is defining objects of interest by a simple descriptor (color content, object texture, etc.). The constrained substitute performs these simplified analyses, and if the object of interest is not identified (with a given probability), no further analysis is required (and so no transmission to the cloud system is needed).

The Predictive Selection pattern (Section 6.2.4) uses a decisioning process similar to that illustrated in Figure 10.2. The comparison of image histogram entropy and, separately, image segmentation complexity, to an appropriate threshold is a simplified form of predictive selection. The decision is predicting that images with low entropy should be selected for processing locally, while images with high entropy should be selected for processing on the cloud system.

Building on the previous paragraph, the Predictive Selection with Secondary Engine pattern introduced in Section 6.3.4 may be applied to the same image histogram entropy-based decisioning. Here, however, there may be two thresholds for comparison. If the entropy is greater than the higher threshold, the image is sent to the cloud system for scene analysis processing. If the entropy is less than the lower threshold, it is processed locally. If, however, the entropy is intermediate to the two thresholds, then a second pattern—for example, the Constrained Substitution pattern—is used to make the decision.

The third-order meta-algorithmic pattern, Feedback (Section 6.4.1), can also be considered in this system. Here, the difference between expected processing time on either the local or cloud-based system (or both) is used to change the threshold for the decision. This can be applied to the image histogram entropy value, the image segmentation complexity value, or both.

Related to the Feedback approach, the Proof by Task Completion pattern of Section 6.4.2 can be readily applied to the scene analysis problem. As described above for the Constrained Substitution pattern, a simplified scene analysis approach can be used to speed up the processing and/or reduce the use of bandwidth under some circumstances. If this approach is deployed, and some bandwidth is available, then from time to time an image will be transmitted and the "full generator" used to verify that the Constrained Substitute algorithm is in fact working properly. The successful match of these two generators—full and "partial"—will be fed back to the personal system as proof by task completion that the partial generator is working appropriately.

Finally, the Regional Optimization pattern introduced in Section 6.4.6 can be used to partially segment an image in the input set and then process the distinct parts of the image with different meta-algorithmic approaches. As but one example among many, low-entropy partitions may be processed locally using a Constrained Substitution pattern while high-entropy partitions may be transmitted to the cloud system and there processed using a Best Engine with Differential Confidence or Second Best Engine (Section 6.3.8) pattern.

This section shows, in abbreviated fashion, how multiple meta-algorithmic approaches can be considered, in parallel or in synchrony, when architecting an intelligent system. In the actual system design (Figure 10.2), an architecture largely based on a Predictive-Selection-like approach was chosen. More than half a dozen other reasonable architectures based on different meta-algorithmic patterns were also outlined. In the end, each of these architectures should be benchmarked against a healthy set of training data, and the best overall architecture—as deduced from the associated cost function—chosen.

10.4   Mobile Systems

In Section 10.3, our concern was whether to move processing to the cloud or process locally, largely as a consequence of transmission bandwidth being the limiting factor in the architecture. In this section, however, one of our considerations is that the resources of the device itself may be the limiting factor. This is more consistent with the mobile systems of today, in which small, multi-threaded devices with limited battery life must decide how best to deploy intelligent systems to accommodate responsiveness, accuracy, battery life, and user data contracts. Essentially, this section is the flip side of the Section 10.3, in which we look at the problem in terms of the device—or the *personal system* in Figure 10.2—rather than the *cloud system.*

In so doing, however, our approach to architecture must be revisited. Mobile devices have redefined not only how the vast majority of people access the internet but also have created a security threat surface unimaginable even as recently as the year 2000, when smart phones emerged. Because of the relentless drive of Moore's law and electronics miniaturization in general, the data storage, processing, and networking capabilities of smart phones continue to lag those of workstations, let alone laptops, by only a few years. This means that smaller and smaller devices are capable of holding more applications, more services, and less obviously more ways to compromise data, access rights, and identity. Some security threats can be relatively well contained; for example, physical biometrics (see Section 3.3.3) such as fingerprints are, in theory, only stored locally and only in encrypted form. However, there is an increased use of location-based services by smart phone owners: the overwhelming majority use one or more GPS services, and in so doing share their information. Combine the convenience of these services with the narcissism of social networking, add in the complicated world of forgotten-password retrieval, and you have created a huge new set of threats.

Taken together, these new and pervasive risks to security mean that the return-on-investment models for the use of mobile devices in conjunction with a cloud system need rethinking in light of (a) the costs of synchronization, (b) the use of multiple passwords, and (c) the greater possibility for direct and indirect social engineering to break down a device's security. This sort of reconsideration is a good opportunity for the application of meta-algorithmics and other forms of parallelism. The fundamental question to ask here is "What should be processed on the mobile device and what should be transmitted off-device for processing, performing of location-based services, and so on?" For example, password federation services can address some of the risks of multiple password systems, but they may also increase the damage should they be compromised, since all of the user's passwords are then lost.

Hence, the security risks, including the long-term risks of allowing the mobile service provider(s) to aggregate too much behavioral, social, and other identity-salient content, must be weighted appropriately. This can alter considerably the decision as to where to process information. In Section 7.3.3, a biometric example was given in which the first-order meta-algorithmic pattern, Constrained Substitute (introduced in Section 6.2.2), was used. This approach has immediacy for mobile systems, where biometric identification applications may be preferably confined to the mobile device itself.

In Section 7.3.3, the equation used to determine whether the costs are higher for processing on the mobile device versus the cloud system is

$$\text{COST} = \left(C_A + C_p t_p + C_B t_B \right) n_p,$$

where n_p is the number of times to run the algorithm, C_A is the cost (in \$) to run the algorithm once, C_B is the cost of the bandwidth for a given time period, t_B is the time to send the information for one run of the algorithm, C_p is the cost of the processing for a given time period, and t_p is the processing time. When we take into account the cost of a security breach, the equation becomes

$$\text{COST} = p_{\text{Breach}} C_{\text{Breach}} + \left(C_A + C_p t_p + C_B t_B \right) n_p,$$

where p_{Breach} is the probability of a security breach and C_{Breach} is the cost of the breach. Using this modified equation, then, we can apply the Constrained Substitute pattern as before.

10.5   Scheduling

Scheduling was considered earlier (Section 2.2.3), where it was noted that the most important concerns of a scheduler are (1) throughput of processes, (2) response time latency, or the time between task submission and system response, (3) turnaround latency, or the time between task submission and completion, and (4) real-time adaptation to provide fairness when unanticipated strains on resources—from system "hangs" to a large influx of new tasks—occur.

In Section 2.2.3, scheduling is considered in light of parallelism by task. As stated in that section, among the many scheduling approaches, there is (1) FIFO (first in, first out, or "queue"), (2) multi-level queue, (3) round robin, (4) fixed priority preemptive, and (5) shortest remaining time (or "shortest job first"). It was also noted that it may well be that using multiple scheduling algorithms in parallel and intelligently combining them may result in a more robust, more effective scheduling approach.

That last comment segues to the use of parallelism by meta-algorithmics for optimizing the deployment of scheduling. It is clear that many of the first-order meta-algorithmic patterns can be directly applied to selecting the scheduler from the list of five above, among others. For example, the Sequential Try pattern of Section 6.2.1 is analogous to the fixed priority preemptive scheduler, where the "try" is the attempt to finish the scheduled task in a given time window. In a looser, more trivial analogy, the Sequential Try pattern is analogous to the FIFO queue where each task is tried sequentially. The Constrained Substitute pattern (Section 6.2.2) has much in common with the multi-level queue scheduler, where a task that has a relatively relaxed completion time target can be assigned to a lower-priority queue that substitutes for the higher-priority queue(s) with the constraint being the required completion-by time. The Voting and Weighted Voting patterns of Section 6.2.3 are analogous to the shortest job first queue in which the voting is the expected processing time of the task. The Predictive Selection pattern (Section 6.2.4) is an excellent analog for the shortest job first queue wherein the job is interrupted if the predicted time for completion that was used to select its location in the queue is disproven during run-time, and the new expected run-time exceeds that of another task waiting in the queue. Finally, the round-robin queue, which incorporates multiple FIFO queues, can also use a form of predictive selection for assignment of tasks to queues. From these simple analogs, then, we can see that parallelism by meta-algorithmic has much in common with traditional parallelism by task as incorporated by queues.

Given the similarity between parallelism by meta-algorithmic and by task with regard to many commonplace queues, it is logical to ask if other forms of parallelism by meta-algorithmic can be used to create new types of queues. Indeed, the implications are that they can. For example, let us consider the Confusion Matrix and the concept of *maximum confusion* as introduced in Section 8.2, which is the foundation of many second-order meta-algorithmic patterns. We define the maximum confusion as

$$\text{Maximum confusion} = \text{argmax} \left(\{i, j \neq i\} \in 1, \ldots, N[C_{ij} + C_{ji}] \right).$$

In this equation, the set of paired off-diagonal elements of the confusion matrix, C_{ij} and C_{ji}, with the highest summed weight, can be targeted for organization of their two classes $\{i,j\}$ as a single aggregate class, which allows the reduction of the order of the confusion matrix. As applied to queues, the classes with maximum confusion can be pooled for scheduling considerations with, presumably, the minimum impact on actual versus expected throughput statistics.

The same type of approach can be taken using the Sensitivity Analysis third-order meta-algorithmic pattern introduced in Section 6.4.5. Here, the correlation between classes of the data can be used to dictate the pooling of classes for scheduling. Like maximum confusion, this approach is used to create a pooled version of any of the traditional queues, which allows us to match the number of classes to the specific aspects of the queue—for example, number of processors or number of queues—required for optimization.

10.6   Classification

Many of the examples in this book have focused on the application of meta-algorithmics to classification problems. With only a few exceptions, these examples highlighted an improvement—often a very significant improvement—in overall classification accuracy after meta-algorithmics are applied. However, there are circumstances in which meta-algorithmics do not improve classification accuracy. One circumstance is when the pattern applied does not take advantage of the proper partitions of the input data space, such that there are differential results for the individual classifiers for these subclasses of input data. Improper, or simply unfortunate, partitioning can significantly reduce the otherwise almost assured efficacy of the several Predictive Selection and the several Confusion Matrix patterns. Another circumstance in which meta-algorithmics can be ineffectual is when there is not a sufficient plurality of classifiers. Applying meta-algorithmic patterns such as Voting and Weighted Voting to a set of only two classifiers, for example, will result in classification accuracy intermediate to that of the two classifiers without differential partitioning. Finally, when the amount of training data is small, the meta-algorithmic patterns will generally not be able to improve accuracy reliably.

In this section, an example suffering from all of these deficiencies is illustrated. The input data set is tiny, comprising the high temperatures in degrees Fahrenheit for Fort Collins, Colorado, USA, for the first 28 days of both January and February, 2012, assigned randomly to four groups {A,B,C,D} of 7 days out of the 28. The goal is to correctly classify the day as belonging to either class [January] or class [February] based only on the single temperature metric. Table 10.9 provides the input data set along with the days in the month corresponding to these sets.

Next, we define our experimental set-up. Since the data set is very small, we decide to use a 2:1:1 ratio of training:validation:test data. We perform training on all three combinations of two of the sets from {A,B,C} whilst using the third set for validation. Two simple classifiers are used for the problem. The first assumes that each of the two classes represents a Gaussian distribution and determines the critical point (CPt) between the two means as the point that

Table 10.9 High temperatures in Fort Collins, Colorado, USA, for the first 28 days of January and February, 2012, assigned randomly to four groups {A,B,C,D} as described in text

Group	Days of the Month	High Temperatures (January)	High Temperatures (February)
A	1, 7, 10, 16, 19, 22, 26	36, 40, 55, 39, 61, 47, 53	48, 27, 36, 41, 43, 59, 39
B	2, 6, 12, 13, 17, 23, 28	44, 51, 43, 52, 29, 44, 43	47, 39, 23, 39, 41, 46, 44
C	3, 8, 11, 14, 20, 21, 25	55, 54, 41, 55, 57, 52, 64	30, 28, 19, 43, 43, 51, 63
D	4, 5, 9, 15, 18, 24, 27	52, 59, 50, 54, 49, 41, 45	32, 31, 39, 34, 39, 43, 50

is the same number of standard deviations of the two populations from the corresponding pair of means. This method, designated the Gaussian model, is described in Simske, Li, and Aronoff (2005). The second classifier is even simpler. Here, the nearest integer or mean of two successive integers is found that classifies the maximum number of correct classifications of the training data. As such, this CPt is termed the value of best separation (VBS), and the method designated the VBS classifier. On the training data, the VBS is guaranteed to have the maximum accuracy of any single CPt-based classifier. The CPt of the Gaussian and VBS classifiers are given in Table 10.10, along with the mean and standard deviations of the high temperature values in each of the three sets of 14 days used for training sets.

In Table 10.11, the two classifiers are validated against the third, left-out, set of 7 days of high temperatures. Here, it observed that the Gaussian model significantly outperforms the VBS model, correctly classifying 28 of 42 days (66.7%) compared to 24 of 42 (57.1%) days for the VBS model. As expected, both the Voting and Weighted Voting meta-algorithmic patterns—having no input data partitions, only two classifiers, and sparsely populated training sets—provided intermediate accuracy to that of the two individual classifiers, correctly classifying 26 of 42 (61.9%) days.

The validation step (Table 10.11) therefore identifies the Gaussian classifier approach to be superior to that of the VBS or the two, in this case both handicapped and hapless, meta-algorithmic approaches. Indeed, when the three training and validation sets are then used to define the CPt for the Gaussian classifier, the January data is 48 ± 9, the February data is 40 ± 11, the CPt is 44.8, and the samples in set {D} are correctly classified 12 out of 14 times (85.7%). This higher value of accuracy is arguably obtained because the training and validation sets have been combined, providing a better estimate of the Gaussian distributions underlying the two populations. Of course, a more modest interpretation is that we are using

Table 10.10 Critical point (CPt) of the binary classification using the Gaussian model outlined in Simske, Li, and Aronoff (2005) and the value of best separation (VBS) as described in the text. Two of the four temperature sets are used for training, as identified. The January and February temperature statistics for the training sets are given in the last two columns (mean $\pm$ standard deviation)

Training Sets	CPt, Gaussian	CPt, VBS	January ($\mu \pm \sigma$)	February ($\mu \pm \sigma$)
A, B	43.24	42	46 ± 8	41 ± 9
A, C	46.54	45	51 ± 9	41 ± 12
B, C	44.93	40	49 ± 9	40 ± 12

Table 10.11 Classification accuracy of the validation sets for the three training sets of Table 10.10. Four classifiers are chosen: Gaussian, VBS, Voting of Gaussian + VBS, and Weighted Voting of Gaussian + VBS, as described in the text

Validation Set	Gaussian	VBS	Voting	Weighted Voting
C	0.786	0.643	0.786	0.786
B	0.571	0.500	0.500	0.500
A	0.643	0.571	0.571	0.571

75% of the overall data sets for training, so accuracy is expected to be higher than when using 50% as in Table 10.11.

As expected, then, meta-algorithmics did not help in this simple, small classification problem. Does this mean that there is no meta-algorithmic pattern that could further improve the accuracy of the classification? For a CPt-based classifier using only the high temperature data, this is indeed the case. The VBS *for the test data*, which gives the best possible performance on the test data (but is of course a cheat), has a CPt of 44 and also has an accuracy of 85.7%. So, we are required to analyze the data in Table 10.9 a little differently in order to improve the classification accuracy. One meta-algorithmic pattern that can be used is the Expert Feedback pattern (Section 6.4.4). The expert feedback can be in the form of input range, co-occurrences of data, and ratios. In the case of the monthly temperature data, suppose we associate each temperature with the most recent peak high temperature and the next low, or vice versa. That is, we keep the temperature along with the co-occurrence of the previous and subsequent local optima. In this way, we can readily reassign one of the two misclassifications and obtain an overall 92.9% accuracy for the data set {D}. Again, this is a cheat (and such co-occurrences are usually unreliable for such small data sets) but one that illustrates how meta-algorithmic patterns of higher complexity can be deployed when the classification problem itself is otherwise compromised by the types of deficiencies described in this section.

10.7 Summary

This chapter, in abbreviated form, considered the way in which having the set of system architecture tools comprising parallelism by meta-algorithmics leads to the possibility of more robust systems. A relatively deep consideration of summarization shows how meta-algorithmics can lead to a change in how ground truth is collected in the first place, and how the output of one meta-algorithmic system can be fed back into another system. Section 10.3 showed how meta-algorithmic patterns can be used in coordination with a hybrid personal system/cloud architecture to create an overall architecture that can be more robust and/or accurate. Section 10.4 included revisiting the cost model for processing information using a Constrained Substitution pattern, now accounting for the impact of a security breach. Scheduling, normally associated with parallelism by task, was shown to align well with first-order meta-algorithmic patterns. Higher-order meta-algorithmic patterns were considered for their ability to create simplified queues for problems involving a surfeit of input classes. Finally, a very simple classification problem was exhaustingly elaborated in order to show some of the considerations that must be traded off when deploying meta-algorithmics.

Combined, then, the approaches outlined in this chapter are intended to give the reader some perspective on how to connect meta-algorithmics to bigger system architecture issues than the more algorithm-specific examples of Chapters 7–9. Combined with those previous chapters, however, these examples should provide credible support for the following *eight principles of meta-algorithmics*:

Principle 1: No single generator—algorithm, system, or engine—encapsulates the complexity of the most challenging problems in artificial intelligence, such as advanced machine learning, machine vision and intelligence, and dynamic biometric tasks.

Principle 2: It makes sense for a system designer to optimize an algorithm for a portion of the input range and then leave the rest of the range to another algorithm.

Principle 3: Patterns of usage are often more powerful—more accurate, more robust, more trainable, more reusable, and less costly—than a single, highly trained algorithm, system, or engine.

Principle 4: Ground truthing, or the labeling of training data, is extremely expensive—in general, we have to assume that there will be a relative sparseness of training data.

Principle 5: First-, second-, and third-order meta-algorithmics are used to create a highly trainable system with minimum cost of on-ramp. Commercial off-the-shelf, open source, and targeted generators can be deployed together in these patterns. The targeted generators are designed to provide good results for partitions of the input space on which the existing generators do not perform acceptably.

Principle 6: The juxtaposition of bottom-up and top-down algorithms, and the combination of targeted and broad algorithms can be used to generate systems that are resilient to changes in the input data, and highly adaptable to subsystem deployment.

Principle 7: Weighting and confidence values must be built throughout the system in order to provide the means to use meta-algorithmics on multiple classifiers at a time; hybridize the multiple classifiers in a plurality of ways; and/or as an alternate to parallelism by task and parallelism by component. Meta-algorithmics, in this way, provide the means for simultaneously combining, learning, and parallel processing.

Principle 8: The goals of modern algorithm designers will be, increasingly, indistinguishable from the goals of modern system designers. Architecting for intelligence is becoming a primary need for architecting any system, with the increasing pervasiveness of *big data*, analytics, voice search, and other machine-to-machine intelligent systems.

Combined, these principles lead to a different way of approaching design. This may have relevance to fields other than intelligent systems, as will be addressed in the final chapter next.

Reference

Simske, S.J., Li, D., and Aronoff, J.S. (2005) *A Statistical Method for Binary Classification of Images*. ACM Symposium on Document Engineering, pp. 127–129.

11

The Future

There are only two truly infinite things, the universe and stupidity. And I am unsure about the universe.

—Albert Einstein

The future is uncertain but the end is always near.

—Jim Morrison

11.1 Recapitulation

Is it stupid to assume that one can architect, implement, and deploy an intelligent system in which one treats some of the world's most powerful intelligence generators—algorithms, systems, and engines—as black boxes? Perhaps so. But, as stupidity is truly infinite, somewhere in its limitless capacity this stupidity may achieve results that are of unexpectedly high quality. The unprecedented processing, parallelism, and analytics capabilities of the modern computing world have changed the way in which intelligent systems can be optimized. The relentless simultaneous improvements in storage, caching, data processing algorithms and architecture, and parallel processing are providing an exponential corollary to Moore's law in which the movement towards full-scale exhaustive search is proceeding even for the largest scale machine intelligence problems.

Exhaustive search will never be practical for many problems—for example, a 100-node "single visit" path search, termed a traveling salesman problem, requires $(99!)/2$, or 4.7×10^{155}, pathways to be searched, a problem likely to be unwieldy for a long, long time. However, intelligent reduction of next-state to a mean of just three options quickly reduces the number of pathways to be explored to no more than 3^{98}, or 5.7×10^{46}, which is more likely to be achieved in our lifetimes. This can be viewed as a form of near-exhaustive search, which is extremely inefficient but could, understandably, be important for extremely valuable optimization problems such as remote sensing, space travel, and the like. Such an approach, incorporating a form of expert input, is readily amenable to parallelism, wherein the number of queues is a multiple of the number of allowed next-states.

Meta-algorithmics: Patterns for Robust, Low-Cost, High-Quality Systems, First Edition. Steven J. Simske.
© 2013 John Wiley & Sons, Ltd. Published 2013 by John Wiley & Sons, Ltd.

Such a near-exhaustive search, however, should be the last resort. The various forms of parallelism outlined in this book provide a large set of possible approaches, with considerable overlap, to perform a much more efficient near-exhaustive search.

For parallelism by task, six primary patterns were described: (1) queue-based, (2) variable sequence-based, (3) latent introspective parallel-separable operation (PSO), (4) unexploited introspective PSO, (5) emergent introspective PSO, and (6) recursively scalable task parallelism. These patterns can be used to expedite throughput of different parts of a problem space. Queues are simple sequential schedulers, or pipelines. Variable-sequence-based parallelism allows a pipeline to be broken up into parallelizable parts where the order of operations is not important. The three types of PSOs look inside specific generators (algorithms, systems, services, engines) and identify separable operations that are either initially hidden, not taken advantage of, or not possible until the generator itself is restructured. Finally, the recursively scalable task parallelism approach connects the final architecture chosen for parallelism with a cost function that allows the deployed architecture to be optimal for the specific concerns of the system designer.

For parallelism by component, eight approaches were overviewed: (1) structural reframing, (2) model down-sampling, (3) componentization through decomposition, (4) temporal parallelism, (5) overlapped parallelism, (6) scaled-correlation parallelism, (7) variable element parallelism, and (8) search parallelism. The most general is that of structural reframing: the preparation of a task for parallelism and the incumbent advantages obtained just by this restructuring. Even when downstream parallelism is not employed, the reframing itself leads to significant increase in throughput for a serial pipeline. Model down-sampling is used to improve processing through, effectively, compressing the data. Componentization through decomposition allows us to separate information into separate data sets, smaller than the original, and process these independently. Temporal parallelism deconstructs a task into two or more parallel processing branches, each sampling the same data but with distinct sampling rates. Overlapped parallelism, on the other hand, provides spatial parallelism through partial overlap of adjacent data sets; for example, subregions of images. Scaled-correlation parallelism uses multiple scales of the input data in parallel. Variable element parallelism and search parallelism are more traditional forms of parallelism in which operations such as image interpretation or database look-up are performed in parallel.

All fourteen of these approaches, or *patterns of design*, for parallelism by task or parallelism by component share in common (a) their ability to speed up the overall processing, reducing a wide variety of system costs, and (b) their general inability to improve the overall accuracy, robustness, resiliency, or convergence efficiency of the system. One exception to this might be the scaled-correlation parallelism, under certain conditions. Regardless, the field of parallel processing, I hope, will benefit from the elaboration of a large set of useful design patterns for system optimization for accuracy, convergence efficiency, and adaptability to changes in the input space.

To address these needs, a third form of parallelism—meta-algorithmics—was introduced, and subsequently became the main theme of this book. The main goal in this book was to provide a relatively exhaustive set of patterns, along with practical examples of their usage, for developing intelligent systems. The 21 main meta-algorithmic patterns were assigned to three categories: first, second, and third order. The first-order patterns are the primitive statistical algorithms, related in several cases to ensemble methods (Zhou, 2012), which use an algorithm to incorporate the output of two or more generators into a single, improved output. The second-order algorithms include those based on the confusion matrix, which often is used to simplify the input space; first-order patterns that meet certain criteria; and the conditional combinations of two of the first-order patterns. These meta-algorithmics introduce decision pipelines, and the set of nine patterns described in Chapters 6 and 8 are certainly nonexhaustive. The generalized pipelining of meta-algorithmic patterns is instead left for the final third-order meta-algorithmic pattern, that of Generalized Hybridization (Section 6.4.7). The other third-order meta-algorithmic patterns are different from the second-order patterns in that they usually involve feedback, explicit optimization over time, and/or sensitivity analysis.

A large, but certainly nonexhaustive, set of examples were then provided in Chapters 7–10. Some of these were trivial, in order to show how to incorporate meta-algorithmic parallelism into a system architecture. Others were more complicated, emphasizing that the use of meta-algorithms is both an art and a science. Finally, several examples incorporated the comparison among many different meta-algorithmic patterns. It is these that are the norm for the intelligent system architect, since the optimal meta-algorithmic pattern to incorporate cannot always be deduced before experimentation.

The meta-algorithmic pattern that will perform best on a given problem is dependent on many factors, including the nature of the input data. If the input data consists of several Gaussian distributions, for example, a clustering-based algorithm will generally perform best. When the input data is less well behaved, a boundary-based approach such as a support vector may be more useful. However, for the huge input sets that modern intelligence systems typically must accommodate, it is unlikely that a single approach will optimally handle all of the various subclasses of data. This is where meta-algorithmics come in. If the system designer does not know where to start, perhaps the easiest way is to use the Predictive Selection algorithm (Section 6.2.4) to determine the extent to which the relative performance of the individual generators varies with the partitions of the input space. If the variance is high, then the Predictive Selection pattern will be used to select the configuration for each partition as part of the system architecture.

11.2   The Pattern of All Patience

More generally, however, the system designer can investigate several meta-algorithmic patterns simultaneously for consideration in the final deployment of the system. The choice is dependent on more than just the behavior of the input data. It can also be based on organizational expertise, licensing and anticipated licensing arrangements, future availability of the components, and so on.

This is a novel approach to system design. It requires open-mindedness as to which intelligence generators will be implemented in the final system. As new intelligent algorithms, systems, and engines are created or prepared for deployment, each of the meta-algorithmic patterns already considered for the solution architecture can be reconsidered for the incorporation of the new generator. As was shown in many of the patterns—Weighed Voting and Confusion Matrix being prominent among them—the addition of a new generator may provide statistical support to omit one or more of the previously implemented generators. The voting weight, for example, may become negative for one or more of the generators (see, e.g., Table 8.11). Overall accuracy, deduced from the sum of the diagonals of the confusion matrix, may be improved when one or more generators are removed, as illustrated in Section 8.3.4. Other meta-algorithmic patterns, such as the Constrained Substitute and Proof by Task Completion, can suggest the removal of a generator on one or more cost bases. This does not prove that the generator left out is inferior in some way to the other generators that are incorporated in the final deployed system. It shows, rather, that for the given input set and the given population of collaborating generators, this generator is simply not required.

What happens to make an otherwise valuable generator superfluous? The easiest answer is that other generators in the system simply outperform this generator. However, experience with meta-algorithmic patterns—including examples shown in Chapters 7–9—highlights the manner in which far less accurate generators can still improve the system performance. For example, the results presented in Section 8.3.1, originally presented in Simske, Wright, and Sturgill (2006), show the substantial reduction in error contributed by the Bayesian engine, even though it is highly correlated with the other two, more accurate engines. The easiest answer, therefore, does not always hold. Instead, it makes more sense for us to explore several patterns of parallelism and deploy the pattern providing the best overall system results.

Thus, like Shakespeare's Lear, *we will be the pattern of all patience*, and unweary-ingly evaluate multiple patterns to identify those that perform optimally for the variety of intelligent systems. The argument is relatively simple. The primary costs involved in meta-algorithmics are in creating the input and output structures and—for a distributed or cloud-based architecture—transmitting the salient data and meta-data from the sensing loca-tion to the processing location. Once the data and meta-data are transmitted, there is effectively unlimited processing capability "in the cloud" and thus every parallel processing pattern that is potentially useful should be explored.

As a general rule, selecting a few meta-algorithmic patterns that *a priori* are expected to cover the input space differently—and in combination completely or nearly completely—should be deployed. A major focus of this book has been parallel processing, with more traditional forms shown as belonging to two broad classes: parallelism by task and parallelism by component. Parallelism by meta-algorithmics underpins a third form of parallelism, supported by the following:

1. *Partitioning of the input space*: Meta-algorithmics tuned to specific subdomains of the input space can be used to provide parallel means of analyzing each subdomain. Predictive-Selection-, Regional-Optimization-, and Confusion-Matrix-based patterns are excellent choices for this partitioning.
2. *Generator independence*: A set of algorithms, intelligent systems, or intelligence engines that are poorly correlated are often excellent candidates for using such multiple generators for a collective decision. The meta-algorithmic patterns based on Voting or Tessellation and Recombination are excellent choices for enabling such collective decisioning, as they allow different—parallel—approaches to analysis to be effectively used together.
3. *Combination of learning approaches*: Meta-algorithmic patterns such as Feedback, Ex-pert Feedback, Sensitivity Analysis, and Generalized Hybridization provide different ap-proaches to learning. Feedback-related patterns provide learning by changing the weight of connections between output and input. The Sensitivity Analysis pattern can provide the fine-tuning of the weighting among multiple learning generators. Generalized Hybridization, finally, allows different learning engines to be combined in useful ways.

4. *Selection of suitable engine*: Meta-algorithmics can be used to provide an advanced form of parallelism by task. The Sequential Try and Constrained Substitute patterns can be significantly sped up when each of the generators to be investigated is run in a separate parallel branch.

Overall, meta-algorithmics provide a different, often substantially more complicated, form of parallelism than parallel processing by task or by component. Because of the ability to perform Generalized Hybridization, in fact, meta-algorithmics can be used to create a type of machine intelligence analogous in architecture to that of human consciousness as proposed by Gerald Edelman (Edelman, 1992). Edelman's theory of Neural Darwinism rests on three tenets: (1) developmental selection, (2) experiential selection, and (3) reentry.

Development selection can be viewed as analogous to the first-order meta-algorithmics, since the more efficient generators are selectively tried, voted for, allowed to substitute for other generators, and so on. More importantly, however—and akin to an important element of how the human nervous system develops—poorly performing generators or meta-algorithmic patterns are selected against. The Sequential Try pattern, for example, can be used to eliminate ineffective generators, while the Weighted Voting pattern can be used to differentially emphasize one generator over another.

Experiential selection can be viewed as analogous to the second-order meta-algorithmics. Here, two or more first-order generators can be used in a cooperative pattern. There are many options for combinations—far more than the nine patterns introduced in Section 6.3. But, there is a commonality among all of these patterns. This commonality is analogous to experiential selection, in which past events have reduced the plasticity while simultaneously increasing the speed of decision-making in a conscious neural system. From experience, we may know when one or more factors in a decision need to be calibrated relative to the other factors. In Section 6.3.2 this is the role of the output space transformation. From experience, we may have learned some rules or other expertise that can be used to guide a decision. This is the role of the expert decisioner in Section 6.3.3. Experience may also guide us to readily identify when a decision has been made incorrectly. This is the analog of the secondary engines described in Section 6.3.4. We also can judge the likelihood of a correct decision from past experience: this is the required precision of Section 6.3.5. Experience also leads us to have more confidence in a decision when multiple experts agree—this is the role of majority voting in Sections 6.3.6 and 6.3.7—or more confidence in a decision when the decision-making generators themselves have more confidence (Sections 6.3.8 and 6.3.9). Physiologically, higher-confidence neuronal paths are carved through reinforcement and the concomitant improved synaptic efficiencies in the decision-making path. Meta-algorithmically, this reinforcement is achieved through a variety of weights assigned to the decision.

Reentry is a powerful concept (Edelman, 1992) that describes how thoughts progress through time through the remapping of intact and altered sets of neuronal activity to the same neuronal tissue. This concept has strong ideological commonality with the third-order meta-algorithmic patterns. Feedback patterns provide revisiting of the input after an output has been computed—reentry through inputting anew. Proof by Task Completion (Section 6.4.2) is also a form of reentry. Sensitivity Analysis (Section 6.4.5), meanwhile, is a means of remapping the system with slight alteration. Each of these patterns addresses the temporal aspect of reentrancy. Regional Optimization (Section 6.4.6), on the other hand, addresses the spatial aspect of reentrancy, allowing for independent remapping of each of the distinct regions.

These are powerful analogies. Taken together, the implication is that a system designed from the ground up using meta-algorithmics is an intelligent—though not yet anyway a conscious—system. The flip side of this analogy, of course, is that the system must be intelligently designed. The need to perform pattern selection during system design is familiar to security system architects. The upfront investment will be more than recuperated downstream, when the whole system does not need to be rearchitected. The fundamentals of system architecture through meta-algorithmics are relatively straightforward. Each knowledge generator must be configured to provide the same output type—data, meta-data, and statistics—as its peers. With this done, the system has a high degree of flexibility in terms of later meta-algorithmic pattern deployment.

11.3 Beyond the Pale

This book already reaches into a large number of domains for its relatively modest length. At the risk of overstretching, I would like to comment on how opening up system architecture to a meta-algorithmics-centered approach may have relevance in other domains outside of the machine intelligence area.

I hope to have shown how meta-algorithmics require a different mind-set for system architecture than traditional systems. Meta-algorithmics is intended to provide a set of patterns for making systems innately as combinations of other systems—system that work together without compromise, but rather collaboration. Meta-algorithmics do not, generally, simply combine systems—although some of the simplest examples, for example, using the Voting pattern, may be argued to provide no more than this. In general, however, meta-algorithmics allow multiple generators to participate as partners equal to their ability to improve the common lot. The relative ability of each meta-algorithmic participant—that is, pattern—to contribute to the overall decision-making will usually change through time. This makes the system robust, flexible, and collaborative. Today, the most reasonable task might be simply to select the single best engine, since it provides both high accuracy and confidence. Tomorrow, however, the input data may drift substantially, making a Confusion-Matrix- and/or Sensitivity-Analysis-based pattern outperform the best engine. This means that every reasonable generator can and should be kept in the system architecture—especially in a cloud environment where the storage and processing requirements of using extra generators and the associated plethora of meta-algorithmic patterns is relatively irrelevant.

This is a collaborative, "blue ocean," approach to intelligent system design. The implications are simple, but nevertheless powerful. A blue ocean approach to product innovation has been shown to offer significant short- and long-term advantages to a company (Kim and Mauborgne, 2005). Surely, a meta-algorithmic blue ocean approach can be applied to other fields.

Many other fields require intelligent analysis of a large set of possible decisions. Consider, for example, politics. Multiple parties bring their set of concerns to the bargaining table, each with a different relative importance, or weighting, assigned to the decisions. These are typical input for meta-algorithmic patterns. As the sets of needs change over time, different patterns of behavior, comprising the political decisions, may be needed. I am not implying that a set of patterns will ever replace the complexity of negotiation, since interpersonal communication results in a fluid weighting of the decisions due to the powers of persuasion, eloquence, and reason. However, meta-algorithmic approaches to negotiation may be useful in showing parties how far they are from the overall optimum consensus. In this way, meta-algorithmics may prove useful in the area of game theory. It is certain that meta-algorithmic patterns different from the ones described in this book will be required for politics. That being said, a Tessellation and Recombination (Section 6.2.5) approach, wherein the broader needs are broken into their composite needs and then recombined to provide an optimal overall set, is of potential value.

Another interesting area for the application of meta-algorithmic approaches is education. There are several different end goals in education, including fact retention, general reasoning ability, linguistic abilities, ability to abstract, and so on. Many different approaches to learning have been discovered that provide excellent results, but often at the emphasis of one type of learning over another. In order to become a true proficient in an area, all of these learned skills must be garnered. What is the best way to combine multiple learning approaches to achieve the overall best education? The answer here is constrained by the amount of time a person can devote to the learning, by the relative amount of time each of the learning approaches requires, and by the minimum amount of time required for each approach to have a positive effect, among other factors. A variety of meta-algorithmic patterns—including Tessellation and Recombination with Expert Decisioner (Section 6.3.3.), Proof by Task Completion (Section 6.4.2), Confusion Matrix for Feedback (Section 6.4.3), Expert Feedback (Section 6.4.4), and Sensitivity Analysis (Section 6.4.5)—are potentially useful patterns for addressing educational optimization. There may be other, education-specific, meta-algorithmic patterns awaiting discovery as well.

A final area for the potential application of meta-algorithmics also deserves mention. The general area of hypothesis testing—especially for relatively comparing two or more theories on the basis of relevant experimental data. Depending on the complexity of the theories, the basic precepts can be tessellated and then recombined (complex theories) or simply incorporate a Weighted Voting approach to decide which theory has the highest score based on the relative importance of each of the precepts. This application of meta-algorithmics is nothing more than an extension of how they are already used, since of course clustering, classification, signal and image understanding, text understanding, and most of the other examples in this book are in fact testing hypotheses.

As an alternative perspective to the blue ocean described above, meta-algorithmics can be viewed as providing the means for Architectural Darwinism. In a complex, high-value, distributed and/or multi-user system, the use of meta-algorithmic patterns provide a dynamic means of generating an architecture that adapts with changes in its environment. With such a system, what thrives survives; that is, what has utility in the overall problem space is used more often. If the creators of the meta-algorithmic generators are paid by use, then indeed the measure of fitness is the money the generator makes. The truly unfit generator, in never being used, will be selected against and perish. It is that simple, that infinite in its stupidity. As with all survival of the fittest, it is a bit of a tautology. Nevertheless, we arrive at this Architectural Darwinism by two routes: (a) the analogy to Edelman's Neural Darwinism and (b) the relative utility of generators in a meta-algorithmic system.

In one way, then, meta-algorithmic architectures can be viewed as collaborative; in another, competitive. In my opinion, the competitive perspective is less compelling, since it ignores the potential value of less effectual generators. It is important to note that some generators that are not effectual on their own are nevertheless effectual when participating in one or more meta-algorithmic patterns. These generators may work together with their more accurate and/or robust counterparts to optimize the system response to the overall set of inputs. The meta-algorithmic pattern, and not the performance of the individual generators, decides the overall *system value* of each generator. This results in a collaborative architecture, since the way the generators work together is more important than how they work alone.

11.4 Coming Soon

I doubt many authors come to the end of their books and feel that they have nothing left to say. Indeed, in my case, I feel that in many ways I have only scratched the surface on meta-algorithmics. I felt it would be relatively painless to show how meta-algorithmics borrowed from the great foundation produced by the last several decades of intelligent system design. The fact that I had already built some rather large meta-algorithmic systems using the first-order, several of the second-order, and two of the third-order patterns also contributed to my underestimating the scope and scale of the possibilities for a meta-algorithmic approach to system design.

In fact, I have not—and could not—properly acknowledge all of the great research in the many fields touched on by this book. I quite likely have missed critical meta-algorithmic patterns that would further benefit the reader. I have used simplified examples in order to keep the flow of the book at a reasonable pace. This book has not sufficiently broken out the generic versus specific aspects of the meta-algorithmic patterns, except perhaps by example—illustrating their utility across a dozen domains should at least allow the reader to get a good appreciation for the difference in the art and science of meta-algorithmic architecture design. The applications of meta-algorithmics to mobile system architecture, cloud system architecture, and scheduling have also much future work to be done.

Given these limitations, what are the primary contributions of this book? I have hoped to provide the following:

1. An exhaustive—or at least relatively exhaustive—collection of the primary patterns for meta-algorithmics, affording a superset to ensemble and other combinatorial patterns.
2. A well-structured, detailed set of design patterns that are organized by their complexity.
3. No-frills, no-obfuscation mathematical description of the patterns, their application, and examples. The focus was on linear systems and statistics, and where additional mathematical complexity would not add to the value, it was eschewed.
4. Detailed instructions on how to use the meta-algorithmic patterns together, and how to build out rather complicated overall systems with as simple of building blocks as possible.
5. A broad set of examples, exploring enough different research fields to make the methods accessible to most computer science, mathematics, science, and engineering professionals; but, not exploring so many different fields that the book was a medley of disconnected examples. Chapter 3 illustrated the great breadth in the fields covered, while Chapters 7–9 illustrated the commonality of meta-algorithmic approach that can be used across these fields.
6. A path forward. I have hopefully conveyed to any reader who has made it this far that they can begin exploring and deploying meta-algorithmics immediately. In some cases, they will not find a huge advantage to do so; for example, with certain expensive image, text, and speech recognition engines the cost of using two or more such engines may not merit the increase in accuracy, robustness, or system architecture flexibility. In other cases, however, the reader is likely to find them highly valuable.

7. The impetus for further research on the patterns and applications of meta-algorithmics. In addition to the fields addressed in this book, there are many other intelligent systems that I believe will benefit from such a patterned, collaborative approach. I hope readers will advance those fields in some way using the information in this book.

   In addition to these specific goals, I had several other forward-looking goals for the book. One of the critical areas for intelligent system design and analysis is—by definition—sensitivity analysis. However, in some ways, this field is still in its adolescence, if not infancy. Through the Sensitivity Analysis pattern (Section 6.4.5) and the Confusion Matrix theory (Section 8.2) presented herein, I hoped to stimulate further, more sophisticated work on sensitivity analysis.

   I also hoped to stimulate the development of more work on dynamic system architectures. With the explosion of parallel processing and mobile access to cloud computing resources, it is clear to me that dynamic system architecture is the future. As such, the collaborative approach enabled by meta-algorithmics seems a very reasonable means of supporting this dynamism.

11.5   Summary

The future is uncertain, but the end is definitely near. The end of this book, that is. This is, therefore, an appropriate time to look back and make sure that I have written what I promised to write. Hopefully, this book has helped connect the broad set of technologies and design patterns comprising meta-algorithmics to the larger world of parallel processing, algorithms, intelligent systems, and knowledge engines. As an interdisciplinary field, meta-algorithmics owes so much to so many that it is certain salient previous work has been left out in these brief 11 chapters. While unintentional, it is also hopefully not a barrier to further elaboration of the field of meta-algorithmics, which I feel can be a universal acid, helping to eat away preconceived concepts of finding a single best algorithm, system, or engine for a task. Instead, we wish to find the single best generator or pattern of generator collaboration for the specific task, fully knowing it may not be the best collaboration on another day. The readiness is all.

References

Edelman, G.M. (1992) *Bright Air, Brilliant Fire: On the Matter of the Mind*, Basic Books, 304 pp.
Kim, W.C. and Mauborgne, R. (2005) *Blue Ocean Strategy*, Harvard Business School Press, 240 pp.
Simske, S.J., Wright, D.W., and Sturgill, M. (2006) *Meta-algorithmic Systems for Document Classification*. ACM DocEng 2006, pp. 98–106.
Zhou, Z.-H. (2012) *Ensemble Methods: Foundations and Algorithms*, CRC Press, 236 pp.

Index